NEAR-RINGS

NORTH-HOLLAND
MATHEMATICS STUDIES **23**

Near-Rings

The Theory and its Applications

GÜNTER PILZ

Institut für Mathematik
Johannes-Kepler-Universität Linz

1977

NORTH-HOLLAND PUBLISHING COMPANY
AMSTERDAM – NEW YORK – OXFORD

North-Holland ISBN: 0 7204 0566 1

PUBLISHERS:
NORTH-HOLLAND PUBLISHING COMPANY
AMSTERDAM, NEW YORK, OXFORD

SOLE DISTRIBUTORS FOR THE U.S.A. AND CANADA:
ELSEVIER / NORTH HOLLAND, INC.
52 VANDERBILT AVENUE, NEW YORK, N.Y. 10017

Library of Congress Cataloging in Publication Data

Pilz, Günter.
Near-rings.

(North-Holland mathematics studies ; 23)
Bibliography: p.
Includes index.
1. Near-rings. I. Title.
QA251.5.P54 1977 512'.4 76-48297
ISBN 0-7204-0566-1

PRINTED IN THE NETHERLANDS

To my
beloved wife
Gerti

INTERDEPENDENCE GUIDE

The numbers indicate the ones of the paragraphes; 7a is §7, section a), and so on. Full lines mean heavy, dotted lines slight dependencies.(§9f is a mere collection of results.)

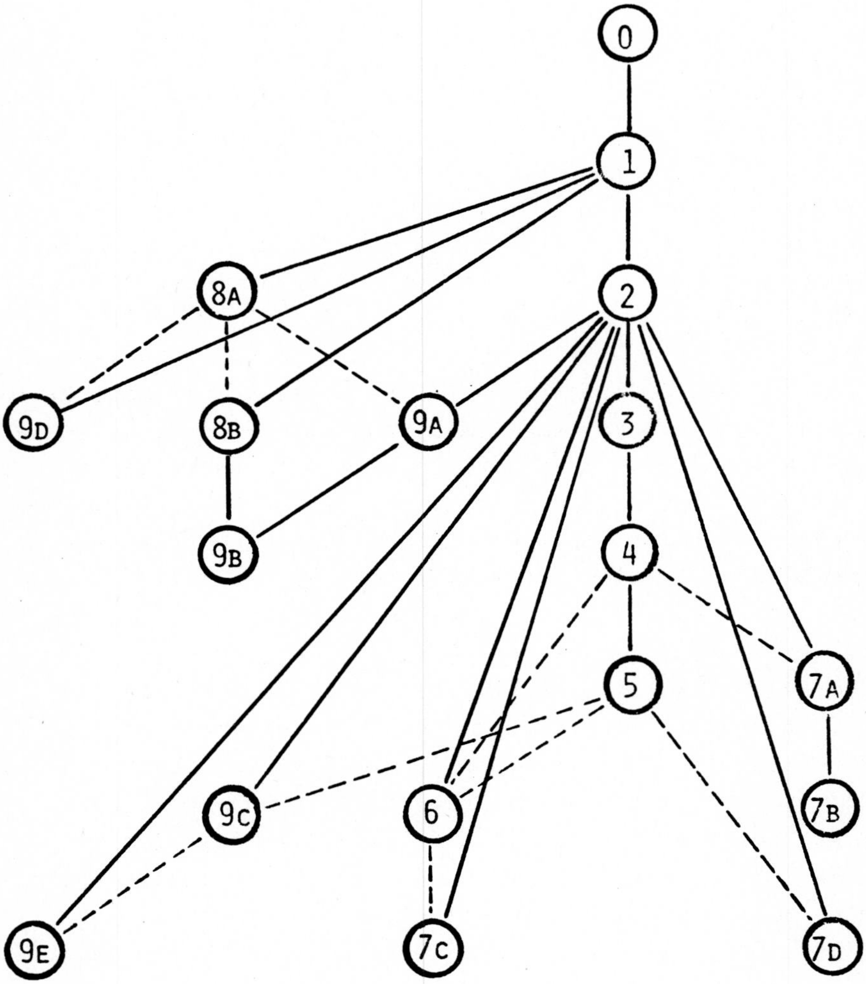

If you want to use this book as a textbook, choose any path you want.

PREFACE

Near-rings are generalized rings. Roughly spoken, a near-ring is "a ring $(N,+,\cdot)$, where + is not necessarily abelian and with only one distributive law".
Near-rings arise in a natural way: take the set $M(\Gamma)$ of all mappings of a group $(\Gamma,+)$ into itself, define addition + pointwise and $\circ$ as composition. Then $(M(\Gamma),+,\circ)$ is a near-ring. Even if Γ is abelian, only one distributive law is always fulfilled: $(f+g)\circ h = f\circ h+g\circ h$ holds by the definition of $f+g$, while for $f\circ(g+h) = f\circ g+f\circ h$ we would have to assume that f is a homomorphism. Another example is supplied by the polynomials w.r.t. addition and substitution.
A well-known result in ring theory says that every ring can be embedded into the ring $E(\Gamma)$ of all endomorphisms of some abelian group Γ. For near-rings we prove (1.86) that every near-ring can be embedded into $M(\Gamma)$ for some group Γ. Hence one might view ring theory as the "linear theory of group mappings", while near-rings provide the "non-linear theory". Surprisingly, a lot of "linear results" can be transferred to the general case after suitable changes. For instance, the "atoms" of ring theory, the primitive rings, were described by the famous density theorem of N. Jacobson (for rings with minimum condition: Wedderburn-Artin theorem on simple rings). For near-rings, similar results concerning primitive near-rings were obtained via the work of several authors (but the proofs are totally different): the rôle of $Hom_D(V,V)$ for rings is played by $M(\Gamma)$ or some related types in the near-ring case (4.52, 4.54).
Historically, the first step toward near-rings was an axiomatic research done by Dickson in 1905. He showed that there do exist "fields with only one distributive law" (= near-fields). Some years later these near-fields showed up again and proved to be useful in coordinatizing certain important

classes of geometric planes (recall that Descarte's method of coordinatizing the "usual" plane by the field of real numbers was one of the most successful steps in geometry). It was Zassenhaus who was able to determine all finite near-fields (8.34). Nowadays, near-fields are a mighty tool in characterizing doubly transitive groups (8.44), incidence groups (8.68) and Frobenius groups (8.81).
Since the sum of two endomorphisms of a non-abelian group $(\Gamma,+)$ is no endomorphism any more, the sets $E(\Gamma)$ of all finite sums and differences of endomorphisms of Γ were considered. With respect to addition and composition, these $E(\Gamma)$'s are near-rings belonging to the class of the "distributively generated" near-rings.
Many parts of the well-established theory of rings were transferred to near-rings and new near-ring-specific features were discovered, building up a theory of near-rings step by step. Up to now, about 550 papers on near-rings (and near-fields) with about 8000 pages appeared in print, but there exists no book on this subject.
This book tries to unify the theory and its terminology and to give a systematic and well-assorted account of the present state of the theory.
Some remarks are to be made:

(a) At the 1972-conference on near-rings in Oberwolfach (Germany), speakers using right near-rings were reviled (for fun - but it happened) by many participiants - including the full-of-shame-author who announces after several "sleepless nights" to use right near-rings. But there are several good reasons for this choice (see 1.2).

(b) Generally, I avoided to give proofs for theorems which are either not along the main stream of discussion or are long ones which contain special methods seemingly applicable only in this context, cannot be simplified by previous results and involve many other (e.g. geometrical) details, but are readily accessible in the literature.

(c) Several results following from universal algebra or from the theory of groups with multiple operators are cited, yet not proved extensively in order to devide between results which are specific for near-rings and those which are not.

Near-rings are already far away of being a mere collection of trivial results concerning some "pathological" system without any application to other branches of mathematics. Apart from the applications concerning axiomatics and geometry mentioned above, special classes of finite near-rings (the "planar" near-rings) give new and highly efficient classes of balanced incomplete block designs already with small parameters (8.117-8.124). Moreover, these planar near-rings can be used to characterize Frobenius groups, hence also finite groups with fixed-point-free automorphism groups (8.96, 8.97). If Γ is a finite, invariantly simple non-abelian group, $E(\Gamma)$ is "primitive" and therefore every self-map of Γ fixing zero is the "sum of endomorphism" (exact formulation in 7.47). Another version of the density theorem 4.52 shows that the density property is (in the near-ring-case) something like an interpolation property, giving the result that if a near-ring N (with some additional properties) of mappings on a group Γ "interpolates" at zero and at two other places then N "interpolates" already at arbitrary (finitely) many points (4.65). Also, near-rings might be the appropriate tool to develop a "non-abelian homological algebra" (9.168) and show up again in algebraic topology (9.167), functional analysis (9.166) and in categories with group objects (9.169). Finally, the author hopes that near-rings and "near-ring modules" (= N-groups) will prove to be useful for a number of theories which try to generalize "linear" results to the "non-linear case", for instance the general systems theory ("If you try to non-linearize, you will find the near-rings nice."). From the ring-theoretical point of view, many bizarre situations arise in near-rings. For example, not every left ideal **is** a subnear-ring. However, there are several implicit applications of near-rings to ring theory (for the near-ring-results show what is

particular to rings and what is not) and to universal algebra (because a high percentage of definitions and results of the near-ring theory carry over to universal algebra).
On quotations: References to other sections are done e.g. by "2d)" meaning "§2, section d)" or by "2d3)" abbreviating "§2, section d), number 3)". Numbers following names of authors refer to the bibliography at the end of the book. If only the author's name is given, all papers of this author cited in the bibliography are meant. This bibliography should be fairly up to date and complete. It was compiled in former years by J. Clay, G. Betsch, J. Malone, H. Heatherly and the author.
Names in brackets refer to the list of "Supplementary works" which contains the non-near-ring-papers cited in this book.

Several results in this monograph are new or in a new (and hopefully improved) form without special notice.
In the beginning of proofs there is no repetition of the assumptions (to save space). "=>" and "<=" mean that the direction indicated is treated at moment (in proofs of equivalences).
Near-rings have the AMS-classification number 16A76, near-fields also 12K05.
It is a pleasure to thank Mr. E. Fredriksson of the editorial staff of the North-Holland Publishing Company and the reviewer for a pleasant cooperation and a lot of useful suggestions. Many thanks go also to Mrs. E. Hospodar for her excellent typing job and to G. Betsch, Y.-S. So, H.E. Bell, J.D.P. Meldrum and to M.L. Holcombe for reading parts of the manuscript and providing useful hints and important comments. Most of all I have to thank my wife for her patience and endurance in living with an absent-mind-husband in the past years.

And now good luck and much fun with near-rings!

Linz, Austria; August 1976 Günter Pilz

CONTENTS

PART II: STRUCTURE THEORY

PART III: SPECIAL CLASSES OF NEAR-RINGS

§0 PREREQUISITES

For the concept of sets we can use any one of the usual set theories with the axiom of choice and using classes. In order to avoid logical difficulties as much as possible, we use statements about classes only as abbreviations of "less obscure ones". For instance, if $\mathcal{F}$ denotes the class of all finite sets, "$F \in \mathcal{F}$" is only an abbreviation for "F is a finite set".

"$\exists x \in A$" stands for "there exists an $x \in A$", "$\exists! x \in A$" for "there exists exactly one $x \in A$" and "$\forall x \in A$" for "for all $x \in A$".

Inclusion will be denoted by $\subseteq$, strict inclusion by $\subset$. $\emptyset$ will denote the (an) empty set and 2^A the power set of A; if A_i ($i \in I$) is a collection of sets, we will write the elements of $\bigtimes_{i \in I} A_i$ as $(\ldots, a_i, \ldots)$, where $a_i \in A_i$. If all $A_i \subseteq A$ then $\bigtimes_{i \in \emptyset} A_i := A$ and also $\bigcap_{i \in \emptyset} A_i := A$. $A \setminus B$ is the set-theoretic difference. If $\sim$ is an equivalence relation on the set A, $A/\sim$ will be the factor set of A w.r.t. $\sim$ and $\pi: A \to A/\sim$ will be the canonical projection.

The sets of all natural numbers will be denoted by $\mathbb{N}$, the natural numbers together with 0 by $\mathbb{N}_0$, the prime numbers by $\mathbb{P}$, the integers by $\mathbb{Z}$, the rationals by $\mathbb{Q}$, the reals by $\mathbb{R}$ and the complex numbers by $\mathbb{C}$.

If f is a function from A to B and if $A_1 \subseteq A$ then $f/_{A_1}$ will be the restriction of f to A_1 and $f(A_1)$ will denote the image of A_1 under f. B^A will be the set of all maps from A to B. If $B \subseteq A$, $\iota: B \to A$ will be reserved for the inclusion map.

If A is any set containing something like a "zero element" 0, A^* will denote $A \setminus \{0\}$. So e.g. $\mathbb{N}_0^* = \mathbb{N}$, while $\mathbb{P}^*$ is not defined.

"Field" will always mean "skew-field". The symmetric (alternating) group on n symbols will be denoted by S_n (A_n, respectively). The integers modulo n will be written as $\mathbb{Z}_n$ and represented by $\mathbb{Z}_n = \{0,1,\ldots,n-1\}$.

We need an abstract version of "generated objects":

0.1 DEFINITION $\mathcal{M} \subseteq 2^A$ is called a Moore-system (Dubreil-Dubreil-Jacotin) on A if
(a) $A \in \mathcal{M}$.
(b) $\mathcal{M}$ is closed w.r.t. arbitrary intersections.

0.2 PROPOSITION If $\mathcal{M}$ is a Moore-system on A and if $B \subseteq A$ then $[B]_{\mathcal{M}} := \bigcap_{\substack{M \in \mathcal{M} \\ M \supseteq B}} M$ is the smallest element of $\mathcal{M}$ (w.r.t. $\subseteq$) containing B.

0.3 DEFINITION Let the notation be as above.
(a) $[B]_{\mathcal{M}}$ is called the element of $\mathcal{M}$ which is generated by B.
(b) $A \in \mathcal{M}$ is called finitely generable (f.g.) if there is a finite subset B of A with $[B]_{\mathcal{M}} = A$.

0.4 DEFINITION A Moore-system $\mathcal{M}$ is called inductive if $\mathcal{M}$ contains the union of every chain of elements of $\mathcal{M}$.

0.5 EXAMPLES
(a) 2^A is an inductive Moore-system on A.
(b) The set of all subgroups of a group Γ is an inductive Moore-system on Γ.
(c) The set of all closed subsets of a topological space T is a Moore-system on T which is not inductive in general.

We now turn to chain conditions.

0.6 DEFINITION A (partially) ordered set $(A,\leq)$ is said to fulfill the minimum condition if every non-empty subset contains (at least) one minimal element.

0.7 PROPOSITION For a partially ordered set $(A,\leq)$, the following conditions are equivalent:

(a) The minimum condition.

(b) The descending chain condition (DCC): every strict chain $a_1 > a_2 > \ldots$ of elements of A terminates after finitely many steps (or, equivalently, for each chain $a_1 \geq a_2 \geq \ldots \ \exists n \in \mathbb{N}: \ a_n = a_{n+1} = \ldots$).

0.8 DEFINITION Linearly ordered sets with the minimum condition are called well-ordered.

0.9 REMARK In replacing $\leq$ by $\geq$, we get the concepts of "maximum condition", "ascending chain condition" (ACC) and "inverse well-order".
Every non-empty subset of an ordered set with the minimum (maximum) condition has the same property.

0.10 PROPOSITION Let $\mathcal{M}$ be a Moore-system on the set A.
$(\mathcal{M},\subseteq)$ fulfills the ACC $\Rightarrow$ every element of $\mathcal{M}$ is f.g..
If $\mathcal{M}$ is inductive, the converse also holds.

Proof. Let $(\mathcal{M},\subseteq)$ have the ACC and assume that some $M\in\mathcal{M}$ is not f.g. and generated by $B\subseteq A$. Take some arbitrary $b_1 \in B$. $[\{b_1\}]_{\mathcal{M}} =: B_1 \neq M$. Take some $b_2 \in M \setminus B_1$ and form $B_2 := [\{b_1, b_2\}]_{\mathcal{M}} \neq M$. Continuing this process, one gets an infinite chain $B_1 \subset B_2 \subset B_3 \subset \ldots$ of elements of $\mathcal{M}$, a contradiction.
Now let $\mathcal{M}$ be inductive and suppose that every element M of $\mathcal{M}$ is f.g.. Assume moreover that $M_1 \subset M_2 \subset M_3 \subset \ldots$ is a strict infinite chain of elements of $\mathcal{M}$. Let $M := \bigcup_{i\in\mathbb{N}} M_i \in \mathcal{M}$ be generated by (say) $\{a_1,\ldots,a_n\}$. But there is some $k\in\mathbb{N}$ with the property that $\{a_1,\ldots,a_n\}\subseteq M_k$, so we get $M_k = M$, which is again a contradiction.

Finally, it should be remarked that in general we use small letters for elements, capitals for sets and script letters for collections of sets.

PART I

NEAR-RINGS FOR BEGINNERS

§ 1 THE ELEMENTARY THEORY OF NEAR-RINGS

a) FUNDAMENTAL DEFINITIONS AND PROPERTIES

Near-rings are generalized rings: addition has not to be commutative and (more important) only one distributive law is postulated.

Examples of near-rings are

(a) the set $M(\Gamma)$ of all mappings on an (additively written) group Γ with pointwise addition and composition;

(b) the polynomials $R[X]$ (R a commutative ring with identity) under addition and substitution;

(c) an arbitrary additively written group with zero multiplication;

as well as many others.

Similar to ring theory, "modules over a near-ring N" ("near-modules" or "N-groups") will be introduced. They play an important rôle in the theory of near-rings. This section contains the basic definitions, examples and properties of near-rings and N-groups, and of substructures and ideal-like objects in these kinds of algebras.

Since near-rings and N-groups (with a zero-symmetric N) are special classes of Ω-groups (groups with multiple operators), a whole bunch of concepts and results is "a priori" available. Compared with ring theory, some complications arise: an element multiplied by 0 is not 0 in general, the characterization of ideals is a little bit more complicated, ideals are not always subalgebras, and so on.

1.) NEAR-RINGS

1.1 DEFINITION A near-ring is a set N together with two binary operations "+" and "." such that

(a) (N,+) is a group (not necessarily abelian)

(b) (N,.) is a semigroup

(c) $\forall\ n_1,n_2,n_3 \varepsilon N$: $(n_1+n_2).n_3 = n_1.n_3+n_2.n_3$ ("right distributive law").

1.2 REMARKS In view of (c), one speaks more precisely of a "right near-ring". Postulating

(c') $\forall\ n_1,n_2,n_3 \varepsilon N$: $n_1.(n_2+n_3) = n_1.n_2+n_1.n_3$

instead of (c), one gets "left near-rings". The theory runs completely parallel in both cases, of course; so one can decide to use just one version.

Although left near-rings are more frequently used in the literature up to now, we will use right near-rings:

- The left distributive law is in some way unnatural in near-rings of functions (the most important examples) and especially unmotivated in near-rings of polynomials and formal power series.
- An ad-hoc-test done by the author showed up that about 80% of the books in which ring-modules play an important rôle use left-modules, which are also more familiar from the theory of vector spaces. In 1.18, we will see that choosing left N-groups forces one to use right near-rings.
- The right distributive law is exclusively used in papers on the closely related concept of composition rings (which were systematically studied prior to near-rings!).

1.3 NOTATION Near-rings will usually be denoted by N,N',N_1 or similar symbols. We abbreviate (N,+,.) by N. Multiplication will in most cases be indicated by juxtaposition; so we write n_1n_2 instead of $n_1.n_2$. In dealing with general near-rings the neutral element of (N,+) will be denoted by 0. $|N|$ will be the order of the near-ring N.

The term "near-ring" will often be abbreviated by "nr.". Throughout this monograph, the class of all near-rings will be denoted by $\mathcal{N}$. If "N" appears, it will always be a near-ring, without further notice.

1.4 EXAMPLES

(a) Let Γ be an additively written (but not necessarily abelian) group with zero o ("omykron"). Then the following sets of mappings from Γ into Γ are nr.'s under pointwise addition and composition:

$M(\Gamma) := \{f : \Gamma \to \Gamma\} = \Gamma^\Gamma$.

$M_o(\Gamma) := \{f : \Gamma \to \Gamma \mid f(o) = o\}$.

$M_c(\Gamma) := \{f : \Gamma \to \Gamma \mid f \text{ is constant}\}$.

$M_c^o(\Gamma) := \{f_\delta : \Gamma \to \Gamma \mid \delta \varepsilon \;\wedge\; f_\delta(\gamma) = \begin{cases} o & \text{if } \gamma = o \\ \delta & \text{if } \gamma \neq o \end{cases}\}$.

(Evidently, Γ, $M_c(\Gamma)$ and $M_c^o(\Gamma)$ are isomorphic groups).

$M_{cont}(\Gamma) := \{f : \Gamma \to \Gamma \mid f \text{ is continuous}\}$ (Γ a topological group).

Another related example is

$M_{diff}(\mathbb{R}) := \{f : \mathbb{R} \to \mathbb{R} \mid f \text{ is differentiable}\}$, while the real functions having an indefinite integral do not form a near-ring (they are not closed w.r.t. composition).

For $S \subseteq End(\Gamma)$ define

$M_S(\Gamma) := \{f : \Gamma \to \Gamma \mid \forall s \in S:\; f \circ s = s \circ f\}$.

Evidently, $M_{\{id\}}(\Gamma) = M(\Gamma)$ and $M_{\{\bar{o}\}}(\Gamma) = M_o(\Gamma)$, where $\bar{o}$ is the zero endomorphism.

These $M_S(\Gamma)$'s will become very important in §4.

(b) Let Γ be as above. Near-rings on Γ are e.g.

$(\Gamma, +, \circ)$ with $\gamma \circ \delta = o$ for all $\gamma, \delta \in \Gamma$;

$(\Gamma, +, *)$ with $\gamma * \delta = \gamma$ for all $\gamma, \delta \in \Gamma$.

More generally, take some subset Δ of Γ and define $\gamma \cdot_\Delta \delta := \begin{cases} \gamma & \text{if } \delta \in \Delta \\ o & \text{if } \delta \notin \Delta \end{cases}$. Then $(\Gamma, +, \cdot_\Delta)$ is a near-ring if $o \notin \Delta$.

(Multiplications of this type are called the "trivial ones" in Malone (3), because they are exactly those ones which can be defined on any group, making this group into a near-ring.)

Now let G be a fixed-point-free automorphism group on Γ (i.e. $\forall g \varepsilon G \, \forall \gamma \varepsilon \Gamma$: $g(\gamma) = \gamma \Rightarrow (\gamma = o \vee g = id)$). Choose any subset $\{B_i | i \varepsilon I\}$ of the set of all non-zero orbits of G on Γ (Betsch called these orbits "<u>1-orbits</u>" and the other ones "<u>0-orbits</u>"); moreover, choose any set of representatives $\{b_i \varepsilon B_i | i \varepsilon I\} =: B$ and define $\gamma \cdot_B \delta$ to be o if $\delta \notin \bigcup_{i \varepsilon I} B_i$ and to be $= g_\delta(\gamma)$ if δ is in some B_i, where g_δ is the unique automorphism in G sending b_i into δ:

Γ

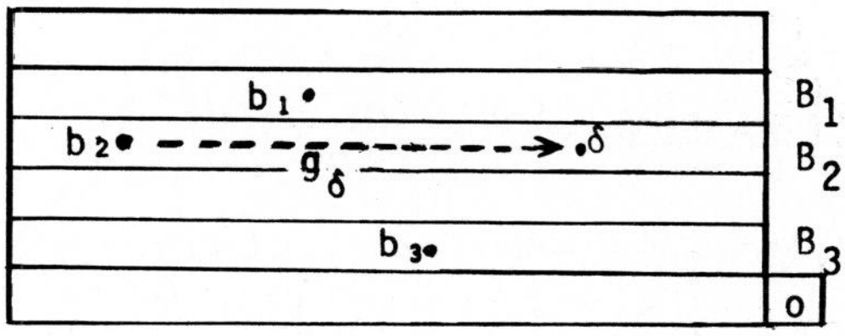

Then $(\Gamma, +, \cdot_B)$ is a near-ring as one sees by looking at the different possible cases. These types of nr.'s were introduced by Ferrero (5) and will prove useful in the theory of planar and integral near-rings. Anyhow, one sees that every group can be made into a near-ring in various ways.

(c) Let V be a vector space over some field F. Call as usual a map $V \to V$ an <u>affine map</u> if it is the sum of a linear and a constant one. The set $M_{aff}(V)$ of all affine maps is again a near-ring (operations as in (a)).

(d) Let R be a commutative ring with identity. Near-rings are $(R[x], +, \circ)$ and $(R[[x]], +, \circ)$, where $\circ$ means substitution. Another near-ring is formed by the set $\overline{R[x]}$ of all polynomial functions on R with the operations as in (a) (see §7d)).

(e) Of course, every ring is a near-ring.

<u>1.5 PROPOSITION</u> $\forall n, n' \varepsilon N$: $On = 0 \wedge (-n)n' = -nn'$.

Proof: as for rings.

1.6 REMARK As most of our examples show, $n0 = 0$ and $n(-n') = -nn'$ do not hold in general. For instance, in $M(\Gamma)$ $f\circ 0 = 0$ means that "f goes through the origin" and $f\circ(-f') = -f\circ f'$ means that "f is an odd function".

One therefore defines for a near-ring N:

1.7 DEFINITION

(a) $N_0 := \{n\in N \mid n0 = 0\}$ is called the zero-symmetric part of N.

(b) $N_c := \{n\in N \mid n0 = n\} = \{n\in N \mid \forall n'\in N: \; nn' = n\}$ is called the constant part of N.

This is motivated by (notation as above)

1.8 EXAMPLES $(M(\Gamma))_0 = M_0(\Gamma)$; $(M(\Gamma))_c = M_c(\Gamma)$.

1.9 DEFINITION $N\in\mathcal{N}$ is called zerosymmetric (constant) if $N = N_0$ ($N = N_c$, respectively).
$\mathcal{N}_0$ ($\mathcal{N}_c$) stand for the classes of all zerosymmetric (constant) near-rings.

1.10 EXAMPLES Elements of $\mathcal{N}_0$ are (notation as in 1.4) $M_0(\Gamma)$, $M_S(\Gamma)$ if $\bar{o}\in S$, $(\Gamma,+,\cdot_B)$, every ring.
$M_c(\Gamma)\in\mathcal{N}_c$, while $M(\Gamma)$ or $R[x]$ are neither in $\mathcal{N}_0$ nor in $\mathcal{N}_c$.

1.11 DEFINITIONS The following concepts are defined as in ring theory: left (right,-) identities, left (right,-) invertible elements, left (right,-) cancellable elements, left (right,-) zero divisors, idempotent and nilpotent elements. Moreover, call $d\in N$ distributive if $\forall n,n'\in N$: $d(n+n') = dn+dn'$. Let $N_d := \{d\in N \mid d$ is distributive$\}$.
Let $\mathcal{N}_1$ be the class of all near-rings with identity (usually denoted by 1).

1.12 EXAMPLES The identity function serves as an identity in $M(\Gamma)$ and $M_0(\Gamma)$. Invertible in these near-rings are exactly the bijective functions. $2x$ is an example of a nilpotent element in $\mathbb{Z}_4[x]$. Cartan (1) characterized all invertible elements in $(F[[x]])_0$, F a field: $\sum_{i=1}^{\infty} a_i x^i$ has an inverse in $(F[[x]])_0$ (w.r.t.$\circ$) iff $a_1 \neq 0$.
If $N = M_{aff}(V)$ then $N_d = \mathrm{Hom}_F(V,V)$. If N is a ring then $N = N_d$. It is clear that $N_d \subseteq N_0$. If N has an identity 1 then $1 \in N_0$.

The next assertion stems from Berman-Silverman (1):

1.13 PROPOSITION If $e \in N$ is idempotent then we get a "Peirce-decomposition":
$\forall n \in N \ \exists x_0 \in \{x \in N \mid xe = 0\} \ \exists x_1 \in Ne$: $\quad n = x_0 + x_1$.
Taking $e = 0$ one gets
$\forall n \in N \ \exists n_0 \in N_0 \ \exists n_c \in N_c$: $\quad n = n_0 + n_c$.
Hence $(N,+) = (N_0,+) + (N_c,+)$ and $N_0 \cap N_c = \{0\}$.
Proof. $n = (n-ne)+ne$ will do the decomposition job.
If $n = x_0+x_1 = x_0'+x_1'$ with $x_0, x_0' \in \{x \in N \mid xe=0\}$ and $x_1 = y_1 e$, $x_1' = y_1' e \in Ne$ then $ne = x_1 e = x_1' e$. But $x_1 e = y_1 ee = y_1 e = x_1$ and $x_1' e = x_1'$. It follows that $x_1 = x_1'$ and $x_0 = x_0'$.

1.14 DEFINITIONS Let N be a near-ring.
If $(N,+)$ is abelian we call N an abelian near-ring; if $(N,.)$ is commutative we call N itself a commutative near-ring. If $N = N_d$, N is said to be distributive. If all non-zero elements of N are left (right,-) cancellable, we say that N fulfills the left (right,-) cancellation law. N is integral if N has no non-zero divisors of zero.
If $(N^* = N \setminus \{0\},.)$ is a group, N is called a near-field (abbreviation: nf.). A near-ring which is not a ring will be referred to as a non-ring. Similarly, a non-field is a nf. which is no field. A near-ring with the property that N_d generates $(N,+)$ is called a distributively generated near-ring (dgnr.).

1.15 EXAMPLES (Notation as in 1.4) $M(\Gamma)$ is abelian iff Γ is abelian. $(\Gamma,+,\circ)$ serves as an example of a commutative and distributive non-ring, while $(\Gamma,+,*)$ is integral. In the language of 1.4(b), $(\Gamma,+,\bullet_B)$ is integral iff all non-zero orbits are "1-orbits". $(\mathbb{Z}_2,+)$ with $0\cdot 0=0\cdot 1=0$, $1\cdot 0 = 1\cdot 1 = 1$ is a nf. All other nf's are zero-symmetric. Let Γ be a group. If Γ is not abelian, the sum of two endomorphisms is not necessarily an endomorphism any more. But the set of all (finite) sums and differences of endomorphisms of Γ is closed under addition and composition and forms a dgnr. $E(\Gamma)$.

1.16 HISTORIC REMARKS Near-fields were the first nr's considered in the literature. In 1905, Dickson (1),(2) changed the multiplication in a field in order to get examples of "one-sided distributive fields" (= nf's) showing that the second distributive law does not follow from the remaining axioms for a (skew-)field. His "changed fields" are called "Dickson nf's" (see §8(a)4).

A couple of years later Veblen and Wedderburn (1) started to use nf's to coordinatize certain kinds of geometric planes.

In 1936, Zassenhaus (1) determined all finite nf's: they have order p^n ($p\in\mathbb{P}$, $n\in\mathbb{N}$) and are (up to 7 exceptional cases) Dickson nf's. In (2) he showed up the connection between nf's and fixed-point free permutation groups.

Ore (1), Furtwängler-Taussky (1) and Taussky (1) started axiomatic considerations in the thirties for what we now call near-rings.

A first name for these structures was proposed in 1938 by Wielandt (1): "Stamm" (=tribe) ("stem" is still used in the Italian literature). Wielandt also announced structure-theoretic results in this note.

The first ones to use the name "near-ring" were Zassenhaus in 1936 and Blackett and P.Jordan in 1950.

In 1932 Fitting (1) characterized those automorphisms of (non-abelian) groups, whose sum is an automorphism, too, thereby implicitely starting to consider dgnr's.

Finally, the fifties brought the start of a rapid development of the theory of near-rings.

Now we are going to define the analogue of the concept of a module in ring theory: certain operator groups.

2.) N-GROUPS

1.17 DEFINITIONS Let $(\Gamma,+)$ be a group with zero o and let $N \in \mathfrak{N}$. Let $\mu: N\times\Gamma \longrightarrow \Gamma$, $(n,\gamma) \to n\gamma$. (Γ,μ) is called an N-group ("near-module over N" (but cf. the different meaning e.g. in Karzel-Pieper (1))) if

$\forall\ \gamma\in\Gamma \quad \forall\ n,n'\in N : (n+n')\gamma = n\gamma+n'\gamma \wedge (nn')\gamma = n(n'\gamma)$.

If the meaning of μ is clear we write ${}_N\Gamma$ for the N-group above. Let ${}_N\mathfrak{g}$ be the class of all N-groups. To simplify formulations, ${}_N\Gamma$ stands for N-groups throughout, without further notice.

1.18 EXAMPLES

(a) Let N be a nr. Then $\mu: N\times N \longrightarrow N$, $(n,n') \to nn'$ makes $(N,+)$ into an N-group, denoted by ${}_NN$.

(b) Each (left) module M over a ring R is an R-group.

(c) Let Γ be a group. Then Γ is an $M(\Gamma)$-group ${}_{M(\Gamma)}\Gamma$ with $\mu: M(\Gamma)\times\Gamma \to \Gamma$, $(f,\gamma) \to f(\gamma)$.

1.19 PROPOSITION Take ${}_N\Gamma \in {}_N\mathfrak{g}$.

(a) $\forall\gamma\in\Gamma;\ 0\gamma = o$;

(b) $\forall\gamma\in\Gamma \quad \forall n\in N: (-n)\gamma = -n\gamma$;

(c) $\forall n\in N_0: no = o$;

(d) $\forall\gamma\in\Gamma \quad \forall n\in N_c: n\gamma = no$.

Proof. (a) and (b): as for (ring-) modules.

(c): $no = n0o = 0o = o$.

(d): $n\gamma = n0\gamma = no$.

<u>1.20 DEFINITION</u> ${}_N\Gamma \in {}_N\mathcal{G}$ is called <u>unitary</u> if $N \in \mathcal{N}_1$ and $\forall\ \gamma \in \Gamma$: $1\gamma = \gamma$.

Since $\mathcal{N}$, $\mathcal{N}_0$, $\mathcal{N}_c$, and all ${}_N\mathcal{G}$ are varieties in the sense of universal algebra it makes sense to speak about a lot of things:

3.) SUBSTRUCTURES

<u>1.21 DEFINITION</u>

(a) A subgroup M of a nr. N with $M.M \subseteq M$ is called a <u>subnear-ring</u> of N (notation: $M \leq N$).

(b) A subgroup Δ of ${}_N\Gamma$ with $N\Delta \subseteq \Delta$ +) is said to be an <u>N-subgroup</u> of Γ ($\Delta \leq_N \Gamma$).

<u>1.22 EXAMPLES</u>

(a) N_0 and N_c are subnear-rings of N. Hence it follows from 1.13 that $(N,+)$ is a "splitting extension" of its subgroups $(N_0,+)$ and $(N_c,+)$. See Pilz (9),(10) for the converse problem of constructing near-rings out of a zero-symmetric and a constant one.

(b) If ${}_N\Gamma$ is a (ring-) module then the N-subgroups are just the submodules of Γ.

Later on we will see that the subnear-rings of the $M(\Gamma)$'s are in a certain sense already all near-rings. We know already one procedure to get subnear-rings of $M(\Gamma)$: the $M_S(\Gamma)$'s of 1.4. Two more methods are:

<u>1.23 EXAMPLES</u>

(a) Take a subgroup Δ of Γ. $M_\Delta(\Gamma) := \{f \in M(\Gamma) \mid f(\Delta) \subseteq \Delta\}$ is a subnear-ring of $M(\Gamma)$.

(b) Take a normal subgroup Δ of Γ. $M_{\Gamma/\Delta}(\Gamma) := \{f \in M(\Gamma) \mid \forall\ \gamma \in \Gamma: f(\gamma+\Delta) \subseteq f(\gamma)+\Delta\}$ is a subnear-ring of $M(\Gamma)$ (cf. Betsch (3)).

+) The term "<u>N-subgroup of N</u>" refers to ${}_N N$.

1.24 REMARK Wielandt (3) proposed a construction method for subnear-rings of $M(\Gamma)$ which gives the 3 kinds of subnear-rings mentioned above as special cases.

The method is as follows:

Take any cardinal number α, form the direct product Γ^{α} and a subgroup Δ of Γ^{α}. Each $f \in M(\Gamma)$ can be considered to be $\in M(\Gamma^{\alpha})$ if it is defined component-wise.

Let $M_{\alpha,\Delta}(\Gamma) := \{f \in M(\Gamma) \mid f(\Delta) \subseteq \Delta\} \leq M(\Gamma)$. Then

(a) $M_{\Delta}(\Gamma) = M_{1,\Delta}(\Gamma)$

(b) $M_{\{s_1,\ldots,s_\alpha\}}(\Gamma) = M_{\alpha+1,\Delta}(\Gamma)$ with $\Delta = \{(\gamma, s_1(\gamma), \ldots, s_\alpha(\gamma) \mid \gamma \in \Gamma\}$

(c) $M_{\Gamma/\Delta}(\Gamma) = M_{2,E}(\Gamma)$ with $E = \{(\gamma,\gamma') \mid \gamma - \gamma' \in \Delta\}$

4.) HOMOMORPHISMS AND IDEAL-LIKE SUBSETS

1.25 DEFINITION Let N, N' be $\in \mathcal{N}$ and ${}_N\Gamma, {}_N\Gamma' \in {}_N\mathcal{G}$.

(a) $h: N \to N'$ is called a (near-ring) homomorphism if
$\forall\, m,n \in N$: $h(m+n) = h(m) + h(n) \wedge h(mn) = h(m)h(n)$.

(b) $h: {}_N\Gamma \to {}_N\Gamma'$ is called an N-homomorphism if
$\forall\, \gamma,\delta \in \Gamma \quad \forall\, n \in N$: $h(\gamma+\delta) = h(\gamma) + h(\delta) \wedge h(n\gamma) = nh(\gamma)$.

There seems to be no need for explicit definitions of nr.-monomorphisms ($N \rightarrowtail N'$), $\mathrm{Hom}(N,N')$, $\mathrm{Hom}_N(\Gamma,\Gamma')$, $\Gamma \cong {}_N\Gamma'$, Ker h, Im h, and so on. If there exists a monomorphism $N \rightarrowtail N'$ we say that N is embeddable in N' and write $N \hookrightarrow N'$. A similar convention applies to N-groups.

1.26 EXAMPLE For all $\gamma \in {}_N\Gamma$: $h_\gamma: \begin{array}{l} N \to \Gamma \\ n \to n\gamma \end{array} \in \mathrm{Hom}_N(N,\Gamma)$

1.27 DEFINITION Let $N \in \mathcal{N}$ and ${}_N\Gamma \in {}_N\mathcal{G}$.

(a) A normal subgroup I of $(N,+)$ is called ideal of N ($I \trianglelefteq N$) if

α) $IN \subseteq I$

β) $\forall\, n,n' \in N \quad \forall\, i \in I$: $n(n'+i) - nn' \in I$.

Normal subgroups R of $(N,+)$ with α) are called right

<u>ideals</u> of N (R $\trianglelefteq_r$ N), while normal subgroups L of (N,+) with β) are said to be <u>left ideals</u> (L $\trianglelefteq_\ell$ N).

(b) A normal subgroup Δ of Γ is called <u>ideal of $_N\Gamma$</u> ($\Delta \trianglelefteq_N \Gamma$) if $\forall\, \gamma\in\Gamma\ \forall\, \delta\in\Delta\ \forall n\in N$: $\overline{n(\gamma+\delta)-n\gamma\in\Delta}$. Other names: N-kernel or submodule (cf. 1.33!). The term "ideal" is motivated by (Kurosh) and is very handy in formulating simultaneous statements about N-groups and near-rings.

<u>1.28 REMARKS</u> The left ideals of N coincide with the ideals of $_NN$. Moreover, one easily sees that a subgroup I of N (Δ of Γ) is an ideal iff

$$\left.\begin{array}{l} n_1 \equiv n_1' \pmod I \\ n_2 \equiv n_2' \pmod I \end{array}\right\} \Rightarrow n_1+n_2 \equiv n_1'+n_2' \pmod I \wedge n_1n_2 \equiv n_1'n_2' \pmod I$$

$$\left(\begin{array}{l} \gamma_1 \equiv \gamma_1' \pmod \Delta \\ \gamma_2 \equiv \gamma_2' \pmod \Delta \end{array}\right. \Rightarrow \gamma_1+\gamma_2 \equiv \gamma_1'+\gamma_2' \pmod \Delta \wedge$$

$\wedge\ \forall n\in N$: $n\gamma_1 \equiv n\gamma_1' \pmod\Delta$, respectively).

So ≡ (mod I or mod Δ) is a "congruence relation" (cf. (Grätzer)) if I (Δ) is an ideal. If $I \trianglelefteq N$ and $I \neq N$, we write $I \triangleleft N$, etc.

In 1.27, (a)β) and (b) can also be written as

$\forall n,n'\in N\ \ \forall i\in I$: $n(i+n')-nn'\in I$ and

$\forall \gamma\in\Gamma\ \ \forall\delta\in\Delta\ \ \forall n\in N$: $n(\delta+\gamma)-n\gamma\in\Delta$.

<u>Factor nr's</u> N/I ($I \trianglelefteq N$) and <u>factor N-groups</u> Γ/Δ ($\Delta \trianglelefteq_N \Gamma$) are defined as usual (cf. any book on universal algebra). If $L \trianglelefteq_\ell N$, then N/L is meant in the sense of N-groups.

Clearly {0} and N are ideals of N as well as {o} and Γ are ones of $_N\Gamma$. These ideals are called the <u>trivial ideals</u>.

1.29 THEOREM ("Homomorphism theorem").

(a) If $I \trianglelefteq N$ then the canonical map $\pi: N \to N/I$ is a nr-epimorphism. So N/I is a homomorphic image of N.

(b) Conversely, if $h: N \twoheadrightarrow N'$ is an epimorphism then $\operatorname{Ker} h \trianglelefteq N$ and $N/_{\operatorname{Ker} h} \cong N'$.

The corresponding statements hold for N-groups.

The proof is analogous to the one for groups, rings or universal algebras, and hence omitted.

So ideals are just the kernels of (N-) homomorphisms.
As usual for "sophisticated" algebraic structures we get with the usual proof:

1.30 THEOREM (so-called "2nd isomorphism theorem")

Let $h: N \twoheadrightarrow N'$ be an epimorphism. Then h induces a 1-1-correspondence between the subnear-rings (ideals) of N containing Ker h and the subnear-rings (ideals) of N' by $A (\subseteq N) \to h(A)$:

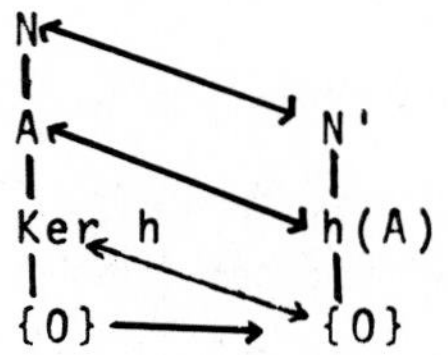

Moreover, for all ideals I of N containing Ker h we get

$$N/I \cong h(N)/h(I).$$

If $\pi: N \to N/I$ is the canonical epimorphism, we therefore get for all ideals J of N containing I

$$N/I\big/_{J/I} \cong N/J .$$

Again the analogous statements hold for N-groups. Observe in this case that for the last formula we have to assume that J is also an N-group to make J/I meaningful (cf. 1.33, 1.34).

<u>1.31 DEFINITION</u> A subnear-ring M of N is called <u>invariant</u> if $MN \subseteq M$ and $NM \subseteq M$.

Invariant subnear-rings and ideals coincide in rings, but not in near-rings:

<u>1.32 PROPOSITION</u>

(a) $N_o \trianglelefteq_\ell N$, but not generally $N_o \trianglelefteq N$.

(b) N_c is an invariant subnear-ring of N, but in general neither a right nor a left ideal.

<u>Proof</u>. (a) N_o is a left ideal: for all $n,n' \in N$ and $n_o \in N_o$ we have $(n+n_o-n)0 = n0+n_o0-n0 = 0$, so $n+n_o-n \in N_o$, and $[(n(n'+n_o)-nn')]0 = n(n'0+n_o0)-nn'0 = 0$, hence $n(n'+n_o)-nn' \in N_o$.

N_o is not necessarily an ideal: $N := M(\mathbb{R})$, $id_{\mathbb{R}} \in N_o = M_o(\mathbb{R})$, $\underline{1}: \begin{matrix}\mathbb{R} \to \mathbb{R} \\ x \to 1\end{matrix} \in M(\mathbb{R})$, but $id \circ \underline{1} = \underline{1} \notin M_o(\mathbb{R})$.

(b) N_c is an invariant subnear-ring:

$\forall\, n_c \in N_c \quad \forall\, n \in N$: $(nn_c)0 = nn_c$ and $(n_c n)0 = n_c 0 = n_c n$, which implies that $nn_c \in N_c$ and $n_c n \in N_c$.

N_c is not a left or right ideal in general, since N_c is not always a normal subgroup of $(N,+)$:
Take a non-abelian group Γ and $\gamma,\delta \in \Gamma$ with $\gamma+\delta \neq \delta+\gamma$. $f_\gamma: \begin{matrix}\Gamma \to \Gamma \\ x \to \gamma\end{matrix} \in M_c(\Gamma)$. Now $(id+f_\gamma-id)(o) = \gamma$, but $(id+f_\gamma-id)(\delta) = \delta+\gamma-\delta \neq \gamma$ implying that $id+f_\gamma-id \notin M_c(\Gamma)$. So $M_c(\Gamma)$ is normal iff Γ is abelian.

<u>1.33 REMARK</u> In general there is no direct connection between N-subgroups and left ideals, as we have seen above. This is the reason for avoiding the terms "near-modules" and "submodules": submodules would be no near-modules in general, for ideals of N-groups are not necessarily N-subgroups. So in general N-groups are no "Ω-groups" ("groups with multiple operators") in the sense of (Kurosh) or (Higgins). This does not happen for zero-symmetric near-rings:

1.34 PROPOSITION

(a) $L \trianglelefteq_\ell N \Rightarrow N_0L \subseteq L$

(b) $N = N_0 \Leftrightarrow$ each left ideal of N is an N-subgroup of N.

(c) $N = N_0 \Rightarrow (\Delta \trianglelefteq_N \Gamma \Rightarrow \Delta \leq_N \Gamma)$.

Proof. (a) $L \trianglelefteq_\ell N \Rightarrow \forall \ell \in L \quad \forall n_o \in N_o : n_o \ell = n_o(0+\ell) - n_o 0 \in L$.

(b) $\Rightarrow$: by (a)

$\Leftarrow$: $\{0\} \trianglelefteq_\ell L \Rightarrow \{0\} \leq_N N \Rightarrow N0 = \{0\} \Rightarrow N = N_0$.

(c) is settled similarly.

1.35 PROPOSITION

(a) $\forall \gamma \in {}_N\Gamma: \quad N\gamma \leq_N \Gamma$.

(b) $\forall \Delta \leq_N \Gamma : \quad No = N_c o \subseteq \Delta$.

So No is the smallest under all N-subgroups of ${}_N\Gamma$. Throughout this monograph we will write

$$No = N_c o =: \Omega .$$

Of course, $N = N_0$ implies $\Omega = \{o\}$. By 1.19(d), $\forall \gamma \in \Gamma: \Omega = N_c \gamma$.

1.36 DEFINITION

(a) $N({}_N\Gamma)$ is simple: $\Leftrightarrow$ $N({}_N\Gamma)$ has no non-trivial ideals.

(b) ${}_N\Gamma$ is called N-simple: $\Leftrightarrow$ ${}_N\Gamma$ has no N-subgroups except Ω and Γ (cf. 1.35).

1.37 PROPOSITION If $N({}_N\Gamma)$ is simple then all (N-) homomorphic images are (N-) isomorphic either to $\{0\}$ or to $N(\{o\}$ or $\Gamma)$.

Proof: by 1.29.

1.38 EXAMPLES

(a) In §7 we will see that $M(\Gamma)$ ($|\Gamma| > 2$) and $M_o(\Gamma)$ are simple nr.'s (7.30, 7.33).

(b) See Blackett (4) for some more examples of simple nr.'s of real functions.

(c) If $N = N_o$ then N-simplicity implies (by 1.34(c)), simplicity for each ${}_N\Gamma \epsilon {}_N\mathcal{G}$.

Since {0} is always minimal in the set of all ideals of N, we define more interesting ones to be minimal:

1.39 DEFINITION A minimal ideal of N is an ideal which is minimal in the set of all non-zero ideals. Similarly, one defines minimal right ideals, left ideals, N-subgroups (minimal under all N-subgroups $\neq \Omega$), etc. . Dually, one gets the concepts of maximal ideals etc. .

1.40 PROPOSITION $I \lhd N$ is maximal in N ($\Delta \lhd_N \Gamma$ is maximal in Γ) iff N/I (Γ/Δ) is simple.

Proof: 1.30.

5.) ANNIHILATORS

We will need the "noetherian quotients" quite frequently:

1.41 DEFINITION Let Δ_1, Δ_2 be subsets of ${}_N\Gamma \epsilon {}_N\mathcal{G}$.
$(\Delta_1 : \Delta_2) := \{n \epsilon N \mid n\Delta_2 \subseteq \Delta_1\}$.
Abbreviations: $(\{\delta\} : \Delta_2) =: (\delta : \Delta_2)$, similarly for $(\Delta:\delta)$, $(\delta:\Delta)$.
$(o:\Delta)$ is called the annihilator of Δ.
If necessary, we indicate the nr. N involved by writing $(\Delta_1 : \Delta_2)_N$.

1.42 PROPOSITION Notation as above.

If Δ_1 is a subgroup (normal subgroup, N-subgroup, ideal), the same applies to $(\Delta_1 : \Delta_2)$.

The proof is easy and therefore omitted.

1.43 COROLLARY

(a) $\forall\ \gamma \in \Gamma : (o:\gamma) \trianglelefteq_{\ell} N$

(b) $\forall\ \Delta \leq_N \Gamma : (o:\Delta) \trianglelefteq N$

1.44 PROPOSITION Let Δ, Δ_i $(i \in I)$ be subsets of ${}_N\Gamma$. Then $\bigcap_{i\in I} (\Delta_i : \Delta) = (\bigcap_{i\in I} \Delta_i : \Delta)$ and $\bigcup_{i\in I} (\Delta_i : \Delta) \subseteq (\bigcup_{i\in I} \Delta_i : \Delta)$.

For $n \in \bigcap_{i\in I} (\Delta_i : \Delta) \Longleftrightarrow \forall i \in I : n\Delta \subseteq \Delta_i \Longleftrightarrow n\Delta \subseteq \bigcap_{i\in I} \Delta_i \Longleftrightarrow$

$\Longleftrightarrow n \in (\bigcap_{i\in I} \Delta_i : \Delta)$ and similarly for the union.

1.45 PROPOSITION Let Δ be a subset of ${}_N\Gamma \in {}_N\mathcal{G}$.

(a) $(o:\Delta) = \bigcap_{\delta\in\Delta} (o:\delta)$

(b) ${}_N\Gamma \cong_N {}_N\Gamma' \Rightarrow (0:\Gamma) = (0:\Gamma')$

Proof: straightforward.

Consider the h_γ's of 1.26.

1.46 PROPOSITION $\operatorname{Ker} h_\gamma = (o:\gamma)$, so $N\gamma \cong N/(o:\gamma)$.

Proof: homomorphism theorem.

1.47 DEFINITION ${}_N\Gamma$ is called faithful if $(o:\Gamma) = \{0\}$.

1.48 PROPOSITION ${}_N\Gamma$ faithful $\Rightarrow N \hookrightarrow M(\Gamma)$.

Proof: Consider for each $n \in N$ the map $f_n : \begin{array}{l}\Gamma \to \Gamma \\ \gamma \to n\gamma\end{array}$. $f_n \in M(\Gamma)$.

Then $h : \begin{array}{l} N \to M(\Gamma) \\ n \to f_n \end{array}$ turns out to be a near-ring homomorphism with $\operatorname{Ker} h = \{n \in N \mid f_n = \bar{o}\} = \{n \in N \mid n \in (o:\Gamma)\} = \{0\}$.

1.49 PROPOSITION Let ${}_N\Gamma$ be faithful.

(a) If Γ is abelian then so is N.

(b) If $\forall\, n\in N \quad \forall\, \gamma,\delta\in\Gamma : n(\gamma+\delta) = n\gamma+n\delta$ then $n\in N_d$.

Proof: (a) by 1.48 and (b) by a straightforward calculation.

More generally one can prove that, if ${}_N\Gamma$ is faithful, every "identity which holds in Γ" (cf. (Grätzer)) also "holds in N".

1.50 PROPOSITION Let ${}_N\Gamma$ be faithful. We assume that $N \subseteq M(\Gamma)$ (by 1.48).

(a) $\Omega = \{o\} \iff N_c = \{0\} \iff N\in\mathfrak{N}_0$;

(b) $\Omega = \Gamma \iff N_c = M_c(\Gamma) \iff \Gamma \cong_N N_c$.

Proof: (a) If $\Omega = \{o\}$ then $\forall\, \gamma\in\Gamma \quad \forall\, n_c\in N_c$: $n_c(\gamma) =$ $= n_c(o) = o = \bar{o}(\gamma)$, so $n_c = \bar{o}$.

If $N_c = \{0\}$ then $N = N_0\in\mathfrak{N}_0$.

If $N\in\mathfrak{N}_0$ then $\Omega = \{o\}$ by 1.19(c) .

(b) If $\Omega = \Gamma$, take some $m\in M_c(\Gamma)$. $m(o) =: \gamma_0$.

Then $\exists\, n_c\in N_c : n_c(o) = \gamma_0$.

So $\forall\, \gamma\in\Gamma : m(\gamma) = m(o) = \gamma_0 = n_c(o) = n_c(\gamma)$, hence $m = n_c$ and $N_c = M_c(\Gamma)$.

If $N_c = M_c(\Gamma)$ then the map $h: \begin{matrix} N_c \to \Gamma \\ n_c \to n_c(o) \end{matrix}$ is an N-isomorphism.

Finally, if $\Gamma \cong_N N_c$ by some N-isomorphism h, take an arbitrary $\gamma\in\Gamma$. $h(\gamma) =: n_c\in N_c$. Then $h(\gamma) =$ $= n_c = n_c n_c = n_c h(\gamma) = h(n_c(\gamma)) = h(n_c(o))$.

So $\gamma = n_c(o) \in \Omega$ and $\Omega = \Gamma$.

6.) GENERATED OBJECTS

1.51 PROPOSITION

(a) The sets of all ideals (right ideals, left ideals, N-subgroups, invariant subnear-rings) form inductive Moore-systems on N.

(b) The sets of all ideals (N-subgroups) of an N-group ${}_N\Gamma$ form inductive Moore-systems on Γ.

Hence it makes sense to speak about the "ideal (...) generated by a subset".

1.52 PROPOSITION (Scott (5)) Let $R \subseteq N$ with $RN \subseteq R$. Then the left ideal L_R generated by R is an ideal.

Proof: $RN \subseteq R \subseteq L_R$, so $R \subseteq (L_R:N)$. Since $(L_R:N) \trianglelefteq_\ell N$ by 1.42, $L_R \subseteq (L_R:N)$. Therefore $L_R N \subseteq L_R$ showing that $L_R \trianglelefteq N$.

See also 2.16.

1.53 THEOREM (Beidleman (1))

(a) If N is fg. (0.3) (e.g. if $N \in \mathcal{N}_1$) then each ideal (right ideal, N-subgroup) different from N is contained in a maximal one.

(b) If ${}_N\Gamma$ is fg. ($N \in \mathcal{N}_0$) then every ideal (N-subgroup) of ${}_N\Gamma$, but unequal to Γ, is contained in a maximal one.

Proof (for ideals I of N). Let N be generated by (say) $x_1, \ldots, x_k$. $\mathcal{I} := \{I \mid I \triangleleft N\}$. Take some chain $I_1 \subseteq I_2 \subseteq \ldots$ of elements of $\mathcal{I}$. $I := \bigcup_{n \in \mathbb{N}} I_n \trianglelefteq N$ by 1.51(a).

If $I = N$ then all of $x_1, \ldots, x_k \in I$. Hence there is some $s \in \mathbb{N}$ with $x_1, \ldots, x_k \in I_s$. But then $I_s = N$, a contradiction.

So $(\mathcal{I}, \subseteq)$ fulfills the hypothesis of Zorn's Lemma (unless $N = \{0\}$, a trivial case) and consequently contains a maximal element.

If $N \in \mathcal{N}_1$, one proceeds as in ring theory.

b) CONSTRUCTIONS

1.) PRODUCTS, DIRECT SUMS AND SUBDIRECT PRODUCTS

For 1.54 - 1.60 cf. each book on groups, rings or universal algebra. We cite e.g. from (Grätzer).

1.54 DEFINITION Let $(N_i)_{i\in I}$ be a family of near-rings. $\bigtimes_{i\in I} N_i$ with the component-wise defined operations "+" and "•" is called the direct product $\prod_{i\in I} N_i$ of the near-rings N_i $(i\in I)$.

1.55 DEFINITION The subnear-ring of $\prod_{i\in I} N_i$ consisting of those elements where all components - except a finite number $\in \mathbb{N}_0$ - equal to zero, is called the (external) direct sum $\bigoplus_{i\in I} N_i$ of the N_i's.

More generally, every subnear-ring N of $\prod_{i\in I} N_i$ where all projection maps π_i $(i\in I)$ are surjective (in other words, $\forall\, i\in I$ $\forall\, n_i\in N_i$: n_i is the i-th component of some element of N) is called a subdirect product of the N_i's.

The definitions of products, direct sums and subdirect products of N-groups should be clear now (for direct sums you need $N = N_0$).

1.56 NOTATION If the N_i $(i\in I)$ are as above, let $\bar{N}_i$ be given by $\bar{N}_i := \{(\ldots,0,n_i,0,\ldots) \mid n_i\in N_i\}$.

1.57 PROPOSITION

(a) $\forall\, i \in I: N_i \cong \overline{N}_i \wedge \overline{N}_i \trianglelefteq \bigoplus_{j \in I} N_j \wedge \overline{N}_i \trianglelefteq \prod_{j \in I} N_j \wedge$

$\wedge\, N_i \hookrightarrow \bigoplus_{j \in I} N_j \wedge N_i \hookrightarrow \prod_{j \in I} N_j$;

(b) $\bigoplus_{i \in I} N_i \trianglelefteq \prod_{i \in I} N_i$;

(c) $J \subseteq I \Rightarrow \bigoplus_{j \in J} N_j \hookrightarrow \bigoplus_{i \in I} N_i \wedge \prod_{j \in J} N_j \hookrightarrow \prod_{i \in I} N_i$.

1.58 REMARKS (cf. (Grätzer)). If N is a subdirect product of near-rings N_i $(i \in I)$ then the N_i's are homomorphic images of N (under the projection maps π_i). If Ker $\pi_i =: K_i$ we get a family $(K_i)_{i \in I}$ of ideals of N with zero intersection.

Conversely, if a family of ideals $(K_i)_{i \in I}$ of some nr. N with $\bigcap_{i \in I} K_i = \{0\}$ is given then N is isomorphic to a subdirect product of the near-rings $N_i := N/K_i$.

Of course, 1.56 - 1.58 can be transferred to N-groups in the obvious way.

1.59 DEFINITION A subdirect product N of near-rings N_i $(i \in I)$ is called trivial if $\exists\, i \in I : \pi_i$ is an isomorphism.

$N \in \mathcal{N}$ is called subdirectly irreducible if N is not isomorphic to a non-trivial subdirect product of near-rings.

The same is defined for N-groups.

1.60 THEOREM ((Grätzer), Fain(1)). The following conditions for a nr. $N \neq \{0\}$ are equivalent:

(a) N is subdirectly irreducible;

(b) If $(I_\alpha)_{\alpha \in A}$ is a family of ideals of N with $\bigcap_{\alpha \in A} I_\alpha = \{0\}$ then $\exists\, \alpha \in A : I_\alpha = \{0\}$;

(c) $\bigcap_{\{0\} \neq I \trianglelefteq N} I \neq \{0\}$;

(d) N contains a unique minimal ideal, contained in all other non-zero ideals.

Replacing "N" by "$_N\Gamma$" yields an analogous theorem for N-groups.

1.61 COROLLARY Each simple nr. (N-group) is subdirectly irreducible.

1.62 THEOREM ((Grätzer), p.124).

(a) Each near-ring is isomorphic to a subdirect product of subdirectly irreducible near-rings.

(b) Each N-group is N-isomorphic to a subdirect product of subdirectly irreducible N-groups.

2.) NEAR-RINGS OF QUOTIENTS

1.63 DEFINITION Let N be a nr. and S a subsemigroup of $(N,\cdot)$. A near-ring N_S is called a <u>near-ring of left (right) quotients</u> of N w.r.t. S if

(a) $N_S \in \mathcal{N}_1$

(b) $N \hookrightarrow N_S$ (by h, say)

(c) $\forall\ s \in S$: $h(s)$ is invertible in $(N_S,\cdot)$

(d) $\forall\ q \in N_S\ \ \exists\ s \in S\ \ \exists\ n \in N : q = h(n)h(s)^{-1}\ \ (q = h(s)^{-1}h(n))$.

Of course there arise the questions about existence and uniqueness of such near-ring of quotients. We will settle these questions after the following

1.64 DEFINITION N is said to fulfill the <u>left (right) Ore condition</u> (Ore (1)) w.r.t. a given subsemigroup S of $(N,\cdot)$ if

$$\forall\ (s,n) \in S\times N\ \ \exists\ (s_1,n_1) \in S\times N :\ ns_1 = sn_1 \quad (s_1 n = n_1 s)$$

<u>1.65 THEOREM</u> (Graves-Malone (1)). Let S be a subsemigroup of $(N,\cdot)$. N has a nr. of left quotients w.r.t. S $\iff$

$\iff$ (a) $S \neq \emptyset$

(b) $\forall\, s \in S$: s is (left and right) cancellable

(c) N satisfies the left Ore condition w.r.t. S.

<u>Proof</u>. $\Rightarrow$: Assume that N has a nr. N_S of left quotients w.r.t. S, and let h be as in 1.63.

(a): By 1.63(d)

(b): $\forall\, s \in S\ \forall\, m,n \in N : ms = ns \Rightarrow h(m)h(s) = h(n)h(s) \Rightarrow$
$\Rightarrow h(m) = h(n) \Rightarrow m = n$.
Similarly, $sm = sn \Rightarrow m = n$.

(c): Take $n \in N$, $s \in S$. $q := h(s)^{-1}h(n) \in N_S$, so by 1.63(d)
$\exists\, n_1 \in N\ \exists\, s_1 \in S : h(s)^{-1}h(n) = h(n_1 h(s_1)^{-1}$;
Therefore $h(ns_1) = h(sn_1)$, whence $ns_1 = sn_1$.

$\Leftarrow$: Similar to ring theory (cf. (N. Jacobson), p. 262):
Define an equivalence relation $\sim$ on $N\times S$ by
$(n,s)\sim(n',s') :\iff \exists\, n_1 \in N\ \exists\, s_1 \in S : (ss_1 = s'n_1 \Rightarrow$
$\Rightarrow ns_1 = n'n_1)$.

$N\times S/\sim\ =: N_S$. Call the equivalence class containing (n,s) "$\frac{n}{s}$" . Define on N_S two operations:

$\frac{n}{s}+\frac{n'}{s'} := \frac{ns_1+n'n_1}{ss_1}$ where $(n_1,s_1) \in N\times S$ fulfills $ss_1 = s'n_1 \in S$

$\frac{n}{s}\cdot\frac{n'}{s'} := \frac{n'n_2}{ss_2}$ where $((n_2,s_2) \in N\times S$ fulfills $ns_2 = s'n_2 \in S$

+ and $\cdot$ are shown to be well-defined and $(N_S,+,\cdot)$ turns out to be a nr. with identity $e = \frac{s}{s}$ (s any element of S). If $t \in S$, the map $h: N \to N_S : n \to \frac{nt}{t}$ is a monomorphism and every $h(s)=\frac{st}{t} \in h(S)$ has the inverse $\frac{t}{st}$. Every $\frac{n}{s}$ can be written as $h(n)h(s)^{-1}$.

<u>1.66 THEOREM</u> If N has nr!s of left quotients N_S, N'_S w.r.t. S then

$$N_S \cong N'_S .$$

(We may therefore speak about "<u>the</u> nr. of left quotients w.r.t. S".)

<u>Proof</u>. N fulfills the left Ore condition by 1.65.

Let h,h' be as in 1.63(b). Define $\bar{h}: N_S \to N'_S$ by $h(n)h(s)^{-1} \to h'(n)h'(s)^{-1}$. $\bar{h}$ is well defined: if $h(n)h(s)^{-1} = h(m)h(t)^{-1}$ then $\exists (t_1,n_1) \varepsilon S\times N : tn_1 = st_1$. So $h(t)^{-1}h(s) = h(n_1)h(t_1)^{-1}$ and $h'(t)^{-1}h'(s) = h'(n_1)h'(t_1)^{-1}$. Therefore

$h(n)h(s)^{-1} = h(m)h(t^{-1}) \Rightarrow h(n) = h(m)h(t)^{-1}h(s) =$
$= h(m)h(n_1)h(t_1)^{-1} \Rightarrow h(n)h(t_1) = h(m)h(n_1) \Rightarrow$
$\Rightarrow nt_1 = mn_1 \Rightarrow h'(n) = h'(m)h'(n_1)h'(t_1)^{-1} =$
$= h'(m)h'(t)^{-1}h'(s) \Rightarrow h'(n)h'(s)^{-1} = h'(m)h'(t)^{-1}$.

Clearly, $\bar{h}$ is an isomorphism.

<u>1.67 REMARKS</u> As Maxson (1) and Graves-Malone (1) pointed out, 1.65 does not hold for near-rings of right quotients, because addition in N_S (as in 1.65) is not necessarily well-defined.

Tewari (1) even showed that there exist near-rings having nr!s of left (right) quotients but no nr!s of right (left) quotients.

<u>1.68 DEFINITION</u> If $S = \{s \varepsilon N | s \text{ cancellable}\}$ then N_S (if it exists) is called the <u>left (right) quotient near-ring</u> of N.

In the section on near-integral domains (§9b2)) we will consider the case that N_S is a near-field.

<u>1.69 REMARK</u> In Chew-Chan (1) a characterization of left quotient near-rings by means of "semi-N-homomorphism" is given.

3.) FREE NEAR-RINGS AND N-GROUPS

For this number, we again use Grätzer's terminology and results. Let $\mathcal{V}$ be any variety of near-rings (e.g. all near-rings, all abelian near-rings, all near-rings with unity or all distributive near-rings.
Let X be any non-empty set.

1.70 DEFINITION A nr. $F_X \in \mathcal{V}$ is called a free-near-ring in $\mathcal{V}$ over X if $\exists\ f: X \to F\ \forall\ N \in \mathcal{V}\ \forall\ g: X \to N\ \exists!\ h \in Hom(F_X, N): h \circ f = g$
(in diagram notation: 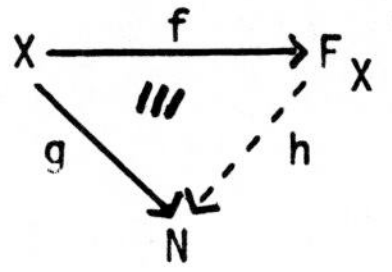).

From (Grätzer) we deduce

1.71 THEOREM In this case, f is injective while X can be regarded as a subset of F_X and generates F_X. F_X is (up to isomorphisms) uniquely determined by $\mathcal{V}$ and $|X|$ and has the form indicated in (Grätzer), p. 163. One therefore is able to speak about "the free nr. in $\mathcal{V}$ determined by some cardinal α".
After several glasses of wine one would describe F_X loosely as the "set of all sums and products of elements of $X \cup \{0\}$ (and possibly more 0-ary symbols such as 1) where one can calculate according to the laws which hold in $\mathcal{V}$".

1.72 DEFINITION If $\mathcal{V} = \mathcal{N}$, we simply speak about the free nr. on X. A near-ring is called free if it is free over some set X.

1.73 REMARK In the same way we define free N-groups in some variety of N-groups, free N-groups over some cardinal number and free N-groups. 1.71 can be transferred to N-groups by making the obvious alterations.

1.74 EXAMPLES If $N \in \mathcal{N}_1$ then ${}_N N$ is free over $\{1\}$.

It is harder to be a free nr. (N-group) than a free ring (module). This fact is rewarded by

1.75 THEOREM If $F, F' \in \mathcal{N}$ (${}_N\Phi, {}_N\Phi' \in {}_N\mathcal{G}$) are free over sets X, X' then $F \cong F'$ ($\Phi \cong_N \Phi'$) $\Longleftrightarrow |X| = |X'|$.

Proof. $\Longleftarrow$ is settled by 1.71.

$\Longrightarrow$: $\mathcal{N}$ (${}_N\mathcal{G}$) contain finite structures with more than one element (for nr!s e.g. the field $\mathbb{Z}_2$, for N-groups e.g. $(\mathbb{Z}_2,+)$ with $n0 = n1 = 0$ for all $n \in N$). Now apply (Grätzer), p. 197.

1.76 REMARK Note that the theorem above also holds e.g. for a free nr. (N-group) in the variety of abelian nr!s (N-groups) ("free abelian nr!s (N-groups)").

1.77 REMARK Let ${}_N\Phi$ be free over X. The usual characterization (in the case of unitary (ring-) modules) of X as a base ("linearly independent generating set") does not carry over to the case of N-groups directly: N-groups do not have to be unitary, the lack of commutativity in ${}_N\Phi$ causes "linear combinations" (defined as usual) to be influenced by the order of the summands (as Maxson (1) pointed out, one has to define linear combinations in terms of ordered sets of elements of ${}_N\Phi$), and - most troublesome of all - ${}_N\Phi$ usually consists of more than the set of all linear combinations, since in general $n(n_1\gamma_1+\ldots+n_k\gamma_k)$ is no linear combination any more.
Anyhow, generalizing the concept of linear independence gives something like a base: let W_n ($n \in \mathbb{N}_0$) be the set of all n-ary words over some set X in a variety $\mathcal{V}$ of N-groups and (for ${}_N\Gamma \in \mathcal{V}$) w_Γ the induced function $\Gamma^n \to \Gamma$. Define in W_n $w \sim_{\mathcal{V}} w' :\Longleftrightarrow \forall \Delta \in \mathcal{V}: w_\Delta = w'_\Delta$.

<u>1.78 DEFINITION</u> A subset B of $_N\Gamma \in \mathfrak{V}$ is called <u>independent</u> if $\forall\ n \in \mathbb{N}_0$ $\forall\ w = w(x_1,\dots,x_n) \in W_n$ $\forall\ \beta_1,\dots,\beta_n \in B$, $\beta_i \neq \beta_j$ for $i \neq j$: $(w_\Gamma(\beta_1,\dots,\beta_n) = o \Rightarrow w \sim_{\mathfrak{V}} \bar{o})$.

<u>1.79 REMARK</u> Let $_RM$ be a unitary ring-module with $\forall\ r \in R$: $rM = \{o\} \Rightarrow r = 0$ (otherwise $_RM$ would have no linearly independent subset at all).
Then each subset of $_RM$ is linearly independent iff it is independent in the sense of 1.78.

<u>1.80 DEFINITION</u> $B \subseteq {_N\Gamma}$ is called a <u>base</u> for $_N\Gamma$ if

(a) B generates $_N\Gamma$

(b) B is independent.

As usual, the following questions arise:
(a) Which N-groups have a base ?
(b) Are different bases equipotent ?

<u>1.81 THEOREM</u> $B \subseteq {_N\Gamma}$ is a base for $_N\Gamma$ iff the inclusion map $\iota: B \to \Gamma$ can be extended to an N-isomorphism $\Phi \to \Gamma$, where Φ is the free N-group on B.

<u>Proof</u>. $\Rightarrow$: Let B be a base for $_N\Gamma$. Consider the diagram

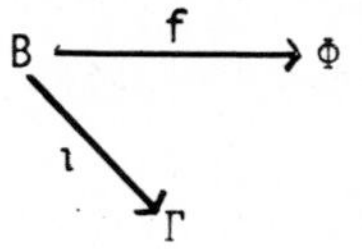

Since Φ is free on B, there is exactly one $h \in \mathrm{Hom}_N(\Phi,\Gamma)$ with $h \circ f = \iota$. We have to show that h is an N-isomorphism:

(a) $B \subseteq h(\Phi) \wedge B$ generates $\Gamma \Rightarrow h(\Phi) = \Gamma$

(b) Let $\phi \in \Phi$ be in Ker h. Represent ϕ by some word $w(f(\beta_1),\dots,f(\beta_{\bar{n}}))$ over $f(B)$ ($\beta_i \neq \beta_j$ for $i \neq j$). Now
$o = h(\phi) = h(w(f(\beta_1),\dots,f(\beta_n))) =$
$= w_\Gamma(h(f(\beta_1),\dots,h(f(\beta_n))) = w_\Gamma(\iota(\beta_1),\dots,\iota(\beta_n)) =$
$= w_\Gamma(\beta_1,\dots,\beta_n)$. B is independent, so $\phi = o$.

<=: Suppose that ι extends to an N-isomorphism $h:\Phi\to\Gamma$. By the construction of ${}_N\Phi$, f(B) is independent and generates ${}_N\Phi$. So f(B) is a base for ${}_N\Phi$. Hence $B = \iota(B) = h(f(B))$ is a base for $h(\Phi) = \Gamma$.

From this theorem we immediately deduce

1.82 THEOREM

(a) ${}_N\Gamma$ has a base <=> ${}_N\Gamma$ is free.

(b) ${}_N\Gamma$ has a base B <=> each map f from B to some N-group ${}_N\Delta$ can uniquely be extended to an N-homomorphism ${}_N\Gamma \to {}_N\Delta$.

And from 1.75 we get

1.83 THEOREM Let ${}_N\Gamma$ be a non-zero N-group possessing a base. Then all bases are equipotent.

1.84 EXAMPLE If N is in $\mathcal{N}_1$ then ${}_NN$ has a base (namely {1}) and all other bases consist of one single element.

1.85 REMARK See more on free products etc. in Fröhlich's paper (4), in Meldrum (2),(3) and in Rao (1). Cf. also Fröhlich (4) and Maxson (1) for a characterization of a base in terms of free products.

See Zeamer (1) for an "arithmetics" in free near-rings.

c) EMBEDDINGS

1.) EMBEDDINGS IN $M(\Gamma)$

The reader might be wondering if all near-rings are near-rings of functions on some group Γ. This is true, although near-rings are also considered under totally different aspects.
The main result is

1.86 THEOREM $\forall N \in \mathcal{N} \; \exists \Gamma \in \mathcal{G} : N \hookrightarrow M(\Gamma)$.

Proof. (Heatherly-Malone (1)). Let Γ be any group properly containing $(N,+)$.

For $n \in N$, define $f_n : \Gamma \to \Gamma$, $\gamma \to \begin{cases} n\gamma & \gamma \in N \\ n & \gamma \notin N \end{cases}$. As one can easily see, $\forall\, n,n' \in N : f_n + f_{n'} = f_{n+n'} \wedge$ $\wedge\, f_n \circ f_{n'} = f_{nn'}$. Thus the map $h : N \to M(\Gamma)$, $n \to f_n$ is a homomorphism.

If $h(n) = h(n')$ then $f_n = f_{n'}$. In particular, $\forall\, \gamma \in \Gamma \setminus N : n = f_n(\gamma) = f_{n'}(\gamma) = n'$.
This implies that h is in fact a monomorphism and an embedding map, as desired.

1.87 REMARK There are several proofs for 1.86. See e.g. Berman-Silverman (3), Nöbauer (8), Heatherly-Malone (1). While Nöbauer embeds in $M(\Gamma)$ with $\Gamma := (M((N,+)),+)$, Heatherly-Malone suggest $\Gamma := (N,+) \oplus (\mathbb{Z}_2,+)$.

1.86 and its proof have many interesting corollaries. Some are in

1.88 COROLLARIES

(a) If N is abelian there is an abelian group Γ with $N \hookrightarrow M(\Gamma)$.

(b) If N is finite there is some finite Γ with $N \hookrightarrow M(\Gamma)$.

(c) $\forall N \in \mathcal{N}_0 \quad \exists \Gamma \in \mathcal{G} : N \hookrightarrow M_0(\Gamma)$.

(d) $\forall N \in \mathcal{N}_c \quad \exists \Gamma \in \mathcal{G} : N \hookrightarrow M_c(\Gamma)$.

(e) If $N \in \mathcal{N}_1$ embeds in $M(\Gamma)$ by ϕ and $n \in N$ is invertible then $\phi(n)$ is bijective.

(f) Every near-field is isomorphic to a nf. F of functions on a group Γ, where all $f \in F^*$ are bijective.

1.89 THEOREM Each N-group can be embedded into some faithful N-group. So each nr. N has some faithful N-group.

Proof. If ${}_N\Gamma \in {}_N\mathcal{G}$, take some group $\Gamma' \supseteq \Gamma$ with $N \hookrightarrow M(\Gamma')$. Evidently, Γ' is an N-group in the natural way and moreover a faithful one.

1.90 REMARK Heatherly (1) showed that $N = N_d \Rightarrow$ $\Rightarrow \exists\ \Gamma \in \mathcal{G} : N \hookrightarrow E(\Gamma)$ (1.15).

It is sometimes desirable to look for an embedding of N into $M(\Gamma)$ with a "smaller" Γ as above. Recall that in 1.86 and 1.87 e.g. $M(\Gamma)$ is embedded into the much bigger $M(M(\Gamma) \oplus \mathbb{Z}_2)$!
For doing this, we generalize a concept due to Menger:

1.91 DEFINITION $B \subseteq N$ is called a base (of equality) if $\forall\ n,n' \in N: (\ \forall b \in B : nb = n'b) \Rightarrow n = n'$.

1.92 REMARK Clearly B forms a base iff $(0:B) = \{0\}$, so it would not be necessary to use a special name. But we do it, because it is a very suggestive one.

1.93 EXAMPLES In $M(\Gamma)$ the set $M_c(\Gamma)$ (a group isomorphic to Γ) forms a base. In $M_{cont}(\Gamma)$ (1.4(a)) it suffices to take a dense subset of Γ.

This motivates the interest in the case that the constants N_c form a base. This can be achieved by force:

<u>1.94 PROPOSITION</u> Let π be the natural epimorphism $N \to N/_{(0:N_c)}$. Then $\pi(N_c)$ forms a base for $\pi(N)$.

<u>Proof</u>. If $\bar{n} = \pi(n)$ and $\bar{n}_1 = \pi(n_1)$ are $\in \pi(N)$ then
$(\forall \bar{n}_c \in \pi(N_c) : \bar{n}\bar{n}_c = \bar{n}_1\bar{n}_c) \Rightarrow (\forall n_c \in N_c : nn_c - n_1 n_c \in (0:N_c))$
$\Rightarrow (\forall n_c \in N_c : 0 = (nn_c - n_1 n_c) n_c = nn_c - n_1 n_c = (n-n_1) n_c) \Rightarrow$
$\Rightarrow (n-n_1) \in (0:N_c) \Rightarrow \bar{n} = \bar{n}_1$.

<u>1.95 EXAMPLES</u>

(a) In $M(\Gamma)$, the constants form a base.

(b) In $\mathbb{Z}_p[x]$, the constants $\mathbb{Z}_p$ do not form a base. In fact, $x^p - x \neq o$ (zero polynomial), but $\forall \alpha \in \mathbb{Z}_p : (x^p - x)(\alpha) = \alpha^p - \alpha = 0$. Therefore $x^p - x \in (0:\mathbb{Z}_p)$. $(0:\mathbb{Z}_p)$ consists of all polynomials whose corresponding polynomial function is the zero map.

The following solves the problem stated after 1.90.

<u>1.96 THEOREM</u> If $B \leq_N N$, the following conditions are equivalent:

(a) B is a base (of equality);

(b) B is a faithful N-group;

(c) $N \hookrightarrow M(B)$.

<u>Proof</u>. 1.48 and 1.92.

<u>1.97 COROLLARY</u> If N_c is a base then N can be considered as a near-ring of functions on N_c. In view of 1.95(a), this is "the natural representation of N".

<u>1.98 DEFINITION</u> Let Γ, Δ be groups. $f \in M(\Gamma)$ is called <u>kernel-free</u> if $\forall \gamma \in \Gamma : (f(\gamma) = o \Rightarrow \gamma = o)$. Put $M(\Gamma) \hookrightarrow_k M(\Delta)$ if there is some $h: M(\Gamma) \rightarrowtail M(\Delta)$ such that h sends kernel-free elements of $M(\Gamma)$ into kernel-free ones of $M(\Delta)$, and $M(\Gamma) \cong_k M(\Delta)$ if $M(\Gamma) \hookrightarrow_k M(\Delta)$ by an isomorphism h.

1.99 THEOREM (Heatherly-Malone (1)). Let Γ, Δ be groups. Then $\Gamma \hookrightarrow \Delta \iff M_0(\Gamma) \underset{k}{\hookrightarrow} M_0(\Delta) \iff M(\Gamma) \hookrightarrow M(\Delta)$.

Proof. (a) Let $\Gamma \hookrightarrow \Delta$ by h. If $\Gamma = \{o\}$, the result is obvious. Assume that $\Gamma \neq \{o\}$ and take some arbitrary, but fixed $\gamma \in \Gamma$, $\gamma \neq o$.
If $f \in M_0(\Gamma)$, define $f_\gamma \in M_0(\Delta)$ by

$$f_\gamma : \Delta \to \Delta$$
$$\delta \to \begin{cases} h(f(h^{-1}(\delta))) & \delta \in \text{Im } h \\ h(f(\gamma)) & \delta \notin \text{Im } h \end{cases}$$

If f is kernel-free, the same applies to f_γ and the map $f \to f_\gamma$ embeds $M_0(\Gamma)$ into $M_0(\Delta)$.

(b) If $M_0(\Gamma) \underset{k}{\hookrightarrow} M_0(\Delta)$ by (say) g then take some fixed $\delta \in \Delta^*$ ($\Delta = \{o\}$ is again trivial).

$h : M_c^o(\Gamma) \to M_c^o(\Delta)$ (Notation as in 1.4(a)) is a
$f_\alpha \to f_{g(f_\alpha)(\delta)}$
group homomorphism. If $\alpha \neq 0$, f_α is kernel-free, h is moreover injective:
$h(f_\alpha) = \bar{o} \Rightarrow h(f_\alpha)(\delta) = o$, so $\delta = o$ (a contradiction) or $\alpha = o$, whence $f_\alpha = \bar{o}$.
Hence $M_c^o(\Gamma) \hookrightarrow M_c^o(\Delta)$ (as groups). But $M_c^o(\Gamma)$ and Γ are isomorphic groups, and the same applies to $M_c^o(\Delta)$ and Δ.
So $\Gamma \hookrightarrow \Delta$.

(c) If $\Gamma \hookrightarrow \Delta$ then proceeding as in (a) one sees that $M(\Gamma) \hookrightarrow M(\Delta)$.

(d) If $M(\Gamma) \hookrightarrow M(\Delta)$ then $M_c(\Gamma) \hookrightarrow M_c(\Delta)$.
$M_c(\Gamma) \cong \Gamma$ and $M_c(\Delta) \cong \Delta$ implies that $\Gamma \hookrightarrow \Delta$.

1.100 COROLLARY (Beidleman (5)) $\Gamma \cong \Delta \iff M_0(\Gamma) \underset{k}{\cong} M_0(\Delta) \iff$
$\iff M(\Gamma) \cong M(\Delta)$.

Since each group Γ can be embedded into some $M(S) = S^S$ (S a suitable set) by Cayley's theorem, we get from 1.99

1.101 COROLLARY (Nöbauer (8)). For all $N\varepsilon\mathcal{N}$ there is some set S with $N \hookrightarrow M(M(S))$. (More precisely, for every nr. N there is a set S and a near-ring N' in $M(M(S))$ such that $N \cong N'$.

2.) MORE BEDS.

Since $M(\Gamma)$ contains an identity (id_Γ) we get from 1.86

1.102 COROLLARY Every (finite, abelian, zerosymmetric) near-ring can be embedded into a (finite, abelian, zerosymmetric) near-ring with identity.

1.103 REMARK Despite of this analogy to ring theory the embedding is totally different from the one in ring theory. Moreover, N is not always embedded as an ideal.

1.104 PROPOSITION Each ring (ring with unity, field) can be embedded into a non-ring (non-ring with identity, non-field).

Proof. (a) For rings (rings with identity) it follows from 1.86. See Clay (3) for another proof.

(b) Let F be a field (Maxson (7)). Take $F(x)$ and define for $f = \frac{f_1}{f_2}\varepsilon F(x)$ $d(f) := \deg f_1 - \deg f_2$. For $a,b\varepsilon F$, $a \neq 0$ put $\theta: F(x) \to F(x)$, $y \to ay+b$. Clearly θ is 1-1. Define for $f,g\varepsilon F(x)$ $f *_\theta g := \begin{cases} 0 & f = 0 \\ (\theta^{d(f)} \circ g) \cdot f & f \neq 0 \end{cases}$

Then it is easy to see that $F_\theta(x) := (F(x),+,*_\theta)$ is a near field and a field iff $\theta = id$.

$\iota: F \to F_\theta(x)$ is the desired embedding map.

1.105 COROLLARY (Maxson (7)). Each commutative ring R without zero divisors $\neq 0$ can be embedded into a non-field.

Proof. R can be embedded into an integral domain, which can be embedded into some field. Now apply 1.104.

1.106 REMARK See more on embeddings in the chapter on nf.'s and near integral domains.

d) SOME AXIOMATIC CONSIDERATIONS

In this section we compile some results on the axiomatics of near-rings: conditions for N to be a ring or to be abelian, cancellable and invertible elements and a brief survey of structures which are closely related to near-rings.

1.) MISCELLANEOUS RESULTS

1.107 PROPOSITION Let N be a nr.

(a) N abelian $\wedge$ N commutative $\Longleftrightarrow$ N is a commutative ring;

(b) N abelian $\wedge$ N distributive $\Longleftrightarrow$ N is a ring;

(c) $N^2 = N \wedge N$ distributive $\Longrightarrow$ N is a ring (Taussky (1)).

Proof. (a) and (b) are obvious.

(c): $\forall\ n,n' \in N\ \ \exists\ a,b,c,d \in N$: $n = ab \wedge n' = cd$. Computing $(a+c)(d+b)$ in two different ways yields $ad+cd+ab+cb = ad+ab+cd+cb$, so $cd+ab = ab+cd$, therefore $n+n' = n'+n$. Now apply (b).

From 1.107(c) and the fact that there exist non-abelian distributive near-rings (1.15) we deduce

1.108 COROLLARY Not every distributive nr. can be embedded into a distributive nr. with identity.

1.109 PROPOSITION Each of the following conditions imply a near-ring N with identity to be abelian:

(a) $\forall\ n\in N : n(-1) = -n$;

(b) (Ligh (6)) N finite $\wedge$ $\forall n\in N : n(-1) = n \Rightarrow n = 0$.

(c) (B.H. Neumann (1)) $(\forall n\in N\ \exists h\in N : n = h+h) \wedge$ $\wedge (\forall\ n\in N : n(-1) = n \Rightarrow n = 0)$.

Proof. (a) $\forall\ n,n'\in N : n+n' = (-n)(-1)+(-n')(-1) =$ $= (-n-n')(-1) = -(-n-n') = n'+n$.

(b) Define $\alpha: \begin{array}{l} N \to N \\ n \to n(-1) \end{array}$. Clearly $\alpha\in Aut(N,+)$ and $\alpha^2 = id$. $\alpha(n) = n$ implies $n = 0$. So by a theorem of group theory (e.g.(W.R. Scott), p. 357), N is abelian.

(c) α (as above) is again a fixed-point-free automorphism of order 2. From group theory (B.H. Neumann (1), p. 206) we know that N is abelian.

1.110 REMARK McQuarrie (2) showed that 1.109(b) does not hold in the infinite case.

We now consider cancellable elements.

1.111 PROPOSITION Let N be a nr.

(a) (Maxson (1)) $n\in N$ is right cancellable $\Longleftrightarrow$ n is not a right zero divisor;

(b) (Maxson (1)) $n\in N_0$ is left cancellable $\begin{array}{c}\Rightarrow \\ \not\Leftarrow\end{array}$ n is not a left zero divisor;

(c) (Timm (3)) If $N\in\mathfrak{N}_0$ then the left cancellation law implies the right one.

Proof. (a) is shown as it is done for rings.

(b) If n is left cancellable and if $nn' = 0 = n0$ then $n' = 0$. To see the "$\not\Leftarrow$"-part, consider near-rings of the type $M_c^0(\Gamma)$ (as introduced in 1.4(a)).

(c) If $n'n = n''n$, $n \neq 0$, then $(n'-n'')n = 0 =$ $= (n'-n'')0$; left cancellation results in $n' = n''$.

1.112 REMARKS Heatherly (1) proved that a finite nr. N has either only right zero divisors or a right identity. Ligh (1) showed that if the right identity is unique then N is a nf.

Moreover, Ligh (1) proved that in a finite non-abelian near-ring without non-trivial zero divisors each element has a unique square-root.

An application of the embedding theorem 1.86 is

1.113 PROPOSITION (Heatherly (1)) Let $N \in \mathcal{N}_1$ be finite. If $n \in N$ has a one-sided inverse then this inverse is two-sided.

Proof. By 1.88(b) there is a finite group Γ with $N \overset{h}{\hookrightarrow} M(\Gamma)$. If $n \in N$ has a one-sided inverse then $h(n)$ has the same. Since Γ is finite, $h(n)$ is a 1-1-map and therefore invertible.

1.114 REMARK If N, N' are nr's and h is a homomorphism (an automorphism) from N to N' then h/N_o and h/N_c are also nr.-homomorphisms (nr.-automorphisms).

Conversely one might ask whether each pair $h_o: N_o \to N_o'$ and $h_c: N_c \to N_c'$ of homomorphisms (automorphisms) can be "mated" together to give a nr.-homomorphism (automorphism) from N to N'. As Malone (1), (4) pointed out, this is not the case in general:

$$h: N \longrightarrow N', \quad n_o+n_c \to h_o(n_o)+h_c(n_c)$$ is a near-ring homomorphism iff

$$\forall n_o \in N_o \quad \forall n_c \in N_c : h_c(n_c)+h_o(n_o) = h_o(n_o)+h_c(n_c) \quad \text{and}$$

$$\forall n \in N \quad \forall m \in N_o : h(nm) = h(n)h_o(m).$$

2.) RELATED STRUCTURES

1.115 SEMINEAR-RINGS

A set S together with two binary operations "+" and "•" is called a seminear-ring if (S,+) and (S,•) are semigroups and $\forall s,s',s''\in S$: $(s+s')s'' = ss''+s's''$.

EXAMPLES: The sets of all mappings on an (additively written) semigroup with pointwise addition and composition, e.g. $M(\mathbb{N})$.

REFERENCES: Pilz (5), Van Hoorn, Van Hoorn-Van Rootselaar, Van Rootselaar.

1.116 NEAR-ALGEBRAS

A vector space A over a (skew-) field F together with an additional binary operation "•" is called a near-algebra over F if (A,+,•) is a near-ring and $\forall\ a,b\in A\ \forall\ \lambda\in F$: $(\lambda a)\cdot b = \lambda(a\cdot b)$.

EXAMPLES: Take the sets of all mappings of a vector space ${}_FV$ into itself with pointwise defined addition, composition and forming λ-folds.

REFERENCES: H.D. Brown, Marin, Timm (8), Williams (2), Yamamuro (1) - (4).

ATTENTION Cf. Holcombe (6) for an essential different definition of a near-algebra.

1.117 COMPOSITION RINGS (TRI-OPERATIONAL ALGEBRAS)

A set R together with 3 binary operations "+", "•", "∘" is called a composition ring (tri-operational algebra, TOA) if (R,+,•) is a ring, (R,+,∘) a near-ring and if $\forall\ r,r',r''\in R$: $(r.r')\circ r'' = (r\circ r'')\cdot(r'\circ r'')$.

EXAMPLES: The sets of all mappings of a ring into itself with pointwise addition, multiplication and composition.

REFERENCES: Adler, F.L. Brown, Burke, Clay-Doi (2), Heller, Mannos, Marschoun, Menger, Milgram, Nöbauer (1) - - (9), Pater, Penner (1), Pilz (1),(3), Riedl, Steinegger, Stueben, Suvak.

1.118 A GENERAL PROCEDURE

Take a universal algebra $A = (A,\Omega)$, form the set $M(A)$ of all self-maps of A and define the operations of Ω pointwise on $M(A)$. Adding the binary operation "$\circ$" of composition yields a new algebra $M(A) = (M(A), \Omega\cup\{\circ\})$.

EXAMPLES:

A	M(A)
naked set	semigroup
semigroup	seminear-ring
group	near-ring
module	(-)
vector space	near-algebra
ring	composition ring
near-ring	(-)
linear algebra	(-)
Ω-group	"Ω-composition-group"

(-): there exists no special name.

REFERENCES: Berman-Silverman (3), Hule, Lausch (2), Lausch-Nöbauer, Mitsch, Mlitz, Nöbauer (8),(10),(11), Nöbauer-Philipp, Pilz (4) and Polin.

Such a lot of interesting structures ! Since one might be attempted to start looking at them more thoroughly we switch back to near-rings very quickly.

§2 IDEAL THEORY

In this paragraph we develop an ideal theory similar to that one for rings.

After defining sums and (internal) direct sums of ideals (of N and ${}_N\Gamma$) we note that, unlike to the ring case, internal and external direct sums of ideals are not necessarily isomorphic. We call an internal direct sum with this property a "distributive sum", and prove that for $N = N_o$ each direct sum is distributive.

Also, we consider the lattices of ideals (and left ideals in N_o). Of course, these lattices are complete modular ones. If $N = N_o$ and no non-zero homomorphic image of N is a ring then the lattice of left ideals is even distributive. Chain conditions play an important rôle throughout this monograph. We prove for example that N has some chain condition iff a direct summand I of N and N/I have the same one.

If N has the DCC on ideals then it is a finite direct sum of indecomposable near-rings. N is called completely reducible if it is the direct sum of simple ideals. This is the case iff every ideal is a direct summand, and then each ideal of N has the same properties. N is a finite direct sum of simple ideals iff N is completely reducible and has DCC and ACC on ideals or (equivalently) iff N is completely reducible and has one of the chain conditions on ideals or (again equivalently) iff N is completely reducible and finitely generated which is in term equivalent to the existence of finitely many maximal ideals with zero intersection. Any two such decompositions are isomorphic.

Finally, we develop the theory of (semi-) prime and nil(potent) ideals which runs fairly parallel to ring theory: every near-ring has minimal prime ideals; the intersection of prime ideals is semiprime; if $I \trianglelefteq N$ then N is nil(potent) iff I and N/I are nil(potent); if $N = N_o$ has DCC on left ideals then N is

a prime near-ring iff N has a unique minimal ideal which is not nilpotent.
Many results carry over to N-groups with $N \in \mathcal{N}_0$.

a) SUMS

1.) SUMS AND DIRECT SUMS

2.1 - 2.11 are formulated for ideals of near-rings; but all (except 2.6(b)) can be transferred to N-groups with $N \in \mathcal{N}_0$ by making the usual changes. The proofs in these considerations run parallel to group or ring theory and are therefore omitted. Also, it is pointed out that these results follow from the general theory of "Ω-groups" (see (Kurosh) or (Higgins)) (note that nr!s are Ω-groups, and N-groups are Ω-groups if $N \in \mathcal{N}_0$).

2.1 THEOREM Let $(I_k)_{k \in K}$ be a family of ideals of a nr. N. Then the following sets are equal:

(a) The set of all finite sums of elements of the I_k's;

(b) The set of all finite sums of elements of different I_k's;

(c) The sum of the normal subgroups $(I_k,+)$;

(d) The subgroup of $(N,+)$ generated by $\bigcup_{k \in K} I_k$;

(e) The normal subgroup of $(N,+)$ generated by $\bigcup_{k \in K} I_k$;

(f) The ideal of N generated by $\bigcup_{k \in K} I_k$.

2.2 DEFINITION The set (a) - (f) above is called the sum of the ideals I_k $(k \in K)$ and denoted by $\sum_{k \in K} I_k$ (for $K = \{1,2,...\}$ also by $I_1+I_2+...$).

From 2.1 (d) - (f) we readily deduce

2.3 COROLLARY

(a) The sum of ideals of N is again an ideal of N.

(b) Forming sums of ideals is an associative and commutative operation.

Certain sums are of particular importance:

2.4 DEFINITION Again let $(I_k)_{k\in K}$ be ideals of N. Their sum $\sum_{k\in K} I_k$ is called an (internal) direct sum if each element of $\sum_{k\in K} I_k$ has a unique representation as a finite sum of elements of different I_k's.
In this case we write for the sum $\dot{\sum_{k\in I}} I_k$ (or $I_1 \dot{+} I_2 \dot{+} \dots$ as in 2.2).

2.5 PROPOSITION For each family $(I_k)_{k\in K}$ of ideals of N the following conditions are equivalent:

(a) The sum of the I_k's is direct.

(b) The sum of the normal subgroups $(I_k,+)$ is direct.

(c) $\forall\ k\in K : I_k \cap (\sum_{\substack{\ell\in K\\ \ell\neq k}} I_\ell) = \{0\}$.

2.6 PROPOSITION

(a) If $\sum_{k\in K} I_k$ is direct then elements of different I_k's commute w.r.t. addition.

(b) If $N = N_0$ and $\sum_{k\in I} I_k$ is direct then elements of different I_k's have product 0.

2.7 EXAMPLE In the notation of 1.56, $\dot{\sum_{i\in I}} \overline{N}_i = \bigoplus_{i\in I} N_i$.

2.8 THEOREM ("First isomorphism theorem")

If $I,J \trianglelefteq N$ then $I \cap J \trianglelefteq J \wedge I+J/_I \cong J/_{I\cap J}$.

2.9 REMARK If the reader should have the same difficulties as the author in remembering this formula he might note the alternating appearance of I and J in the isomorphism statement.

2.10 DEFINITION $I \trianglelefteq N$ is called a direct summand (of N) if
$\exists\ J \trianglelefteq N$: $N = I \dot{+} J$.
J is then called a direct complement of I in N.

2.11 PROPOSITION $I \trianglelefteq N$ is a direct summand $\Leftrightarrow$ $\forall \alpha \in \mathrm{Aut}\, I$:
: α can be extended to an epimorphism $N \twoheadrightarrow I$.

The following result will be frequently used.

2.12 THEOREM If $I \trianglelefteq N$ is a direct summand then each ideal of I is an ideal of N.

2.13 COROLLARY If, as in 1.56, $N = \bigoplus_{i \in I} N_i$ and $J_i \trianglelefteq N_i$ then $J_i \trianglelefteq N$.

2.14 REMARK In general, the (group-theoretic) sum of two N-subgroups is no N-subgroup any more. But:

2.15 PROPOSITION (Fain (1)).
If $\Delta \leq_N \Gamma$ and $E \trianglelefteq_N \Gamma$ then $\Delta + E \leq_N \Gamma$.

Proof. $\forall\ \delta \in \Delta\quad \forall \eta \in E\quad \forall n \in N$: $n(\delta+\eta) = n(\delta+\eta) - n\delta + n\delta \in E + \Delta =$
$= \Delta + E$.

2.16 COROLLARY (Mlitz (2)). If $\Delta \trianglelefteq_N \Gamma$ then the N-subgroup of ${}_N\Gamma$ generated by Δ is given by $\Delta + \Omega$.

Proof: 2.15 and 1.35(b).

With this equipment we can consider the relation of ideals of N and of some homomorphic image N' more closely than in 1.30.

2.17 PROPOSITION Let $h: N \twoheadrightarrow N'$ be an epimorphism with $\mathrm{Ker}\, h =: K$. Let A, A' be ideals (left ideals, N-subgroups) of N, N', respectively. Then
(a) $h(h^{-1}(A')) = A'$.
(b) $h^{-1}(h(A)) = A + K \geq A$.

The same applies to ideals or N-subgroups of N-groups.

Proof. (for ideals A, A' of near-rings N, N').

Let $n \in N$, $n' \in N'$.

(a) $n' \in h(h^{-1}(A')) \iff \exists n \in h^{-1}(A'): n' = h(n) \iff n' \in A'$.

(b) $n \in h^{-1}(h(A)) \iff h(n) \in h(A) \iff \exists a \in A: h(n-a) = h(n)-h(a) = 0 \iff \exists a \in A: n-a \in K \iff n \in A+K$.

2.18 PROPOSITION $R \trianglelefteq_r N \Rightarrow R = R \cap (N_0+N_c) = R \cap N_0 + R \cap N_c = R_o + R_c$.

Proof. $\forall r \in R \quad \exists! n_0 \in N_0 \quad \exists! n_c \in N_c : r = n_0 + n_c$.

$R \trianglelefteq_r N \Rightarrow n_c = n_c 0 = (n_0+n_c)0 = r0 \in R$, so $n_0 \in R$, too.

The rest is trivial.

2.19 REMARKS 2.18 does not hold for left ideals L of N. All $\ell \in L$ have the form $\ell = n_0 + n_c$ with $n_0 \in N_0$ and $n_c \in N_c$, but in general $n_0 \notin L$ and $n_c \notin L$:

Consider $N = \mathbb{Z}[x]$ and $L := \{\Sigma a_i x^i \mid \Sigma a_i \in 2\mathbb{Z} = \{0, \pm 2, \pm 4, \ldots\}\}$.

L is a left ideal of N (even a maximal one - see So (1)), but $\ell := x+1 \in L$ decomposes as $\ell = n_0 + n_c$ with $n_0 = x \notin L$.

2.20 THEOREM Under forming sums and intersections, the ideals of N (${}_N\Gamma$ with $N \in \mathcal{N}_0$) form a complete modular lattice.

Proof. follows from (Kurosh), p. 143.

2.21 REMARK These lattices are not necessarily distributive. But cf. the following considerations and 2.18 (and also Scott (3)).

2.22 PROPOSITION (Scott (3)) If $A, B \trianglelefteq_N \Gamma$ and $A, B \leq_N \Gamma$ then $\forall n \in N \ \forall \alpha \in A \ \forall \beta \in B : n(\alpha+\beta) \equiv n\alpha + n\beta \pmod{A \cap B}$.

Proof. $n(\alpha+\beta) - n\beta - n\alpha \in A+A = A$

So $n(\alpha+\beta) \equiv n\alpha + n\beta \pmod{A}$.

Similarly, $n(\alpha+\beta) \equiv n\beta + n\alpha \pmod{B} \equiv n\alpha + n\beta \pmod{B}$, and the result follows.

One can suspect that 2.22 will be particularly important for $A \cap B = \{o\}$: see 2.29.

2.23 PROPOSITION (Wielandt (2)). If $N\varepsilon\mathcal{N}_0$ and $A,B,\Delta \trianglelefteq_N \Gamma$ then

$$\Gamma' := (A+\Delta)\cap(B+\Delta)/_{(A\cap B)+\Delta}$$

is commutative and $\forall\ n\varepsilon N\ \forall\ \gamma_1,\gamma_2\varepsilon\Gamma' : n(\gamma_1+\gamma_2) = n\gamma_1+n\gamma_2$.

Proof. (Betsch (5)). $E := (A\cap B)+\Delta$; $H := (A+\Delta)\cap(B+\Delta)$.
Let $\eta_1,\eta_2\varepsilon H$ and $n\varepsilon N$.
Then $\exists\ \alpha\varepsilon A\ \exists\ \beta\varepsilon B$: $\eta_1 \equiv \alpha(\text{mod } E) \wedge \eta_2 \equiv \beta(\text{mod } E)$.
Now $\alpha+\beta \equiv \beta+\alpha(\text{mod } A\cap B)$ and $n(\alpha+\beta) \equiv n\alpha+n\beta(\text{mod } A\cap B)$ by 2.22.
Since $A\cap B \subseteq E$ we get
$\eta_1+\eta_2 \equiv \alpha+\beta \equiv \beta+\alpha \equiv \eta_2+\eta_1(\text{mod } E)$ and
$n(\eta_1+\eta_2) \equiv n(\alpha+\beta) \equiv n\alpha+n\beta \equiv n\eta_1+n\eta_2(\text{mod } E)$, and the proposition is proved.

2.24 COROLLARY (Betsch (5)). With the assumptions and notations of 2.23, $\overline{N} := N/_{(0:\Gamma')}$ is a ring.

Proof. Γ' can be considered as a faithful $\overline{N}$-group in the obvious way. Now the result follows from 1.49.

2.25 COROLLARY (Betsch (5)). If $N\varepsilon\mathcal{N}_0$ and $N\varepsilon\mathcal{N}_1$ and if no non-zero homomorphic image of N is a ring then the lattice of left ideals of N is distributive.

Proof. Let L_1,L_2,L_3 be left ideals of N. Consider the N-group $\Gamma := (L_1+L_3)\cap(L_2+L_3)/_{(L_1\cap L_2)+L_3}$.
If $\Gamma \neq \{o\}$ then $(o:\Gamma) \neq N$, for $N\varepsilon\mathcal{N}_1$. From 2.24 we know that $N/_{(o:\Gamma)}$ is a ring ($\neq N$), a contradiction. So $\Gamma = \{o\}$ and the lattice of left ideals is distributive.

2.) DISTRIBUTIVE SUMS

2.26 DEFINITION

(a) A direct sum $\sum^{\bullet}_{\alpha\in A} I_\alpha =: I$ of ideals I_α of N $(\alpha\in A)$ is called distributive: <=>

<=> $\forall \sum_{\alpha\in A} i_\alpha, \sum_{\beta\in A} i'_\beta \in I : (\sum_{\alpha\in A} i_\alpha)(\sum_{\beta\in A} i'_\beta) = \sum_{\alpha\in A} i_\alpha i'_\alpha$.

(b) A direct sum $\sum^{\bullet}_{\alpha\in A} \Delta_\alpha =: \Delta$ of ideals Δ_α $(\alpha\in A)$ of $_N\Gamma$ is called distributive: <=>

<=> $\forall \sum_{\alpha\in A} \delta_\alpha \in \Delta \quad \forall n\in N : n(\sum_{\alpha\in A} \delta_\alpha) = \sum_{\alpha\in A} n\delta_\alpha$.

(Note that the sums involved are actually finite ones.)

2.27 EXAMPLES If $N = \bigoplus_{\alpha\in A} N_\alpha$ then N is the distributive sum of the ideals $\overline{N}_\alpha$ (1.56).

The same applies to N-groups. Moreover:

2.28 PROPOSITION Let $(I_\alpha)_{\alpha\in A}$ be a family of ideals of N whose sum is direct. Then $\sum^{\bullet}_{\alpha\in A} I_\alpha \cong \bigoplus_{\alpha\in A} I_\alpha$ <=> $\sum^{\bullet}_{\alpha\in A} I_\alpha$ is distributive.

The analogous result holds for N-groups with $N = N_0$.

Proof. obvious.

2.29 PROPOSITION (Heatherly (3)). Let $(\Delta_\alpha)_{\alpha\in A}$ be a family of ideals of $_N\Gamma$ with $\sum_{\alpha\in A} \Delta_\alpha = \sum^{\bullet}_{\alpha\in A} \Delta_\alpha =: \Delta$. Then

$$\forall n\in N_0 \quad \forall \Sigma\delta_\alpha \in \Delta : n(\Sigma\delta_\alpha) = \Sigma n\delta_\alpha .$$

Conversely, if $_N\Gamma$ is faithful and if for $n\in N$ $\forall \Sigma\delta_\alpha\in\Delta : n(\Sigma\delta_\alpha) = \Sigma n\delta_\alpha$ then $n\in N_0$.

Proof. The first assertion follows from 2.22 and by induction.

If for $n\in N$ and all $\Sigma\delta_\alpha\in\Delta$ $n(\Sigma\delta_\alpha) = \Sigma n\delta_\alpha$ then $n(o+o) = no+no$, hence $no = o$. So

$\forall \gamma\in\Gamma : (n0)\gamma = n(0\gamma) = no = o = 0\gamma$, hence $n0 = 0$.

From 2.29 we get the following satisfactory result (recall that for $N \neq N_0$ there is no chance at all that always $\sum_{\alpha\in A}^{\bullet} \Delta_\alpha \cong_N \bigoplus_{\alpha\in A} \Delta_\alpha$, for the Δ_α's are not necessarily N-groups).

2.30 THEOREM (Betsch (3)). If $N \in \mathcal{N}_0$ then each direct sum of ideals (in N and $_N\Gamma$) is distributive.

Proof. The statement for $_N\Gamma$ is clear from 2.29.

If $\sum_{\alpha\in A}^{\bullet} I_\alpha =: I$ and $\sum_{\alpha\in A} i_\alpha, \sum_{\beta\in A} i'_\beta \in I$ then

$$(\sum_{\alpha\in A} i_\alpha)(\sum_{\beta\in A} i'_\beta) = \sum_{\alpha\in A} i_\alpha(\sum_{\beta\in A} i'_\beta) .$$

Now $i_\alpha(\sum_{\beta\in A} i'_\beta) = \sum_{\beta\in A} i_\alpha i'_\beta = i_\alpha i'_\alpha$ by 2.6 (b)

b) CHAIN CONDITIONS

2.31 REMARKS By 1.51, the ideals form an inductive Moore-system. It makes sense to speak about things like "the ideals fulfill the DCC" etc.
By 0.10, if the ideals fulfill the ACC then each ideal is f.g. .

2.32 CONVENTION If the set of ideals fulfills the DCC we say that "N fulfills the DCC for ideals" or more briefly that "N has the DCCI. To simplify statements, the phrase "Let N have the DCCI" will be abbreviated by "DCCI". Similar conventions apply to right ideals (DCCR), left ideals (DCCL) and N-subgroups (DCCN).
Of course, the same is done for the ACC.

2.33 REMARK Clearly the DCCN implies the DCCI if $N = N_0$; in N, DCCR or DCCL imply the DCCI. If $N = N_0$ then the DCCN implies the DCCL.
The same holds for the ACC.

2.34 EXAMPLES

(a) (Beidleman (1)). Let a group Γ contain only finitely many normal subgroups but an infinite chain $\Gamma = \Delta_1 \supset \Delta_2 \supset \ldots$ of subgroups (such groups are known to exist). $N := \{f \in M(\Gamma) \mid \forall i \in \mathbb{N} : f(\Delta_i) \subseteq \Delta_i\}$. Then it is immediate that ${}_N\Gamma$ has the DCCI but not the DCCN (since all $\Delta_i \leq_N \Gamma$).

(b) Each ring satisfying the ACCI but not the DCCI ($\mathbb{Z}$, for instance) or conversely is of course an example of a nr. with the same properties.

2.35 THEOREM

(a) If $I \trianglelefteq N$ and N has the DCCI (DCCN, DCCL) then the same applies to N/I.

(b) If $I \trianglelefteq N$ and I is a direct summand then N has the DCCI (DCCN, DCCL) <=> I and N/I have the DCCI (DCCN, DCCL).

(c) If $\Delta \trianglelefteq_N \Gamma$ ($N \in \mathcal{N}_0$ is a direct summand then Γ has the DCCI (DCCN) iff Δ and Γ/Δ have this property.

Proof. (for ideals of N and the DCC).

(a) Let $\bar{J}_1 \supseteq \bar{J}_2 \supseteq \ldots$ be a descending chain of ideals of N/I. If $J_i := \pi^{-1}(\bar{J}_i)$ ($i \in \mathbb{N}$) then $J_1 \supseteq J_2 \supseteq \ldots$ by 1.30.
So $\exists\, n \in \mathbb{N}\ \forall\, k \geq n : J_k = J_n$. Since $\forall\, i \in \mathbb{N} : \pi(J_i) = \pi(\pi^{-1}(\bar{J}_i)) = \bar{J}_i$ by 2.14(a), $\bar{J}_k = \bar{J}_n$ for all $k \geq n$.

(b) =>: It remains to show that I has also the DCC. But this follows from the fact that each ideal of I is an ideal of N.

<=: Let I and N/I have the DCC and let $J_1 \supseteq J_2 \supseteq \ldots$ be a chain of ideals of N. The chains $J_1 \cap I \supseteq J_2 \cap I \supseteq \ldots$ and $\pi(J_1+I) \supseteq \pi(J_2+I) \supseteq \ldots$ get constant after some $n \in \mathbb{N}$. Therefore $\forall\, k \geq n : J_k \cap I = J_n \cap I \wedge \pi(J_k+I) = \pi(J_n+I)$.

Since $\pi^{-1}(\pi(J_i+I)) = J_i+I+I = J_i+I$, $J_k+I = J_n+I$ for all $k \geq n$. Now $\forall\ x \in J_n : x \in J_n+I = J_k+I$, so $\exists\ y \in J_n\ \exists\ i \in I : x = y+i$.
Therefore $x-y \in I \cap J_k = J_n \cap I \subseteq J_n$ and so $x \in J_n$.
This shows that $\forall\ k \geq n: J_k = J_n$.

(c) The proof is similar to that one of (b).

2.36 REMARK Lausch (4) showed that if $N \in \mathcal{N}_1$ has the DCCN and $e \in N$ has some $e' \in N$ with $e'e = 1$ then $ee' = 1$.

The "Jordan-Hölder-theory" carries over to near-rings and N-groups with $N \in \mathcal{N}_0$ (but we only formulate it for near-rings). The proofs are nearly word for word the same as in group or ring theory and hence omitted. This omission is again justified by the fact that all of 2.37 - 2.41 is a special case of the Jordan-Hölder-theory of Ω-groups (see e.g. (Kurosh), IV, §2).

2.37 DEFINITION A finite sequence

$$N = N_0 \supseteq N_1 \supseteq N_2 \supseteq \ldots \supseteq N_n = \{0\} \qquad (*)$$

of subnear-rings N_i of N is called a normal sequence of N $\Longleftrightarrow \forall\ i \in \{1,\ldots,n\} : N_i \trianglelefteq N_{i-1}$.
In the special case that all $N_i \trianglelefteq N$ we call the normal sequence (*) an invariant sequence.
n is called the length of the sequence (*) and the near-rings N_{i-1}/N_i $(i \in \{1,\ldots,n\})$ are called the factors of (*).
Another normal (invariant) sequence

$$N = M_0 \supseteq M_1 \supseteq M_2 \supseteq \ldots \supseteq M_m = \{0\} \qquad (**)$$

is called a refinement of (*) if
$\forall\ i \in \{0,\ldots,n\}\ \exists\ j \in \{0,\ldots,m\} : N_i = M_j$.
(*) and (**) are called isomorphic if $n = m$ and the factors of (*) and (**) are (after a possibly necessary re-ordering) isomorphic.

(**) is called a proper refinement of (*) if (*) is not a refinement of (*).
A normal (invariant) sequence (*) is called a composition sequence (principal sequence) if (*) has no proper refinement.

2.38 PROPOSITION (*) is a composition (principal) sequence <=> <=> all factors are simple.

2.39 COROLLARY A sequence isomorphic to a composition (principal) sequence is itself a composition (principal) sequence.

We now state the famous Jordan-Hölder-theorem:

2.40 COROLLARY Let N have a composition (principal) sequence. Then each normal (invariant) sequence can be refined to a composition (principal) sequence and all these sequences are isomorphic.

2.41 THEOREM N has a principal sequence <=> the ideals of N fulfill both chain conditions.

c) DECOMPOSITION THEOREMS

2.42 DEFINITION N ($_N\Gamma$) is called decomposable if it is the direct sum of non-trivial ideals (or, equivalently, if it has a non-trivial direct summand), otherwise indecomposable.

2.43 EXAMPLES Clearly each simple nr. (N-group) is indecomposable. The ring $\mathbb{Z}$ is indecomposable, but not simple.
Between the concepts of simplicity, indecomposability and minimality of an ideal (which is at the same time supposed to be an N-subgroup in the case of N-groups) there are the following relations:

minimal <= simple => indecomposable

If the ideal in question is even a direct summand, we get

minimal <=> simple => indecomposable.

2.44 REMARK The next considerations concern merely N-groups with $N\in\mathfrak{N}_0$. The reason is obvious: in general the ideals of N-groups are not necessarily N-groups again. But we have to speak about "simple ideals" etc. . Cf. also Roth (1) and the remarks preceding 2.1.

2.45 THEOREM Let N ($_N\Gamma$ with $N\in\mathfrak{N}_0$) have the DCCI. Then N ($_N\Gamma$) is the finite direct sum of indecomposable ideals.

Proof (for near-rings). If N is not indecomposable then there are non-trivial ideals I_1, I_2 with $N = I_1 \dot{+} I_2$.
If I_1, I_2 are indecomposable, we are through. If not, I_1 or I_2 decompose again properly, et cetera. By the DCCI, these decompositions stop after finitely many steps thereby proving that N is the direct sum of finitely many indecomposable ideals.

The corresponding assertion in nuclear physics is much harder to prove!

2.46 DEFINITION N ($_N\Gamma$ with $N\in\mathfrak{N}_0$) is called completely reducible if N ($_N\Gamma$) is the direct sum of simple ideals. $I \trianglelefteq N$ ($\Delta \leq_N \Gamma$) is completely reducible if I (Δ) is completely reducible when considered as a near-ring (N-group).

2.47 REMARK Another usual name is "semisimple". However, "semisimple" will have another meaning in §5.

2.48 THEOREM (Roth (1), Beidleman (1)). If $N\in\mathcal{N}$, the following conditions are equivalent:

(a) Every ideal of N is the sum of simple ideals.
(b) N is the sum of simple ideals.
(c) N is the direct sum of simple ideals.
(d) N is completely reducible.
(e) Each ideal of N is a direct summand.
(f) $\forall\ I \trianglelefteq N$: I and N/I are completely reducible.
(g) N is the sum of minimal ideals.

The analogous theorem holds for N-groups with $N\in\mathcal{N}_0$.

Proof. (for near-rings).

(a) => (b): trivial.

(b) => (c): If $N = \sum_{\alpha\in A} I_\alpha$, define $\mathcal{A} :=$
$= \{B\subseteq A \mid \sum_{\beta\in B} I_\beta = \dot{\sum_{\beta\in B}} I_\beta\}$. $\mathcal{A} \neq \emptyset$. By Zorn's Lemma, $\mathcal{A}$ contains a maximal element (w.r.t. $\subseteq$) $\bar{B}$.
$\sum_{\beta\in B} I_\beta =: N'$. $\forall\ \alpha\in A$: $(I_\alpha \cap N' = I_\alpha \vee I_\alpha \cap N' = \{0\})$.
$I_\alpha \cap N' = \{0\}$ is a contradiction to the maximality of $\bar{B}$. So $\forall\ \alpha\in A$: $I_\alpha \subseteq N'$ and hence $N = N' = \dot{\sum_{\beta\in\bar{B}}} I_\beta$.

(c) => (d): by definition.

(c) => (e): If $I \trianglelefteq N$, consider an ideal J maximal (Zorn!) with the property that $J\cap I = \{0\}$. $N' := I \dot{+} J$. If $N \neq N'$, $\exists\ J_0 \trianglelefteq N$: J_0 simple $\wedge\ J_0 \not\subseteq N' \wedge J_0 \neq \{0\}$. Then $J_0 \cap N' = \{0\}$, so $J+J_0 \supset J$. Also, $(J+J_0)\cap I = \{0\}$, since $x = j+j_0 \in (J+J_0)\cap I$ implies that $j_0 =$ $= x-j \in (I+J)\cap J_0 = N'\cap J_0 = \{0\}$. This contradicts the maximality of J.
Therefore $I\dot{+}J = N$ and I is a direct summand.

(e) => (a): If $I \trianglelefteq N$, denote by $\bar{I}$ the sum of all simple ideals of I. Assume that $\bar{I} \neq I$.
$\bar{I} \trianglelefteq I \trianglelefteq N \wedge$ I is direct summand => $\bar{I} \trianglelefteq N$. Hence $\bar{I}$ is itself a direct summand and there is some $J \trianglelefteq N$ with $\bar{I}\dot{+}J = N$.

Consequently each simple ideal of $\bar{I}$ is a simple ideal of N. $\bar{I}\dot{+}(J\cap I) = I$, since each $i\in I$ has the form $i = \bar{i}+j$ with $\bar{i}\in\bar{I}$ and $j\in J$; because of $\bar{I}\subseteq I$ we know that $j\in I$. We now show that $J\cap I$ contains a simple non-zero ideal of N and arrive at a contradiction. By assumption, $J\cap I \neq \{0\}$.
If $J\cap I$ is f.g. then there exists a maximal ideal I^* in $J\cap I$, and each direct complement (existence as before) of I^* in $J\cap I$ is a simple non-zero ideal of $J\cap I$ and of N.
If $J\cap I$ is not f.g., take any fg. ideal $F \neq \{0\}$ of $J\cap I$. Then $F \neq J\cap I$. F contains a maximal ideal $M \lhd F$. As before, each direct complement of M in F is a non-zero simple ideal of F and of N contained in $J\cap I$.

(c) => (f): Since (c) => (a), every $I \trianglelefteq N$ is the sum (and by (c) the direct sum) of simple ideals, implying that I is completely reducible.
If $I \trianglelefteq N$, take some $J \trianglelefteq N$ (again, J is completely reducible) with $I\dot{+}J = N$. But then $N/I \cong J$ by 2.8 and N/I is completely reducible.

(f) => (d): trivial (take $I = N$).

(a) => (g): trivial.

(g) => (e): as in (c) => (e).

<u>2.49 COROLLARY</u> The direct sum of completely reducible near-rings (N-groups with $N\in\eta_0$) is again completely reducible.

Near-rings (N-groups) which decompose into finitely many simple ideals are especially important.
The following theorem will be used frequently throughout this book.

2.50 THEOREM (Beidleman (1), Betsch (3)). Let N be a nr. . Equivalent are:

(a) N is the sum of finitely many simple ideals.

(b) N is the direct sum of finitely many simple ideals.

(c) N is completely reducible and has the DCCI and the ACCI.

(d) N is completely reducible and has the ACCI.

(e) N is completely reducible and has the DCCI.

(f) N is completely reducible and every ideal of N is f.g..

(g) There exist maximal ideals $I_1,\dots,I_n$ of N with zero intersection, but all $J_r := \bigcap_{k \neq r} I_k \neq \{0\}$.

(in this case, $N = \sum_{r=1}^{n}{}^{\bullet} J_r$ and $J_1,\dots,J_n$ are simple).

(h) There exist maximal ideals $I_1,\dots,I_n$ with $\bigcap_{r=1}^{n} I_r = \{0\}$.

The usual changes yield analogous results for N-groups with $N \in \mathcal{N}_0$ (remark also the additional results in Oswald (2)).

Proof. (a) <=> (b): as in 2.48.

(b) => (c): If $N = \sum_{k=1}^{n} I_k$ (all I_k simple) then N is clearly completely decomposable. Moreover, $N = I_1 \dot{+} \dots \dot{+} I_n \supset I_1 \dot{+} \dots \dot{+} I_{n-1} \supset \dots \supset I_1 \supset \{0\}$ is a principal series, so N fulfills both chain conditions by 2.41.

(c) => (d) and (c) => (e) are trivial.

(d) <=> (f): by 0.10.

(d) => (b) and (e) => (b): If $N = \sum_{\alpha \in A}{}^{\bullet} I_\alpha$, the ACC (DCC) forces A to be finite.

(b) => (g): If $N = J_1 \dot{+} \ldots \dot{+} J_n$ (J_i simple ideals), define $I_k := \sum_{r \neq k} J_r$. Because of $N/I_k \cong J_k$, all I_k are maximal ideals. If $x \in \bigcap_{k=1}^{n} I_k$, $x = j_1 + \ldots + j_n$ ($j_i \in J_i$) and if $\exists\, k \in \{1,\ldots,n\}: j_k \neq 0$ then $x \notin I_k$, a contradiction. So $\bigcap_{k=1}^{n} I_k = \{0\}$.

Since $\bigcap_{k \neq r} I_k = J_r \neq \{0\}$, we are through.

(g) => (h): trivial.

(h) => (b): Let $I_1,\ldots,I_n$ be minimal w.r.t. the property that their intersection $= \{0\}$. Then each $J_r := \bigcap_{k \neq r} I_k \neq \{0\}$. Since $\forall\, r \in \{1,\ldots,n\}: J_r \nsubseteq I_r$, but $J_r \cap I_r = \{0\}$, we have $N = J_r \dot{+} I_r$. Hence $J_r \cong N/I_r$ and J_r is simple.

Let for $r \in \{1,\ldots,n\}$ $K_r := I_1 \cap \ldots \cap I_r$.

We claim that $N = J_1 \dot{+} \ldots \dot{+} J_r \dot{+} K_r$ and prove this by induction on r.

If $r = 1$ then $K_r = I_1$ and $J_1 \dot{+} I_1 = N$.

Assume that it is shown for r ($< n$). We show the assertion for $r+1$.

Since $I_{r+1} + K_r = N$ (by maximality),

$$N/I_{r+1} = I_{r+1}+K_r/I_{r+1} \cong K_r/I_{r+1} \cap K_r = K_r/K_{r+1}.$$

Since N/I_{r+1} is simple, the same applies to K_r/K_{r+1} and K_{r+1} is a maximal ideal in K_r.

$$J_{r+1} \cap K_{r+1} = \Big(\bigcap_{k \neq r+1} I_k\Big) \cap \Big(\bigcap_{\ell=1}^{r+1} I_\ell\Big) = \bigcap_{k=1}^{n} I_k = \{0\}.$$

Also $J_{r+1} \subseteq K_r$, but $J_{r+1} \nsubseteq K_{r+1}$

Hence $K_{r+1} \dot{+} J_{r+1} = K_r$ and $N = J_1 \dot{+} \ldots \dot{+} J_r \dot{+} K_r =$

$= J_1 \dot{+} \ldots \dot{+} J_r \dot{+} J_{r+1} \dot{+} K_{r+1}$.

But $K_n = \{0\}$, so $N = \sum_{k=1}^{n} J_k$.

2.51 REMARKS The proof of (h) ⟹ (b) in 2.50 could also be done by using subdirect products and "words generating prime ideals" similar to (McCoy), p. 59. Cf. also (Higgins), §9. At a first glance one might assume that "f.g." implies already "completely reducible". This is not the case: take the zero-nr. N on the dihedral group D_8 on 8 elements. Then normal subgroups and ideals coincide. But D_8 is known to have $G \trianglelefteq D_8$ and $H \trianglelefteq G$, but $H \ntrianglelefteq D_8$. By 2.48(e) and 2.10 N cannot be completely reducible.

2.52 COROLLARIES

(a) If N fulfills one (and hence all) of the conditions in 2.50 and if $I \trianglelefteq N$ then the same applies to I (use 2.48(f), 2.48(e) and 2.35(b)).

(b) If N has the DCCI and is a subdirect product of simple near-rings N_i $(i \in I)$ then $\exists J \subseteq I$, J finite: $N = \bigoplus_{j \in J} N_j$ (apply 2.50(h) and 1.60).

Again, similar statements hold for N-groups with $N = N_0$.

2.53 DEFINITION Two decompositions of N : $N = \sum^{\bullet}_{\alpha \in A} I_\alpha = \sum^{\bullet}_{\beta \in B} J_\beta$ are called isomorphic if $|A| = |B|$ and the I_α's and J_β's are - up to order - isomorphic.

The Krull-Schmidt-Theorem reads as

2.54 THEOREM (Roth (1)). If N (${}_N\Gamma$ with $N \in \mathcal{N}_0$) fulfills one (and hence all) of the conditions of 2.50 then any two decompositions of N (${}_N\Gamma$) into simple ideals are isomorphic.

Proof (for nr.'s) If $N = I_1 \dot{+} \ldots \dot{+} I_n = J_1 \dot{+} \ldots \dot{+} J_m$ (I_k, J_ℓ simple) then $N \supseteq I_1 \dot{+} \ldots \dot{+} I_{n-1} \supseteq \ldots \supseteq I_1 \supseteq \{o\}$ and $N \supseteq J_1 \dot{+} \ldots \dot{+} J_{m-1} \supseteq \ldots \supseteq J_1 \supseteq \{o\}$ are two invariant sequences with simple factors $\sum^{\bullet r}_{k=1} I_k \Big/ \sum^{\bullet r-1}_{k=1} I_k \cong I_r$ and

$$\dot{\sum_{\ell=1}^{s}} J_\ell \Big/ \dot{\sum_{\ell=1}^{s-1}} J_\ell \cong J_s \qquad (2 \le r \le n \quad \text{and} \quad 2 \le s \le m) .$$

By 2.40 these sequences and therefore these decompositions are isomorphic.

Compare the following result with §9 of (Higgins).

2.55 THEOREM

(a) If $N = \dot{\sum_{r=1}^{n}} I_r$ (all I_r simple) and if $I \trianglelefteq N$ then there is a subset S of $\{1,\dots,n\}$ with

$$I \cong \dot{\sum_{s\in S}} I_s .$$

(b) If $I,J \trianglelefteq N$ are such that N/I and N/J are completely reducible then $N/_{I\cap J}$ is completely reducible, too, all of whose simple summands being isomorphic to one of the simple components of N/I or N/J.

Again, the corresponding theorem holds for N-groups with $N = N_0$.

Proof. (a) Let $K_r := I+I_1+\dots+I_r$ $(1\le r\le n)$, $K_0 := I$. Then $K_n = N$. $\forall\, r\in\{1,\dots,n-1\}$: $K_r \trianglelefteq N \wedge$ $\wedge\ K_r \cap I_{r+1} \trianglelefteq I_{r+1}$. Thus we have either $K_{r+1} = K_r$ or $K_{r+1} = K_r \dot{+} I_{r+1}$.

Hence $\exists\ T\subseteq\{1,\dots,n\}$: $N = I \dot{+} \dot{\sum_{t\in T}} I_t$, and so by 2.8

$$I \cong N\Big/\dot{\sum_{t\in T}} I_t = \dot{\sum_{r=1}^{n}} I_r \Big/ \dot{\sum_{t\in T}} I_t \cong \dot{\sum_{s\in S}} I_s \quad \text{with}$$

$S := \{1,\dots,n\}\setminus T$.

(b) Let $K := I+J$. Then $K/I \trianglelefteq N/I$ and $\exists\ M \trianglelefteq N$: $N/I = (K/I) \dot{+} (M/I)$, whence $K+M = N$ and $K\cap M = I$. So $M\cap J = M\cap K\cap J = I\cap J$ and $M+J \supseteq I+J = K$, $M+J \supseteq M+K = N$, hence $M+J = N$. Consequently

$N/_{I\cap J} = M/_{I\cap J} \dotplus J/_{I\cap J}$.

$M/_{I\cap J} = M/_{M\cap J} \cong M+J/_{J} = N/J$ and

$J/_{I\cap J} \cong K/I \trianglelefteq N/I$ are completely reducible, so

$N/_{I\cap J}$ is completely reducible by 2.49.

The rest follows from (a).

d) PRIME IDEALS

1.) PRODUCTS OF SUBSETS.

2.56 NOTATION If $S,T \subseteq N$ then $ST := \{st \mid s\in S \wedge t\in T\}$.
For $n\in\mathbb{N}$, the definition of S^n is then clear.

2.57 PROPOSITION (Maxson (1)).

(a) $\forall\ R,S,T \subseteq N$: $(RS)T = R(ST)$.

(b) If $h: N \to \overline{N}$ then $\forall\ S,T \subseteq N$: $h(ST) = h(S)h(T)$ and $\forall\ \overline{S},\overline{T} \subseteq \overline{N}$: $h^{-1}(\overline{S}\overline{T}) \supseteq h^{-1}(\overline{S})h^{-1}(\overline{T})$.

(c) $\forall\ I \trianglelefteq N$ $\forall S,T \subseteq N$: $(S+I)(T+I) = ST+I$.

Proof. (a) and (b) are immediate.

(c) follows from (b) for $\pi: N \to N/I$.

2.58 REMARK Note that ST has no particular structure in general. Even if S,T are ideals, ST is not even a subsemigroup of $(N,+)$ except in some very special cases.

2.) PRIME IDEALS

2.59 DEFINITION $P \trianglelefteq N$ is called prime if $\forall\ I,J \trianglelefteq N$: $IJ \subseteq P \Rightarrow$ $\Rightarrow I \subseteq P \vee J \subseteq P$.

2.60 NOTATION For $S \subseteq N$, let (S) be the ideal generated by S. $(\{n\}) =: (n)$.

2.61 PROPOSITION (Van der Walt (1)). Let P be an ideal of N. Equivalent are

(a) P is a prime ideal.

(b) $\forall\ I,J \trianglelefteq N$: $(IJ) \subseteq P \Rightarrow I \subseteq P \vee J \subseteq P$.

(c) $\forall\ i,j \in N$: $i \notin P \wedge j \notin P \Rightarrow (i)(j) \not\subseteq P$.

(d) $\forall\ I,J \trianglelefteq N$: $I \supsetneq P \wedge J \supsetneq P \Rightarrow IJ \not\subseteq P$.

(e) $\forall\ I,J \trianglelefteq N$: $I \not\subseteq P \wedge J \not\subseteq P \Rightarrow IJ \not\subseteq P$.

Proof. (a) $\Leftrightarrow$ (b) $\Leftrightarrow$ (e) is trivial.

(a) $\Rightarrow$ (c): If $(i)(j) \subseteq P$ then $(i) \subseteq P$ or $(j) \subseteq P$, so $i \in P \vee j \in P$.

(c) $\Rightarrow$ (d): If $I \supsetneq P \wedge J \supsetneq P$, take $i \in I \setminus P$ and $j \in J \setminus P$. Then $(i)(j) \not\subseteq P$, so $IJ \not\subseteq P$.

(d) $\Rightarrow$ (e): If $I \not\subseteq P \wedge J \not\subseteq P$, take $i \in I \setminus P$ and $j \in J \setminus P$. Then $(i)+P \supsetneq P$ and $(j)+P \supsetneq P$. Then $((i)+P)((j)+P) \not\subseteq P$. So $\exists\ i' \in (i)\ \exists\ j' \in (j)\ \exists\ p,p' \in P$: $(i'+p)(j'+p') \notin P$. Therefore $i'(j'+p')-i'j'+i'j'+p(j'+p') \notin P$. But since $i'(j'+p')-i'j' \in P$ and $p(j'+p') \in P$, $i'j' \notin P$, hence $IJ \not\subseteq P$.

2.62 PROPOSITION Let $(P_\alpha)_{\alpha \in A}$ be a family of prime ideals, totally ordered by inclusion. Then $\bigcap_{\alpha \in A} P_\alpha =: P$ is a prime ideal, too.

Proof. We may assume that A is ordered such that for $\alpha,\beta \in A$ $\alpha \leqq \beta \Rightarrow P_\alpha \subseteq P_\beta$.

Of course, P is an ideal. Let I,J be ideals of N. $IJ \subseteq \bigcap_{\alpha\in A} P_\alpha \Rightarrow \forall\, \alpha\in A: IJ \subseteq P_\alpha$. If $\exists\, \alpha\in A: I \nsubseteq P_\alpha$ then $J \subseteq P_\alpha$. $\forall\, \beta\geq\alpha: J \subseteq P_\beta$. If $\exists\, \gamma<\alpha: J \nsubseteq P_\gamma$ then $I \subseteq P_\gamma$, so $I \subseteq P_\alpha$, a contradiction.
So $\forall\, \alpha\in A: J \subseteq P_\alpha$ and $J \subseteq \bigcap_{\alpha\in A} P_\alpha$.

<u>2.63 PROPOSITION</u> (Maxson (1)). If $I \trianglelefteq N$ is a direct summand and $P \trianglelefteq N$ is prime then $P\cap I$ is a prime ideal in I.

<u>Proof</u>. If $J_1J_2 \subseteq P\cap I$ $(J_1,J_2 \trianglelefteq I)$ then $J_1J_2 \subseteq P$ and $J_1,J_2 \trianglelefteq N$, so $J_1 \subseteq P$ or $J_2 \subseteq P$ and therefore $J_1 \subseteq P\cap I$ or $J_2 \subseteq P\cap I$.

<u>2.64 PROPOSITION</u> If $I \trianglelefteq N$ and $I \subseteq P \trianglelefteq N$ and if $\pi: N \to N/I =: \overline{N}$ is the canonical epimorphism as usual then: P is prime $\iff \pi(P)$ is prime.

<u>Proof</u>. $\Rightarrow$: If $\mathcal{J}_1\mathcal{J}_2 \subseteq \pi(P)$ $(\mathcal{J}_1,\mathcal{J}_2 \trianglelefteq \overline{N})$, let $J_i := \pi^{-1}(\mathcal{J}_i)$ $(i\in\{1,2\})$. By 2.57, $J_1J_2 = \pi^{-1}(\mathcal{J}_1)\pi^{-1}(\mathcal{J}_2) \subseteq$
$\subseteq \pi^{-1}(\mathcal{J}_1\mathcal{J}_2) \subseteq \pi^{-1}(\pi(P)) = P+I = P$.
So $J_1\subseteq P \vee J_2\subseteq P$, hence $\mathcal{J}_1 = \pi(\pi^{-1}(\mathcal{J}_1)) = \pi(J_1)\subseteq\pi(P)$ or $\mathcal{J}_2\subseteq\pi(P)$.

$\Leftarrow$: If $J_1J_2\subseteq P$ then $\pi(J_1)\pi(J_2) = \pi(J_1J_2)\subseteq\pi(P)$.
So $\pi(J_1)\subseteq\pi(P)$ or $\pi(J_2)\subseteq\pi(P)$. This shows that either $J_1\subseteq J_1+I = \pi^{-1}(\pi(J_1))\subseteq\pi^{-1}(\pi(P)) = P+I = P$ or $J_2\subseteq P$.

<u>2.65 DEFINITION</u> Call N a <u>prime near-ring</u> if $\{0\}$ is a prime ideal.

2.66 EXAMPLES

(a) Every integral near-ring is of course a prime near-ring (for $I \cdot J \subseteq \{0\}$ and $I \neq \{0\}$, $J \neq \{0\}$ would guarantee the existence of some $i \in I^*$, $j \in J^*$ (see p. 1) with $ij = 0$).

(b) N is a prime ideal of N, so $\{0\}$ is a prime ring. More generally:

2.67 PROPOSITION If $I \trianglelefteq N$, I is a prime ideal iff N/I is a prime ring.

Proof. Take $P = I$ in 2.64.

2.68 EXAMPLE If $N \in \mathcal{N}_c$ then each normal subgroup of $(N,+)$ is a prime ideal.

2.69 COROLLARY Each constant near-ring is a prime near-ring.

2.70 PROPOSITION N simple $\Rightarrow$ N is prime or N is a zero-near-ring. The proof is trivial. More generally:

2.71 PROPOSITION If $I \triangleleft N$ is a maximal ideal then I is either prime or $N^2 \subseteq I$.

Proof. N/I is simple. By 2.70, N/I is either prime (implying that I is a prime ideal) or N/I is a zero-nr. which causes $N^2 \subseteq I$.

2.72 COROLLARY If $I \triangleleft N \in \mathcal{N}_1$ is maximal then I is prime.

2.73 REMARK If I is prime, I is not necessarily maximal (not even for finite dgnr's.: see Laxton-Machin (1)).

2.74 DEFINITION An ideal minimal in the set of all prime ideals containing some given ideal I is called a minimal prime ideal of I.

Applying 2.66(b) and Zorn's lemma on $(\{P \trianglelefteq N \mid P \supseteq I \wedge P \text{ prime}\}, \supseteq)$ we get

<u>2.75 PROPOSITION</u> For each ideal I there exists a minimal prime ideal of I.

<u>2.76 DEFINITION</u> A minimal prime ideal of $\{0\}$ is called a <u>minimal prime ideal</u> (in N).

<u>2.77 COROLLARY</u>

(a) Each prime ideal contains a minimal prime ideal.
(b) N has a minimal prime ideal.

<u>Proof</u>. Take $I = \{0\}$ in 2.75.

As in ring theory (cf. e.g.(McCoy)), the complements of prime ideals deserve some interest.

<u>2.78 DEFINITION</u> $M \subseteq N$ is called an <u>m-system</u> if $\forall\, a,b \in M\ \exists\, a_1 \in (a)\ \exists\, b_1 \in (b) : a_1 b_1 \in M$.

<u>2.79 EXAMPLES</u>

(a) $\emptyset$ and N are trivial examples of m-systems.
(b) $\forall\, n \in N$: $\{n, n^2, n^3, \ldots\}$ is an m-system.

2.61(c) gives us

<u>2.80 COROLLARY</u> If $P \trianglelefteq N$, P is a prime ideal iff $N \setminus P$ is an m-system.

<u>2.81 PROPOSITION</u> (Van der Walt (1)). Let $M \subseteq N$ be a non-void m-system in N and I an ideal of N with $I \cap M = \emptyset$. Then I is contained in a prime ideal $P \neq N$ with $P \cap M = \emptyset$.

Proof. $\mathcal{I} := \{J \trianglelefteq N : J \cap M = \emptyset\}$. $I \in \mathcal{I}$. By Zorn's Lemma, $\mathcal{I}$ contains a maximal element P. P is an ideal $\neq N$.
P is in fact a prime ideal:
If $J_1 \supsetneq P \wedge J_2 \supsetneq P$ then take some $j_1 \in J_1 \cap M$ and $j_2 \in J_2 \cap M$.
$(j_1)(j_2) \subseteq J_1 J_2$, and $\exists\, j_1' \in (j_1)\ \exists\, j_2' \in (j_2) : j_1' j_2' \in M$.
So $(J_1 J_2) \cap M \neq \emptyset$, $(J_1 J_2) \nsubseteq P$ and $J_1 J_2 \nsubseteq P$.

3.) SEMIPRIME IDEALS

2.82 DEFINITION $S \trianglelefteq N$ is semiprime: $\Longleftrightarrow \forall\, I \trianglelefteq N: I^2 \subseteq S \Rightarrow I \subseteq S$.

Evidently, each prime ideal is semiprime.
Similar to 2.61 we get

2.83 PROPOSITION For an ideal S of N the following conditions are equivalent:

(a) S is semiprime.

(b) $\forall\, I \trianglelefteq N: (I^2) \subseteq S \Rightarrow I \subseteq S$.

(c) $\forall\, n \in N: (n)^2 \subseteq S \Rightarrow n \in S$.

(d) $\forall\, I \trianglelefteq N: I \supsetneq S \Rightarrow I^2 \nsubseteq S$.

(e) $\forall\, I \trianglelefteq N: I \nsubseteq S \Rightarrow I^2 \nsubseteq S$.

2.84 PROPOSITION If $(S_\alpha)_{\alpha \in A}$ is a family of semiprime ideals then $\bigcap_{\alpha \in A} S_\alpha$ is again semiprime.

Proof. If $I \trianglelefteq N$ and $I^2 \subseteq \bigcap_{\alpha \in A} S_\alpha$ then $\forall\, \alpha \in A: I^2 \subseteq S_\alpha$,
so $\forall\, \alpha \in A: I \subseteq S_\alpha$, hence $I \subseteq \bigcap_{\alpha \in A} S_\alpha$.

As in 2.63 and 2.64 we get

2.85 PROPOSITION Let $I \trianglelefteq N$ be a direct summand and $S \trianglelefteq N$ be semiprime then $S \cap I$ is semiprime in I.

2.86 PROPOSITION $I \trianglelefteq N \wedge I \subseteq S \trianglelefteq N$. Then S is semiprime iff $\pi(S) \subseteq N/I$ is semiprime.

2.87 DEFINITION N is called a semiprime near-ring if $\{0\}$ is a semiprime ideal.

2.67, and 2.74 - 2.77 can again be transferred to semiprime near-rings (ideals).

2.88 DEFINITION (Maxson (1)). $S \subseteq N$ is called an sp-system if $\forall\ s \in S\ \exists\ s_1, s_2 \in (s) : s_1 s_2 \in S$.

2.89 PROPOSITION (Maxson (1)).

(a) Each m-system is an sp-system.

(b) $\forall\ S \trianglelefteq N$: S is semiprime $\Leftrightarrow$ $N \setminus S$ is an sp-system.

The proof is trivial.

2.90 PROPOSITION (cf. 2.81). Let S be a non-void sp-system in N. Let I be an ideal of N with $I \cap S = \emptyset$. Then I is contained in a semiprime ideal $\neq N$.

Now we study some relations between prime and semiprime ideals. Since each prime ideal is semiprime we get at once from 2.84

2.91 PROPOSITION Any intersection of prime ideals is a semiprime ideal.

2.92 PROPOSITION (Maxson (1)). Let S be an sp-system and $s \in S$. Then there is some m-system M with $s \in M \subseteq S$.

Proof. $s \in S \Rightarrow \exists\ s_1, s_2 \in (s) : s_1 s_2 \in S \Rightarrow \exists\ s_1', s_2' \in (s_1 s_2) :$ $: s_1' s_2' \in S$.

Continuing this process, one gets a sequence

$$s,\ s_1 s_2,\ s_1' s_2',\ \ldots,\ s_1^{(k)} s_2^{(k)},\ \ldots$$

with $\forall\ k \in \mathbb{N} : s_1^{(k)} s_2^{(k)} \in S$ and $(s) \supseteq (s_1 s_2) \supseteq (s_1' s_2') \supseteq \ldots$.

Take $M := \{s, s_1 s_2, s_1' s_2', \ldots\}$. We show that M is a desired m-system.

If $s_1^{(k)} s_2^{(k)}, s_1^{(\ell)} s_2^{(\ell)} \in M$ (w.l.o.g. $\ell \leq k$) then $(s_1^{(k)} s_2^{(k)}) \supseteq (s_1^{(\ell)} s_2^{(\ell)})$. Take $s_1^{(\ell+1)} s_2^{(\ell+1)} \in$

$\in (s_1^{(\ell)} s_2^{(\ell)}) \subseteq (s_1^{(k)} s_2^{(k)})$; then $s_1^{(\ell+1)} s_2^{(\ell+1)} \in M$.

2.93 DEFINITION If $I \trianglelefteq N$, call $\mathcal{P}(I) := \bigcap_{\substack{P \text{ prime id.} \\ P \supseteq I}} P$ the prime radical of I.

Of course, $\mathcal{P}(I)$ is a semiprime ideal (by 2.91) containing I.

2.94 PROPOSITION $n \in \mathcal{P}(I) \Rightarrow \exists\, k \in \mathbb{N} : n^k \in I$.

Proof. $M := \{n, n^2, n^3, \ldots\}$ is an m-system (2.79(b)).
If $I \cap M = \emptyset$ then by 2.81 there is some prime ideal $P \supseteq I$ with $P \cap M = \emptyset$, a contradiction to $n \in \mathcal{P}(I)$.
Hence $I \cap M \neq \emptyset$ and $\exists\, k \in \mathbb{N} : n^k \in I$.

2.95 REMARK Beidleman (7) also defined prime N-groups ${}_N\Gamma$ as those ones in which

$\forall\, \Delta \leq_N \Gamma\ \forall\, L \trianglelefteq_\ell N : L\Delta = \{o\} \Rightarrow L = \{0\} \vee \Delta = \{o\}$.

N is called strictly prime if ${}_N N$ is prime. He showed that

(a) for zerosymmetric near-rings in $\mathcal{N}_1$, "${}_N\Gamma$ prime" implies "N prime" (but not conversely),

(b) N strictly prime $\Rightarrow$ N prime (but not conversely, even if N is finite),

and some more results related to the structure of ${}_N\Gamma$ (see §4).

e) NIL AND NILPOTENT

2.96 DEFINITION

(a) $n \in N$ is called nilpotent if $\exists\, k \in \mathbb{N}: n^k = 0$.

(b) $S \subseteq N$ is called nilpotent if $\exists\, k \in \mathbb{N}: S^k = \{0\}$

(c) $S \subseteq N$ is called nil if all $s \in S$ are nilpotent.

2.97 REMARKS

(a) $S \subseteq N$ nilpotent $\Rightarrow$ S nil. (In 3.40 we will see that if $N \in \mathcal{N}$ has the DCCN then "nil" and "nilpotent" coincide for N-subgroups.)

(b) $S \subseteq T \subseteq N \wedge T$ nil(potent) $\Rightarrow$ S nil(potent).

2.98 EXAMPLES

(a) In $\mathbb{Z}_4[x]$, $2x$ is nilpotent.

(b) If $n \in N_c$ is nilpotent then $n = 0$.

2.99 COROLLARY If $I \trianglelefteq N$ is nil then $I \subseteq N_0$.

Proof. By 2.18, $I = I_0 + I_c$, so by 2.97(b) $I_c = N_c \cap I$ is nil, hence by 2.98(b) $I_c = \{o\}$ and $I = I_0 \subseteq N_0$.

2.100 THEOREM $I \trianglelefteq N$. N is nil(potent) $\Leftrightarrow$ I and N/I are nil(potent).

Proof (for nilpotence)

$\Rightarrow$: by 2.97(b), I is nilpotent.
If $\exists\, k \in \mathbb{N}: N^k = \{0\}$ then $(N/I)^k = N^k/I = \{I\}$.
$\Leftarrow$: $\exists\, k_1 \in \mathbb{N}: (N/I)^{k_1} = \{I\}$, so $N^{k_1} \subseteq I$ and
$\exists\, k_2 \in \mathbb{N}: I^{k_2} = \{0\}$.
Therefore $(N^{k_1})^{k_2} = \{0\}$ and N is nilpotent.

The proof for "nil" is similar.

<u>2.101 PROPOSITION</u> (Ramakotaiah (1)). Let I_α $(\alpha\varepsilon A)$ be ideals of N.

(a) ($\forall\alpha\varepsilon A$: I_α nilpotent $\wedge$ A finite) $\Rightarrow$ $\sum_{\alpha\varepsilon A} I_\alpha$ is nilpotent.

(b) ($\forall\alpha\varepsilon A$: I_α nil) $\Rightarrow$ $\sum_{\alpha\varepsilon A} I_\alpha$ is nil.

<u>Proof</u>. Let I,J be nil(potent) ideals. $I+J/_I \cong J/_{I\cap J}$. By 2.100 $J/_{I\cap J}$ and by assumption I are nil(potent). Harassing 2.100 again, I+J is nil(potent). By induction we get (a) and (b) for a finite A. In (b), let $i\varepsilon \sum_{\alpha\varepsilon A} I_\alpha$: $i = i_{\alpha_1}+\ldots+i_{\alpha_k}$ (say). Then $i\varepsilon I_{\alpha_1}+\ldots+I_{\alpha_k}$ and i is again nilpotent.

<u>2.102 PROPOSITION</u> (Polin (2)). Let N be isomorphic to a sub-direct product of near-rings N_α $(\alpha\varepsilon A)$ without non-zero nil(potent) N-subgroups, left ideals or ideals. Then N has the same property.

<u>Proof</u> (for N-subgroups of N). Let $M \leq_N N$ be nil(potent). Let $\pi_\alpha: N \twoheadrightarrow N_\alpha$ be the usual epimorphisms (1.58). Then all $\pi_\alpha(M)$ are nil(potent) in N_α, hence $= \{0\}$, so $M \subseteq \bigcap_{\alpha\varepsilon A} \mathrm{Ker}\, \pi_\alpha = \{0\}$ and therefore $M = \{0\}$. (If $N\notin\mathcal{N}_0$ then $\exists\, \alpha\varepsilon A$: $N_\alpha\notin\mathcal{N}_0$. So neither in N nor in N_α there are nil(potent) N-subgroups and the proposition is meaningless in this case).

The following proposition will be useful later on.

<u>2.103 PROPOSITION</u> (Polin (2)). $I,J \trianglelefteq N \wedge I$ nil(potent). Then $I+J/_J$ is nil(potent) in N/J.

<u>Proof</u>: by 2.8 and 2.100.

There are several connections between nil(potent) and (semi)prime ideals:

2.104 PROPOSITION (Maxson (1)). If $I \trianglelefteq N$. Then N/I has no nilpotent ideals iff I is semiprime.

Proof. =>: Assume that N/I has no nilpotent ideals and that $J \trianglelefteq N$, $J^2 \subseteq I$. Then $\pi(J)^2 = \pi(J^2) = \{I\}$ (zero ideal of N/I). So $\pi(J) = \{I\}$ and $J \subseteq I$.

<=: Conversely, assume that $\bar{J} \trianglelefteq N/I$ and $\exists\ k \in \mathbb{N} : \bar{J}^k = \{I\}$. Then $J := \pi^{-1}(\bar{J})$ fulfills (by 2.57(b)) $I = \pi^{-1}(\bar{J}^k) \supseteq (\pi^{-1}(\bar{J}))^k = J^k$.
If I is semiprime, J can be shown to be in I, so $\bar{J} = \{I\}$.

2.105 THEOREM

(a) (Polin (2)). If $I \trianglelefteq N$ then $\mathcal{P}(I)$ contains all nilpotent ideals of N.

(b) $\mathcal{P}(\{0\})$ is nil.

Proof. (a) $\mathcal{P}(I)$ is semiprime. So $N/\mathcal{P}(I)$ has no nilpotent ideals. Assume that J is a nilpotent ideal of N. By 2.103, $J+\mathcal{P}(I)/\mathcal{P}(I)$ is nilpotent in $N/\mathcal{P}(I)$ and therefore zero. Hence $J \subseteq \mathcal{P}(I)$.

(b) follows from 2.94.

2.106 THEOREM (Scott (3)). If $N \in \mathcal{N}_0$ has DCCI then N is a prime near-ring <=> N has a smallest ideal I under all non-zero ideals and I is not nilpotent.

Proof. =>: If N is prime, {0} is a prime ideal. Let I be a minimal ideal (existence guaranteed by the DCCI). I is not nilpotent by 2.104.
If J is another minimal ideal then $\{0\} \neq IJ \subseteq I$, so $(IJ) = I$. Similarly, $(IJ) = J$, so $I = J$.
If K is another non-zero ideal, K contains a minimal ideal of N, so K contains I. Hence I is the smallest of all non-zero ideals.

<=: Conversely, let I be the unique minimal ideal and suppose that I is not nilpotent.
If $J_1, J_2 \trianglelefteq N$, $J_1J_2 = \{0\}$, but $J_1 \neq \{0\}$ and $J_2 \neq \{0\}$ then J_1, J_2 contain a minimal ideal of N by the DCCI and this minimal ideal = I.
Hence $I \subseteq J_1 \wedge I \subseteq J_2$, so $I^2 \subseteq J_1J_2 = \{0\}$ and I is nilpotent, a contradiction.

From 1.60 we get the following

<u>2.107 COROLLARY</u> If $N \in \mathcal{N}_0$ is prime and has the DCCI then N is subdirectly irreducible.

<u>2.108 REMARKS</u> See Oswald (2) for a discussion of "strictly (semi-)prime" near-rings and Holcombe (1) for "0-, 1- and 2-(semi-)prime ideals" and their connection to ν-primitive ideals (4.2 (c), cf. 4.34). Thereby $I \trianglelefteq N$ is called <u>0-(1-,2-) prime</u> if for all ideals (left ideals, N-subgroups) A,B of N: $AB \subseteq I \Rightarrow A \subseteq I \vee B \subseteq I$, and similar for 0-(1-,2-) semiprime ideals. So 0-(semi-)prime ideals are just our (semi-)prime ideals.

PART II

STRUCTURE THEORY

§ 3 ELEMENTS OF THE STRUCTURE THEORY

Irreducible (ring-) modules ${}_RM$ (i.e. simple ones with $RM \neq \{0\}$) play an important rôle in ring theory. They have e.g. the property that $\forall\ m \in M$: $Rm = \{o\} \vee Rm = M$. However, simple N-groups ${}_N\Gamma$ with $N\Gamma \neq \{o\}$ do not enjoy this property. It might also come to mind to use N-simplicity or N_o-simplicity (both equivalent to simplicity in the ring case). So we define 3 types (type 0,1 and 2) of N-groups, all coinciding with irreducibility in the case of modules, with type 2 implying type 1 and this in turn type 0. Monogenic N-groups ($\exists\ \gamma \in N$: $N\gamma = \Gamma$) are particularly important. For instance, we prove that every monogenic N-subgroup of N contains a right identity if $N=N_o$ has the DCCN.
We then study the effect on the type of ${}_N\Gamma$ of changing N into N/I, N_o or N_c.
In c), modular left ideals are introduced in the same way as for rings. Many theorems of ring theory carry over to near-rings: each modular left ideal is contained in a maximal one, modular left ideals are exactly the annihilators of generators of monogenic N-groups, the intersection of two maximal modular left ideals is modular, etc. It is advisable to call a modular left ideal L ν-modular if N/L is an N-group of type ν.
We prove e.g. that if I is a direct summand of $N=N_o$ then every ν-modular left ideal in I is the "trace" of one in N.
We also introduce quasiregularity for abusing it to show that "nil" and "nilpotent" coincide for $M \leq_N N$ if N is a zero-symmetric near-ring with DCCN.
If $N \in \mathcal{N}_o$ has a right identity e and if N is the finite direct sum of left ideals then decomposing e into Σe_i yields "orthogonal" idempotents e_i. Another method (due to S.D. Scott) to get orthogonal idempotents is presented and central idempotents are discussed.

Finally, we consider zero-symmetric near-rings N with minimum condition on N-subgroups and show e.g. that every "minimal non-nilpotent" N-subgroup (left ideal) contains a right identity (a non-zero idempotent, respectively), and that every minimal ideal is a finite direct sum of N-isomorphic minimal left ideals.

a) TYPES OF N-GROUPS

3.1 DEFINITION

(a) ${}_N\Gamma$ is monogenic: $\Leftrightarrow \exists\, \gamma \in \Gamma$: $N\gamma = \Gamma$.
(In this case we say that ${}_N\Gamma$ is "monogenic by γ" and γ is called a generator for ${}_N\Gamma$.)

(b) ${}_N\Gamma$ is strongly monogenic: $\Leftrightarrow$ ${}_N\Gamma$ is monogenic and $\forall\, \gamma \in \Gamma$: $(N\gamma = \{o\} \vee N\gamma = \Gamma)$.

3.2 REMARK Observe that a strongly monogenic N-group ${}_N\Gamma$ has $\Omega = \{o\}$ or $\Omega = \Gamma$.

3.3 EXAMPLES

(a) Each ${}_N\Gamma$ with $\Omega = \Gamma$ is strongly monogenic.

(b) ${}_{M(\Gamma)}\Gamma$ and ${}_{M_o(\Gamma)}\Gamma$ are strongly monogenic.

See also the examples 3.8, 3.9 below.

Now we list some properties of monogenic N-groups which are useful for the sequel.

3.4 PROPOSITION Let ${}_N\Gamma$ be monogenic (by γ_0). Then

(a) $L \trianglelefteq_\ell N \Rightarrow L\gamma_0 \trianglelefteq_N \Gamma$.

(b) If e is a left identity of N then $\forall\ \gamma\in\Gamma$: $e\gamma = \gamma$.

(c) If e is a left identity of N and if ${}_N\Gamma$ is faithful then e is a two-sided identity.

(d) If Γ is N_0-simple (Γ can be considered as an N_0-group!) then either $N\Gamma = \{o\}$ or ${}_N\Gamma$ is strongly monogenic.

(e) ${}_N\Gamma \cong_N N/(o:\gamma_0)$.

(f) ${}_N\Gamma$ is simple $\Leftrightarrow$ $(o:\gamma_0)$ is a maximal left ideal or $= N$.

(g) ${}_N\Gamma$ is N-simple $\Leftrightarrow$ there is no N-subgroup strictly between $(o:\gamma_0)$ and N $\Leftrightarrow$ $(o:\gamma_0)+N_c$ is a maximal N-subgroup or $= N$.

(h) Γ is N_0-simple $\Leftrightarrow$ $(o:\gamma_0)$ is a maximal N_0-subgroup or $= N$.

(i) (Betsch (5)) If ${}_N\Gamma$ is faithful, $N\in\mathfrak{N}_0$, and if $\exists\ L_1,L_2 \trianglelefteq_\ell N$ with $L_1+(o:\gamma_0) = L_2+(o:\gamma_0) = N$, but $L_1\cap L_2 \subseteq (o:\gamma_0)$, then N is a ring.

Proof. (a) $\forall\ \gamma\in\Gamma\ \exists\ n_\gamma\in N$: $\gamma = n_\gamma\gamma_0$. So

$\forall\ \ell\gamma_0\in L\gamma_0\ \forall\ n\in N\ \forall\ \gamma\in\Gamma$: $n(\gamma+\ell\gamma_0)-n\gamma =$

$= n(n_\gamma\gamma_0+\ell\gamma_0)-nn_\gamma\gamma_0 = (n(n_\gamma+\ell)-nn_\gamma)\gamma_0 \in L\gamma_0$.

In the same way one shows that $L\gamma_0$ is a normal subgroup of Γ.

(b) As in (a), $e\gamma = en_\gamma\gamma_0 = n_\gamma\gamma_0 = \gamma$.

(c) $\forall\ \gamma\in\Gamma\ \forall\ n\in N$: $o = n\gamma-ne\gamma = (n-ne)\gamma$, so $n = ne$.

(d) $\forall\ \gamma\in\Gamma$: $N\gamma \leq_{N_0} \Gamma$ implies $(\gamma = o)$ Ω to be $= \{o\}$ or $\Omega = \Gamma$. So each $N\gamma$ equals either $\{o\}$ or Γ.

(e) Consider the N-epimorphism $h\colon \begin{array}{l} N \to \Gamma \\ n \to n\gamma_o \end{array}$ and apply the homomorphism theorem.

(f) - (h) follow from (e), 2.16 and the "second isomorphism theorem".

(i) By (e), $\Gamma \cong_N N/(o:\gamma_0) =$

$= (L_1+(o:\gamma_0))\cap(L_2+(o:\gamma_0))/(L_1\cap L_2)+(o:\gamma_0) =: \Gamma'$.

$(o:\Gamma') = (o:\Gamma) = \{o\}$. So by 2.24, N is a ring.

3.5 DEFINITION A monogenic N-group Γ with $N\Gamma \neq \{o\}$ is said to be of

type 0: <=> ${}_N\Gamma$ is simple

type 1: <=> ${}_N\Gamma$ is simple and strongly monogenic

type 2: <=> Γ is N_0-simple.

The definition of "type 2" cries for

3.6 REMARK Of course it seems more natural to define "type 2" by "N-simple" (see e.g. Fain (1)). But N-simplicity says very little about Γ. For instance, every non-zero sub-near-ring N of $M(\Gamma)$ with $N \supseteq M_c(\Gamma)$ has ${}_N\Gamma$ of "type 2" then. So one can get nearly everything except nice structure theorems. Moreover, we would not get

3.7 PROPOSITION

(a) ${}_N\Gamma$ of type 2 => ${}_N\Gamma$ of type 1 => ${}_N\Gamma$ of type 0.

(b) If ${}_N\Gamma$ is of type 1 or 2 then $\Omega = \{o\}$ or $\Omega = \Gamma$.

(c) If ${}_N\Gamma$ is a unitary $N = N_0$-group then ${}_N\Gamma$ is of type 1 <=> ${}_N\Gamma$ is of type 2. In this case,

$\forall\ \gamma\in\Gamma^*:\ N\gamma = \Gamma$ (see also 3.19(a)!).

Proof. (a): by 3.4(d) and 1.34.

(b) follows from (a) and 3.2.

(c) =>: Let ${}_N\Gamma$ be of type 1. If $\Delta \leq_{N_0} \Gamma$ then $\forall\ \delta\in\Delta$: $N\delta = \{o\}$ or $= \Gamma$. Hence $\Delta = \{o\}$ or $\Delta = \Gamma$, since each $\delta\in N\delta$.

<=: by (a).

<u>3.8 EXAMPLES</u> If $\Gamma = \mathbb{Z}_4$, define (Betsch (3))

$N_0 := \{f \in M_0(\Gamma) \mid f(2) \in \{0,2\}\}$

$N_1 := \{f \in M_0(\Gamma) \mid f(2) = 0\}$

$N_2 := \{f \in M_0(\Gamma) \mid f(3) = 0\}$.

Then ${}_{N_0}\Gamma$ is of type 0, but not of type 1
${}_{N_1}\Gamma$ is of type 1, but not of type 2
${}_{N_2}\Gamma$ is of type 2

(where $n\gamma$ is defined as in 1.18(c)).
This can be seen by simple calculations.

<u>3.9 EXAMPLES</u> If $N = N_c$ then

${}_N\Gamma$ is of type 0 <=> ${}_N\Gamma$ is of type 1 <=> Γ is a simple group with $\Omega = \Gamma$,

${}_N\Gamma$ is of type 2 <=> Γ is a cyclic group of prime order.

This holds since N-kernels (N_0 = {0}-subgroups) in Γ coincide with normal subgroups (subgroups, respectively) of Γ; $\exists\ \gamma_0 \in \Gamma$: $N\gamma_0 = \Gamma$ results in $\Omega = N_0 = N\gamma_0 = \Gamma$.

<u>3.10 PROPOSITION</u> (Betsch (3)). Let ${}_N\Gamma$ be of type 0 (with generator γ) and let $L \leq_N N$ be a minimal left ideal with $L \nsubseteq (0:\gamma)$. Then $L \cong_N \Gamma$.

<u>Proof.</u> By 3.4(a), $L\gamma \trianglelefteq_N \Gamma$. By $L \nsubseteq (0:\gamma)$, $L\gamma \neq \{o\}$.

Since ${}_N\Gamma$ is simple, $L\gamma = \Gamma$.

$h: L \to \Gamma \in \mathrm{Hom}_N(L,\Gamma)$ and $\operatorname{Ker} h = L \cap (o:\gamma) = \{o\}$
$\ell \to \ell\gamma$

(since L is minimal).

<u>3.11 COROLLARY</u> Let $N = \sum_{i \in I} L_i \in \mathcal{N}_0$, where I is some index set and all L_i are minimal left ideals of N. Let ν be $\in \{0,1,2\}$.

(a) Each N-group of type ν is N-isomorphic to some L_i.

(b) I finite => there are only finitely many classes of non-N-isomorphic N-groups of type ν.

Proof. (a) Let ${}_N\Gamma$ be of type ν and generated by γ.
Since $\Gamma = N\gamma \neq \{o\}$, $\exists$ $i\varepsilon I$: $L_i \not\subseteq (0:\gamma)$.
Now apply 3.10.

(b) Follows from (a).

3.12 LEMMA (Scott (4)) Let N have the DCCN and let $M \leq_N N$ be monogenic (by m_o) and $\exists$ $m_1 \varepsilon M$: $(0:m_1) = \{0\}$. Then M contains a right identity and $(0:m_o)_M = \{0\}$.

Proof. Let all annihilators be taken in M.

(a) Since $(0:m_1) = \{0\}$, the map $h: M \to Mm_1$ is an N-isomorphism; moreover, $Mm_1 \subseteq M$. If $Mm_1 \subset M$ then applying h we get $Mm_1{}^2 \subset Mm_1$, and so on, contradicting the DCCN. So $Mm_1 = M$ and $\exists$ $e\varepsilon M$: $em_1 = m_1$. But then $\forall$ $m\varepsilon M$: $mem_1 = mm_1$, so $\forall$ $m\varepsilon M$: $me-m\varepsilon(0:m_1) = \{o\}$.
Therefore e is a right identity in M.

(b) $Mm_o = M \Rightarrow \exists m_2 \varepsilon M$: $m_2m_o = e$. If $mm_2 = 0$ then also $me = mm_2m_o = 0m_o = 0$, so $(0:m_2) \subseteq (0:e) = \{0\}$. So $(0:m_2) = \{0\}$ and - as in the beginning of (a)- $M = Mm_2$.
Suppose that $m_3\varepsilon(0:m_o)$. $\exists$ $m_4\varepsilon M$: $m_3 = m_4m_2$. So $m_4 = m_4e = m_4m_2m_o = m_3m_o = 0$ and $m_3 = m_4m_2 = 0$.
This shows that $(0:m_o) = \{0\}$.

3.13 THEOREM (Scott (4)). If $N = N_o$ has the DCCN and $M \leq_N N$ is monogenic (by m_o) then M contains a right identity and $(0:m_o)_M = \{0\}$.

Proof. Again, all annihilators are to be taken in M. In view of lemma 3.12 we "only" have to show that $\exists$ $m_1\varepsilon M$: $(0:m_1) = \{0\}$.
Suppose that $\exists$ $M' \leq_N N$ $\exists$ $m_o'\varepsilon M'$: M' monogenic by m_o' and $(0:m_o') \neq \{0\}$. W.l.o.g. we may assume that M' is minimal for containing such an m'.
Let $m_1'\varepsilon M'$ be such that $(0:m_1')$ is minimal in $\{(0:m') | M'm' = M'\}$ (so also $M'm_1' = M'$). Therefore

$\exists\ m_2' \varepsilon M' : m_2' m_1' = m_1'$ and $(0:m_2') \subseteq (0:m_1')$ as in 3.12(b).

If m_2' generates M' then minimality of $(0:m_1')$ forces $(0:m_2') = (0:m_1')$. $\forall\ m' \varepsilon M' : m' m_2' m_1' = m' m_1'$, so $\forall\ m' \varepsilon M' : m' m_2' - m' \varepsilon (0:m_1') = (0:m_2')$ and hence $\forall\ m' \varepsilon M' : (m' m_2') m_2' = m' m_2'$ showing that m_2' is a right identity in $M' m_2' = M'$ and so $(0:m_2') = \{0\}$. By 3.12(b), $(0:m_o) = \{0\}$.

If m_2' does not generate M' then $M' m_2' < M'$. Consider the sequence $M' > M' m_2' \geq M' (m_2')^2 \geq \ldots$.
$\exists\ k \varepsilon \mathbb{N} : M' (m_2')^k = M' (m_2')^{k+1} = \ldots$. Thus $(M' (m_2')^k)(m_2')^{k+1} = M' (m_2')^k$ and since $m_3' : = (m_2')^{k+1} \varepsilon M' (m_2')^k$, m_3' generates $M' (m_2')^k$.
By the minimality of M', $(0:m_3') \cap M' (m_2')^k = \{0\}$.
Again using the minimality of M' we see that each generator m_4' of $M' (m_2')^k = M' m_3'$ has $(0:m_4') \cap M' m_3' = \{0\}$.
We shall show that $m_4' : = m_1' m_3'$ violates this statement.

(a) $m_1' m_3'$ generates $M' m_3'$, for $m_1' = m_3' m_1'$ and $M' m_1' = M'$ imply that $(M' m_3')(m_1' m_3') = M' m_1' m_3' = M' m_3'$.

(b) Observe that $(0:m_3') \neq \{0\}$, for otherwise $M' \cong_N M' m_3' \leq M' m_2' < M'$. Take some non-zero $m_5' \varepsilon (0:m_3')$. $\exists\ m_6' \varepsilon M : m_6' m_1' = m_5'$, since m_1' generates M'.
Now $0 = m_5' m_3' = m_6' m_1' m_3' = m_6' m_3' m_1' m_3'$.
Hence $m_6' m_3' \varepsilon (0:m_1' m_3') \cap M' m_3'$, but $m_6' m_3' \neq 0$ since $m_6' m_3' m_1' = m_6' m_1' = m_5' \neq 0$.

So we arrive at a contradiction and the proof is complete.

b) CHANGE OF THE NEAR-RING

Up to now we had an unjust situation: a near-ring keeps an harem of N-groups, but not conversely. Now we let an N-group ${}_N\Gamma$ change into ${}_{N/I}\Gamma$ (for some $I \trianglelefteq N$), ${}_{N_0}\Gamma$, ${}_{N_c}\Gamma$. These changes will be an important tool in later considerations.

3.14 PROPOSITION (Betsch (3)). Let I be an ideal of N, Γ a group and $\nu\varepsilon\{0,1,2\}$.

(a) If Γ is an N-group with $I \subseteq (o:\Gamma)$ then

$$(n+I)\gamma: = n\gamma$$

makes Γ into an N/I-group ${}_{N/I}\Gamma$.

If ${}_N\Gamma$ is of type ν, so is ${}_{N/I}\Gamma$.

If ${}_N\Gamma$ is faithful, the same applies to ${}_{N/I}\Gamma$.

(b) If Γ is an N/I-group then

$$n\gamma: = (n+I)\gamma$$

makes Γ into an N-group ${}_N\Gamma$ with $I \subseteq (o:\Gamma)_N$.

If ${}_{N/I}\Gamma$ is of tpye ν, so is ${}_N\Gamma$.

If ${}_{N/I}\Gamma$ is faithful then $I = (o:\Gamma)_N$.

The proof is a collection of straightforward arguments and therefore omitted.
Observe that $(N/I)_o = \{n_0+I \mid n_0\varepsilon N_0\}$.

Each N-group Γ can be viewed as an N_0-group ${}_{N_0}\Gamma$ and as an N_c-group ${}_{N_c}\Gamma$ in an obvious way (by restriction). In 3.4(d) we already mentioned this fact. We now study the relation between ${}_{N_0}\Gamma$, ${}_{N_c}\Gamma$ and ${}_N\Gamma$:

3.15 PROPOSITION Let Γ be an N-group and Δ a subset of Γ.

(a) ${}_N\Gamma$ is faithful iff ${}_{N_0}\Gamma$ and ${}_{N_c}\Gamma$ are faithful.

(b) $\Delta \trianglelefteq_N \Gamma \iff \Delta \trianglelefteq_{N_0} \Gamma$

(c) $\Delta \leq_N \Gamma \iff \Delta \leq_{N_0} \Gamma \wedge \Omega \subseteq \Delta$.

Proof. (a) If ${}_N\Gamma$ is faithful, the same trivially applies to ${}_{N_0}\Gamma$ and ${}_{N_c}\Gamma$. Conversely, let $n\Gamma$ be $= \{o\}$.

Then (with $n = n_0+n_c$ as in 1.13) $\forall\ \gamma\varepsilon\Gamma$: $n_o\gamma+n_co =$
$= n_o\gamma+n_c\gamma = n\gamma = o$.
Taking $\gamma = o$ yields $n_co = o$. So $\forall\ \gamma\varepsilon\Gamma$: $n_0\gamma = o$ and $n_0 = 0$. But $n_co = o$ gives $\forall\ \gamma\varepsilon\Gamma$: $n_c\gamma = o$, hence $n_c = o$. Therefore $n = n_0+n_c = 0$.

(b) $\Rightarrow$ is trivial. If $\Delta \trianglelefteq_{N_0} \Gamma$ then

$\forall\ \delta\varepsilon\Delta\ \forall\ \gamma\varepsilon\Gamma\ \forall\ n\varepsilon N$: $n(\delta+\gamma)-n\gamma = n_0(\delta+\gamma)+n_c(\delta+\gamma)-$
$-n_co-n_o\gamma = n_o(\delta+\gamma)+n_co-n_co-n_o\gamma\varepsilon\Delta$.

(c) is even more trivial.

The relation between ${}_N\Gamma$ and ${}_{N_0}\Gamma$ is particularly important.

3.16 COROLLARY Let ${}_N\Gamma \in {}_N\mathcal{G}$.

(a) ${}_N\Gamma$ is simple $\iff$ ${}_{N_0}\Gamma$ is simple.

(b) ${}_{N_0}\Gamma$ is monogenic by $\gamma \Rightarrow {}_N\Gamma$ is monogenic by γ.

(c) ${}_{N_0}\Gamma$ is strongly monogenic $\Rightarrow {}_N\Gamma$ is strongly monogenic or $\{o\} \neq \Omega \neq \Gamma$.

(d) Γ is N_0-simple $\Rightarrow$ Γ is N-simple.

3.17 EXAMPLES If $N = M_c(\mathbb{Z}_4)$ then $\mathbb{Z}_4$ is N-simple but not N_0-simple (since $\{0,2\}$ is an $N_0 = \{0\}$-subgroup). So N-simplicity does not imply N_0-simplicity.

Plugging all together yields

<u>3.18 THEOREM</u> Let ${}_N\Gamma$ be an N-group and $\nu\varepsilon\{0,1,2\}$.

(a) ${}_N\Gamma$ is of type $\nu \Rightarrow {}_{N_0}\Gamma$ is of type ν or $N_0\Gamma = \{o\}$.

(b) ${}_{N_0}\Gamma$ is of type ν (for $\nu = 1$ assume that in ${}_N\Gamma$ $\Omega = \{o\}$ or $\Omega = \Gamma$) $\Rightarrow {}_N\Gamma$ is of type ν.

<u>Proof</u>. (a) Anyhow, ${}_N\Gamma$ is simple, therefore also ${}_{N_0}\Gamma$ by 3.16(a).

Let ${}_N\Gamma$ be monogenic by γ. Then $N_0\gamma \trianglelefteq_N \Gamma$ by 3.4(a). Hence $N_0\gamma = \{o\}$ or $N_0\gamma = \Gamma$.

If $N_0\gamma = \Gamma$, ${}_{N_0}\Gamma$ is monogenic, too.

If $N_0\gamma = \{o\}$ then $\Gamma = N\gamma = N_0\gamma+\Omega = \Omega$ implies that $\forall\ \gamma\varepsilon\Gamma$: $N\gamma = \Gamma$.

Again by 3.4(a), $\forall\ \gamma\varepsilon\Gamma$: $N_0\gamma = \{o\}$ or $= \Gamma$.

So either ${}_{N_0}\Gamma$ is monogenic or $N_o\Gamma = \{o\}$.

If ${}_N\Gamma$ is of type 1 then $\Omega = \{o\}$ or $\Omega = \Gamma$.

If $\Omega = \{o\}$ then $\forall\ \gamma\varepsilon\Gamma$: $N\gamma = N_0\gamma+\Omega = N_0\gamma$ and ${}_{N_0}\Gamma$ is again of type 1.

If $\Omega = \Gamma$ then each $\gamma\varepsilon\Gamma$ generates ${}_N\Gamma$ so (again by 3.4(a)) $\forall\ \gamma\varepsilon\Gamma$: $N_0\gamma = \{o\}$ or $N_0\gamma = \Gamma$.

So ${}_{N_0}\Gamma$ is either of type 1 or $N_0\Gamma = \{o\}$.

The assertion for $\nu = 2$ is trivial.

(b) By 3.16.

<u>3.19 REMARKS</u>

(a) 3.18(a) and (b) show that 3.7(c) holds for arbitrary near-rings!

(b) Information about the behaviour of ${}_M\Gamma$ with $M \leq_N N$ can be found in Mlitz (3).

c) MODULARITY

3.20 DEFINITION $L \trianglelefteq_\ell N$ is called modular: $\Longleftrightarrow$
$\Longleftrightarrow \exists\, e \in N\ \forall\, n \in N : n - ne \in L$.
In this case we also say that L is "modular by e" and that e is a "right identity modulo L" (since $\forall\, n \in N: ne \equiv n \pmod L$).

3.21 REMARKS

(a) If $L_1, L_2 \trianglelefteq_\ell N$ with $L_1 \subseteq L_2$ and L_1 modular by e then L_2 is modular by e, too.

(b) $\{0\}$ is modular iff N contains a right identity.

(c) Every normal subgroup of $(N_c,+)$ is a modular left ideal of N_c (by any element of N_c).

(d) If L is modular by e in $N \in \mathcal{N}_0$ then $e \in L$ iff $L = N$.

3.22 PROPOSITION (Betsch (3)). Each modular left ideal $L \neq N \in \mathcal{N}_0$ is contained in a maximal one (which is modular, too).

Proof. Let L be modular by e. Apply Zorn's Lemma to the set of all left ideals $I \supseteq L$ with $e \notin I$ and use 3.21(a).

Proposition 3.22 is not always true if $N \neq N_0$: see 3.21(c).

3.23 PROPOSITION (Betsch (3)). $L \trianglelefteq_\ell N$ is modular $\Longleftrightarrow$
$\Longleftrightarrow \exists\, {}_N\Gamma \in {}_N\mathcal{G}\ \exists\, \gamma \in \Gamma$: ${}_N\Gamma$ monogenic by $\gamma \wedge L = (0:\gamma)$.

Proof. $\Rightarrow$: Let L be modular by e. Then ${}_N\Gamma := N/L$ is monogenic by $\gamma := e+L$, since $N(e+L) = \{ne+L \mid n \in N\} = \{n+L \mid n \in N\} = N/L = \Gamma$. Moreover, $n \in (0:\gamma) = (L:e+L) \Longleftrightarrow n(e+L) = L \Longleftrightarrow ne \in L \Longleftrightarrow n \in L$.

$\Leftarrow$: Let ${}_N\Gamma$ be monogenic by γ. Then $\exists\, e \in N: e\gamma = \gamma$. But then $\forall\, n \in N: ne\gamma = n\gamma$, so $\forall\, n \in N: n - ne \in (0:\gamma) = L$ and L is shown to be modular by e.

Applying 3.4(e) we get

<u>3.24 COROLLARY</u> $L \trianglelefteq_\ell N$ is modular => $L \supseteq (L:N)$.

<u>Proof</u>. Take some (by γ) monogenic N-group Γ with $L = (o:\gamma)$. Then $L = (o:\gamma) \supseteq (o:\Gamma) = (o:N/L) = (L:N)$.

3.22 - 3.24 are similar to the ring case ((Jacobson), pp. 5-6). Looking at $(L:N)$ more closely gives for future use (cf. Ramakotaiah (1)):

<u>3.25 PROPOSITION</u> Let L be modular by e. Then $(L:N) = (L:Ne)$ and this is the greatest ideal of N contained in L.

<u>Proof</u>. $(L:N) \subseteq (L:Ne)$ is clear. If $n \in (L:Ne)$ then $\forall\ n' \in N : nn'e \in L$. But $nn' - nn'e \in L$, hence $\forall\ n' \in N: nn' \in L$. So $n \in (L:N)$ and $(L:N) = (L:Ne)$. By 1.42, $(L:N)$ is a left ideal and it is easy to see that it is even an ideal of N, $(L:N) \subseteq L$ holds by 3.24.
If $I \trianglelefteq N$ with $I \subseteq L$ then trivially $I \subseteq (L:N)$.

<u>3.26 THEOREM</u> (cf. (Kertesz), p. 122). If $N = L_1 + L_2$, where L_1, L_2 are modular left ideals, then $L_1 \cap L_2$ is again modular.

<u>Proof</u>. Let L_1, L_2 be modular by e_1, e_2, respectively. Decompose e_1, e_2:

$$e_1 = \ell_{11} + \ell_{12} \qquad \text{where } \ell_{11}, \ell_{21} \in L_1,\ \ell_{12}, \ell_{22} \in L_2.$$
$$e_2 = \ell_{21} + \ell_{22}$$

We claim that $L_1 \cap L_2$ is modular by $\ell_{21} + \ell_{12} =: e$.
If $n \in N$ then $n - ne = n - n(\ell_{21} + \ell_{12}) = n - n\ell_{12} + n\ell_{12} - n(\ell_{21} + \ell_{12}) = n - ne_1 + ne_1 - n(-\ell_{11} + e_1) + n\ell_{12} - n(\ell_{21} + \ell_{12})$.
But $n - ne_1 \in L_1$, $ne_1 - n(-\ell_{11} + e_1) \in L_1$ and $n\ell_{12} - n(\ell_{21} + \ell_{12}) \in L_1$.
Therefore $\forall\ n \in N: n - ne \in L_1$.
Similarly, $\forall\ n \in N: n - ne \in L_2$, and we are through.

3.27 COROLLARY

(a) If L is a modular and M a maximal modular left ideal then $L \cap M$ is modular.

(b) A finite intersection of maximal modular left ideals is modular.

(c) If N is a direct sum of two modular left ideals then N contains a right identity.

(d) (Betsch (3)). If N contains a finite family of maximal modular left ideals with zero intersection then N contains a right identity.

3.28 DEFINITION Let ν be $\varepsilon\{0,1,2\}$. A left ideal L of N is called ν-modular if L is modular and N/L is an N-group (via $n(n'+L) := nn'+L$) of type ν.
Let $\mathcal{L}_\nu(N)$ be the set of all ν-modular left ideals of N.

3.29 REMARK So a 0-modular left ideal is just a modular maximal one and a 2-modular left ideal L is a modular maximal left ideal with no N_0-subgroup strictly between L and N. (Beidleman calls these left ideals "strictly maximal".)

ν-modular left ideals turn out to be very useful in determining radicals of related near-rings. The next two propositions prepare enough information on ν-modular left ideals to accomplish this (5.18, 5.20).

3.30 PROPOSITION Let $I \trianglelefteq N$ be a direct summand of $N\varepsilon\mathcal{N}_0$.
Then $\forall\ \nu\varepsilon\{0,1,2\}\ \forall\ L\varepsilon\mathcal{L}_\nu(I)\ \exists\ \bar{L}\varepsilon\mathcal{L}_\nu(N)$: $L = \bar{L} \cap I$.

Proof. (Derived from an argument in Fain (1)).
Let $L\varepsilon\mathcal{L}_\nu(I)$. By $i(i'+L) := ii'+L$ $(i,i'\varepsilon I)$, I/L becomes an I-group of type ν (3.28).
If $n\varepsilon N$ and $i\varepsilon I$, let $n(i+L) := ni+L$. Since I is a direct summand, this multiplication is well-defined: if $N = I \dot{+} J$ $(J \trianglelefteq N)$; $n = i+j$ $(i\varepsilon I,\ j\varepsilon J)$ and $i_1+L = i_2+L$ then $\exists\ \ell\varepsilon L$: $i_2 = i_1+\ell$. But then (using 2.6 (b)) $ni_2 - ni_1 = (i+j)i_2 - (i+j)i_1 =$

$= ii_2 + ji_2 - ji_1 - ii_1 = ii_2 - ii_1 = i(i_1+\ell) - ii_1 \equiv$
$\equiv 0 \pmod L$.

In this way, I/L becomes an N-group.

Let $e \in I$ be a right identity modulo L. Then
$\forall i \in I$: $i - ie \in L$, so $\forall i \in I$: $i+L = i(e+L)$.
Therefore $e+L$ is a generator of ${}_I(I/L)$ and of ${}_N(I/L)$.

Define a new left ideal $\bar{L}$ by $\bar{L} := (0:e+L)_N$.
By 3.23, $\bar{L}$ is a modular left ideal of N and by 3.4(e) we get $N/\bar{L} \cong_N I/L$.

We know that ${}_I(I/L)$ is of type ν. In order to show that $N/\bar{L}$ is of type ν, we prove this for ${}_N(I/L)$.

(1) By above, ${}_N(I/L)$ is monogenic by $e+L$.

(2) ${}_N(I/L)$ is simple: if $\Delta \trianglelefteq_N I/L$ then $\Delta \trianglelefteq_I I/L$, so Δ is trivial.

(3) First observe that for $i \in I$ we get from 2.6 (b) that $Ni = (I \dotplus J)i = Ii + Ji = Ii$.
Now suppose that ${}_N(I/L)$ is strongly monogenic.
If $i+L \in I/L$ then either $N(i+L) = \{0+L\}$ or $N(i+L) \neq \{0+L\}$, in which case $Ni+L \neq L$, therefore $Ii+L \neq L$, hence $I(i+L) \neq \{0+L\}$, so $I(i+L) = I/L$, whence $N(i+L) = I/L$, proving that ${}_N(I/L)$ is strongly monogenic.

(4) Finally, let ${}_I(I/L)$ be of type 2. Consider $\Gamma \leq_{N_0} I/L$. Then $\Gamma \leq_{I_0} I/L$, therefore $\Gamma = I/L$ or $\Gamma = \{0+L\}$.

Hence $\bar{L} \in \mathcal{L}_\nu(N)$ and clearly $L = \bar{L} \cap I$.

<u>3.31 EXAMPLE</u> Take the near-ring $N = N_0$ of example 3.8 and consider the ideal $I := (0:2)$. $\mathbb{Z}_4$ is an I-group of type 1 (see 3.8), so $L := \{0\}$ is 1-modular in I.
Examining the 1-modular left ideals of N (by hand or whatever) shows that all of them contain I. Hence there is no 1-modular left ideal $\bar{L}$ of N with $\bar{L} \cap I = \{0\} = L$.
This shows that 3.30 is not true if I is not a direct summand.

3.32 REMARKS

(a) It is not known to the author if one can release the restriction that I is a direct summand (at least for $\nu=2$).
(b) One can deduce from 5.19 (b) that if $I \trianglelefteq N=N_0$ and $\mathcal{L} \in \mathscr{L}_\nu(N)$ then $\mathcal{L} \cap I$ is not in $\mathscr{L}_\nu(I)$ in general.

3.33 PROPOSITION Let $(N_i)_{i\in I}$ be near-rings and N their direct product. Let L_i be a left ideal of N_i for some $i \in I$.

Denote $\prod_{j\in I} M_j$ with $M_j := \begin{cases} N_j & i \neq j \\ L_i & i = j \end{cases}$ by $\mathcal{L}_i \trianglelefteq_\ell N$.

Then for $\nu \in \{0,1,2\}$, L_i is ν-modular in N_i iff $\mathcal{L}_i$ is ν-modular in N.

Proof. (a) If L_i is ν-modular in N_i then ${}_{N_i}(N_i/L_i)$ is of type ν. By $n_i((\ldots,n_i',\ldots)+\mathcal{L}_i) :=$ $= (\ldots,0,n_i n_i',0,\ldots)+\mathcal{L}_i$, $N/\mathcal{L}_i$ becomes an N_i-group and clearly $N/\mathcal{L}_i \cong_{N_i} N_i/L_i$. So $N/\mathcal{L}_i$ is an N_i-group of type ν. If $J_i := \overline{\{0\}}_i$ (notation as in the statement), $N_i \cong N/J_i$, so 3.14(b) shows that $N/\mathcal{L}_i$ is an N-group of type ν (and the multiplication is the same as in 3.28). Hence $\mathcal{L}_i$ is ν-modular in N.

(b) If $\mathcal{L}_i$ is ν-modular in N then $N/\mathcal{L}_i$ is an N-group of type ν. Similar to (a), $N/\mathcal{L}_i \cong_N N_i/L_i$ where $(\ldots,n_i,\ldots)(n_i'+L_i) := n_i n_i' + L_i$. The annihilator of N_i/L_i in N contains J_i (as in (a)), so by 3.14(a) N_i/L_i is an N_i-group of type ν in the sense of 3.28. Therefore L_i is ν-modular in N_i.

3.34 REMARK 3.33 remains valid if $\prod$ is changed to $\bigoplus$.

We now turn to the concept of quasiregularity (cf. e.g. (Jacobson), p.7), although ν-modularity seems to be more important for the radical theory of near-rings then quasi-regularity (in contrast to ring theory). Nevertheless, we get a very important result (3.40) via quasiregularity.

d) QUASIREGULARITY

3.35 NOTATION For $z\in N$, denote the left ideal generated by the set $\{n-nz \mid n\in N\}$ by L_z.
(Note that for $L = N$, $z = 0$, N_0 has still one single meaning.)

3.36 DEFINITION

(a) $z\in N$ is called quasiregular (=: qr) if $z\in L_z$.

(b) $S\subseteq N$ is called quasiregular (=: qr): $\Leftrightarrow \forall s\in S$: s is qr.

3.37 REMARKS

(a) If $N\in\mathcal{N}_0$, $z\in N$ is qr $\Leftrightarrow L_z = N$.

(b) L_z is modular (by z).

(c) Beidleman (1) calls (for a near-ring $N\in\mathcal{N}_0$ with identity 1) $z\in N$ quasiregular if $\exists\, y\in N$: $y(1-z) = 1$. In this case, z is also quasiregular in the sense of 3.36.

3.38 PROPOSITION (Ramakotaiah (1)). Let N be $\in\mathcal{N}_0$.
(a) $z\in N$ nilpotent $\Rightarrow$ z is qr.
(b) Each nil subset of N is qr.
(c) If $L \triangleleft_\ell N$ is modular by e then e is not qr.
(d) If e is a non-zero idempotent then e is not qr.

Proof. (a) If $z^n = 0$, consider any $x\in N$. Then $x-xz\in L_z$, $xz-xz^2\in L_z, \dots, xz^{n-1}-xz^n\in L_z$.
Hence $x-xz^n\in L_z$, so $x\in L_z$ and $L_z = N$.

(b) Follows from (a).

(c) $\forall\, n\in N$: $n-ne\in L$. If e is qr then $L = L_e = N$.

(d) Assume that the idempotent e is $\neq 0$. Consider the N-endomorphism $h_e\colon \begin{array}{l} N \to N \\ x \to xe \end{array}$. $h_e(e) = e^2 = e \neq 0$

shows that $h_e \neq \bar{o}$.
$\forall\ x\in N$: $h_e(x-xe) = xe-xe^2 = 0$, so $L_e \subseteq \mathrm{Ker}\ h_e \neq N$ and e cannot be quasiregular.

3.39 PROPOSITION Each nil ideal I of a near-ring N is quasiregular.

Proof. Proceeding as in 3.38(a) one sees that
$\forall\ i\in I$, $N_o = \{x-x0 \mid x\in N\} \subseteq L_i$.
If $i\in I$ then by 2.99 $i\in N_o$, so $i\in L_i$ and i is qr.

3.40 THEOREM (Ramakotaiah (1)). $N\in\mathcal{N}_0$, DCCN, $M \leq_N N$. Then M is qr <=> M is nilpotent <=> M is nil.

Proof. (a) Let M be qr. For $k\in\mathbb{N}$, let $M^{(k)}$ be the N-subgroup of N generated by M^k (2.56). We get a chain $M \supseteq M^{(2)} \supseteq M^{(3)} \supseteq \ldots$. By the DCCN,
$\exists\ k\in\mathbb{N}$: $M^{(k)} = M^{(k+1)} = \ldots =: P$.
If $P = \{0\}$, we are through.
If not, observe that $P^{(2)} = P \neq \{0\}$, so
$P\in\{K \mid K \leq_N N \wedge K\subseteq P \wedge PK \neq \{0\}\} =: \mathcal{P}$. So $\mathcal{P} \neq \emptyset$.
The DCCN assures the existence of a minimal element K_o in $\mathcal{P}$. Since $PK_o \neq \{0\}$, $\exists\ k_o\in K_o$: $Pk_o \neq \{0\}$.
$Pk_o \leq_N N$, $Pk_o\subseteq K_o\subseteq P$, $P(Pk_o) \neq \{0\}$ (since otherwise $P^2 \subseteq (0:k_o)$, so $P = P^{(2)} \subseteq (0:k_o)$, a contradiction).
These three assertions qualify Pk_o to be $\in \mathcal{P}$.
Since $Pk_o \subseteq K_o$, $Pk_o = K_o$. Therefore $\exists\ p\in P$: $pk_o = k_o$.
So $\forall\ n\in N$: $(n-np)k_o = nk_o-npk_o = 0$.
Hence $\forall\ n\in N$: $n-np\in(0:k_o) \neq N$, so $L_p \neq N$ and p is not quasiregular.

(b) If M is nilpotent, M is trivially nil.

(c) Follows from 3.38(b).

3.41 REMARK 3.40 does not hold for $N\notin\mathcal{N}_0$: take a finite near-ring N with $N = N_c \neq \{0\}$. $M = N$ is quasiregular, but not nilpotent. If $I \trianglelefteq N=N_o$ and $i\in I$ is quasiregular in I then (Holcombe (1)) i is quasiregular in N.

e) IDEMPOTENTS

3.42 DEFINITION A set E of idempotents is called orthogonal if
$\forall\, e,f \in E:\ e \neq f \Rightarrow e \cdot f = 0.$

The standard method to get orthogonal idempotents is to decompose a right identity:

3.43 THEOREM (Beidleman (1)). If $N \in \mathcal{N}_0$ contains a right identity e, if $N = \dot{\sum}_{i=1}^{k} L_i$ ($L_i \trianglelefteq_\ell N$) and if $e = \sum_{i=1}^{k} e_i$ ($e_i \in L_i$), then $e_1,\dots,e_k$ are orthogonal idempotents and each e_i is a right identity in L_i which generates $L_i = Ne_i$.

Proof. If $e = e_1+\dots+e_k$ then $\forall\, n \in N$: $n = ne = n \sum_{i=1}^{k} e_i = \sum_{i=1}^{k} ne_i$ (by 2.30). If $n \in L_i$, the uniqueness of the representation yields $n = ne_i$, so e_i is a right identity for L_i. In particular, e_i is idempotent, while for $i \neq j$ one gets $e_j e_i = 0$ by taking $n = e_j$ above.
Finally, $L_i = Ne_i$, since each $\ell_i \in L_i$ can be written as $\ell_i = \ell_i e_i$.

If one has a right identity in 3.43, but no direct decomposition it is sometimes still possible to get orthogonal idempotents:

3.44 THEOREM (Scott (4)). $N \in \mathcal{N}_0$, DCCN, $M \leq_N N$, $L_1, L_2 \trianglelefteq_\ell N$, $L_1, L_2 \subseteq M$. If M contains a right identity e and if both L_1, L_2 are minimal for the property that $L_1+L_2 = M$ then there exist orthogonal idempotents $e_1 \in L_1$, $e_2 \in L_2$ with

(a) $e_1+e_2 \equiv e \pmod{L_1 \cap L_2}$

(b) $(0:e_1) \cap M = L_2$ and $(0:e_2) \cap M = L_1$.

Proof. If $e = \ell_1+\ell_2$ $(\ell_1 \varepsilon L_1, \ell_2 \varepsilon L_2)$ then $\ell_1 = \ell_1 e =$
$= \ell_1(\ell_1+\ell_2)$.

But $\ell_1(\ell_1+\ell_2)-\ell_1\ell_2-\ell_1^2 \varepsilon L_1$, so $\ell_1(\ell_1+\ell_2) \equiv$
$\equiv \ell_1^2+\ell_1\ell_2 \pmod{L_1}$

and $\ell_1(\ell_1+\ell_2)-\ell_1^2-\ell_1\ell_2 \varepsilon L_2$, so $\ell_1(\ell_1+\ell_2) \equiv$
$\equiv \ell_1\ell_2+\ell_1^2 \pmod{L_2}$.

From 2.22 we conclude that $\ell_1 \equiv \ell_1\ell_2+\ell_1^2 \pmod{L_1 \cap L_2}$,
hence $\ell_1-\ell_1^2 = \ell_1\ell_2 \varepsilon L_1 \cap L_2$, so $\forall\ m \varepsilon \mathbb{N}: \ell_1^m \equiv \ell_1 \pmod{L_1 \cap L_2}$.
Similarly, $\forall\ m \varepsilon \mathbb{N}: \ell_2^m \equiv \ell_2 \pmod{L_1 \cap L_2}$.
Now let i be $\varepsilon\{1,2\}$.
All $L_i\ell_i^m \leq_N N$. By DCC $\exists\ k \varepsilon \mathbb{N}: L_i\ell_i^k = L_i\ell_i^{k+1} = \ldots$
Therefore $(L_i\ell_i^k)\ell_i^{k+1} = L_i\ell_i^k$ and ℓ_i^{k+1} generates $L_i\ell_i^k$. Moreover, $\ell_i^{k+1} \varepsilon L_i\ell_i^k$. We can apply 3.13 and get

$$(0:\ell_i^{k+1}) \cap L\ell_i^k = \{o\}$$

and $L_i\ell_i^k$ has a right identity e_i with
$e_i\ell_i^{k+1} = \ell_i^{k+1}$.

(b) By 1.13, $M = Me_1+(0:e_1) \cap M = L_1+(0:e_1) \cap M$.
From $e_1\ell_1^{k+1} = \ell_1^{k+1}$ we get
$(0:e_1) \cap M \subseteq (0:\ell_1^{k+1}) \cap M \subseteq (L_1 \cap L_2:\ell_1^{k+1}) \cap M$.
By the remarks at the beginning of the proof,
$e \equiv \ell_1^{k+1}+\ell_2^{k+1} \pmod{L_1 \cap L_2}$.
Thus $\forall\ m \varepsilon (L_1+L_2:\ell_1^{k+1}) \cap M$: $m = me \equiv m\ell_1^{k+1}+m\ell_2^{k+1} \equiv$
$\equiv m\ell_2^{k+1} \pmod{L_1 \cap L_2}$ (since $me = m(\ell_1+\ell_2)$), and
$m(\ell_1+\ell_2)-m\ell_2 \varepsilon L_1$, $m(\ell_1+\ell_2)-m\ell_1 \varepsilon L_2$, $m(\ell_1+\ell_2) \equiv m\ell_1+m\ell_2 \equiv$
$\equiv m\ell_1^{k+1}+m\ell_2^{k+1} \pmod{L_1 \cap L_2}$)).
But $\ell_2^{k+1} \varepsilon L_2$, so $m \varepsilon L_2$, hence $(L_1 \cap L_2:\ell_1^{k+1}) \subseteq L_2$
and by the minimality of L_2 we get $(0:e_1) \cap M = L_2$.
By symmetry, $(0:e_2) \cap M = L_1$.

(a) $e_1+e_2 = (e_1+e_2)e \equiv (e_1+e_2)\ell_1^{k+1}+(e_1+e_2)\ell_2^{k+1} \pmod{L_1 \cap L_2}$ (as above). Because of $e_1\ell_1^{k+1} = \ell_1^{k+1}$ and $e_2\ell_1^{k+1} \in L_1 \cap L_2$ (since $(L_1 \cap L_2 : \ell_1^{k+1}) = L_2$) we get $e_1+e_2 \equiv \ell_1^{k+1}+\ell_2^{k+2} \equiv \ell_1+\ell_2 = e \pmod{L_1 \cap L_2}$.

Finally, since $e_1 \in L_1 = (0:e_2) \cap M$, $e_1e_2 = 0$ and by symmetry, $e_2e_1 = 0$. So e_1,e_2 are orthogonal idempotents.

3.45 <u>REMARK</u> See Lausch (5) for applying sets E of orthogonal idempotents to get a decomposition of N into "blocks" "spanned" by some partitions of E (similar to (Artin-Nesbitt-Thrall)).
See Fain (1) (Th. 6.4) and Lyons (3),(4) for more decompositions induced by orthogonal idempotents.

3.46 <u>DEFINITION</u> An idempotent $e \in N$ is called <u>central</u> if it is in the center of $(N,\cdot)$, i.e. if $\forall\, n \in N$: $en = ne$.

3.47 <u>PROPOSITION</u> (Betsch (3)). Let e be a central idempotent with $Ne \trianglelefteq N$. Then N is the direct sum of the ideals Ne and $(0:Ne) = (0:e)$.

<u>Proof</u>. Clearly Ne (by assumption) and $(0:Ne)$ (by 1.43(b)) are ideals. By 1.13, $N = Ne+(0:e)$ and $Ne \cap (0:e) = \{0\}$. But $(0:e) = (0:Ne)$, since e is central.

3.48 <u>PROPOSITION</u> (Fain (1)). Let E be a set of orthogonal central idempotents and Σe_i any sum of distinct elements of E. Then

(a) $E \subseteq N_d$.

(b) Σe_i is idempotent.

(c) $\forall\, n \in N$: $n\Sigma e_i - \Sigma ne_i \in (0:E) \trianglelefteq N$.

(d) $(0 \in E \vee |E| \geq 2) \Rightarrow N = N_0$.

Proof. (a) is trivial.

(b) $(\Sigma e_i)^2 = \Sigma e_j(\Sigma e_i) = \Sigma\Sigma e_j e_i = \Sigma e_i e_i = \Sigma e_i$.

(c) $\forall$ $e \in E$: $(n\Sigma e_i - \Sigma n e_i)e = 0$.

$(0:E) = \bigcap_{e \in E} (0:e) \trianglelefteq_\ell N$ (by 1.43(a)). Moreover,

$(0:E)N \subseteq (0:E)$ since

$\forall$ $n \in N$ $\forall$ $m \in (0:E)$ $\forall$ $e \in E$: $(mn)e = men = 0n = 0$.

(d) If $0 \in E$ then clearly $N = N_0$.
If $|E| \geq 2$, let $e \neq f$ be in E. Then for all $n \in N$, $n0 = nef = efn = 0n = 0$.

3.49 REMARK A ring is called biregular if each principal ideal is generated by an idempotent. In (3), Betsch defined a near-ring to be biregular if there exists some set E of central idempotents with

(a) $\forall$ $e \in N$: $Ne \trianglelefteq N$.

(b) $\forall$ $n \in N$ $\exists$ $e \in E$: $Ne = (n)$ (principal ideal generated by n).

(c) $\forall$ $e,f \in E$: $e+f = f+e$.

(d) $\forall$ $e,f \in E$: $ef \in E \wedge e+f-ef \in E$.

Ramakotaiah (1) showed that each commutative biregular near-ring is isomorphic to a subdirect product of fields and hence a biregular ring.

f) MORE ON MINIMALITY

We conclude this paragraph with some results concerning minimality of non-nilpotent N-subgroups and left ideals of N. As we will see, considering minimality does not imply that the results can be reached by minimal efforts.
However, we first reap the fruits of previous sections.

3.50 DEFINITION $M \leq_N N$ ($L \trianglelefteq_\ell N$) is called a minimal non-nilpotent N-subgroup (left ideal) if it is minimal in the set of all N-subgroups (left ideals) of N which are not nilpotent.

Clearly if $L \trianglelefteq_\ell N = N_0$ and L is a minimal non-nilpotent N-subgroup then L is a minimal non-nilpotent left ideal.

3.51 THEOREM $N \in \mathcal{N}_0$, DCCN.

(a) (Scott (4)). $M \leq_N N$ is a minimal non-nilpotent N-subgroup $\Rightarrow$ M contains a right identity e with $Ne = Me = M$.

(b) If $L \trianglelefteq_\ell N$ is a minimal non-nilpotent left ideal then L contains a non-zero idempotent.

(c) (Beidleman (1)). If $L \trianglelefteq_\ell N$ is a minimal non-nilpotent N-subgroup then L is a direct summand of $_NN$.

Proof. (a) If $m \in M$ is not nilpotent then $m^2 \in Mm \subseteq M$ is not nilpotent, so $Mm = M$ and by 3.13 M contains a right identity e. By the minimality of M, Ne (not nilpotent!) = M = Me.

(b) By the minimum condition in N, L contains a minimal non-nilpotent N-subgroup M. M has a right identity e by (a) and so L has a non-zero idempotent e.

(c) L contains a right identity e by (a) with $Le = L$. By 1.13, $N = L + (0:e)$ and L is a direct summand of $_NN$.

<u>3.52 COROLLARY</u> (Blackett (2)). $N \in \mathcal{N}_0$, DCCN, N without non-zero nilpotent N-subgroups. Then each minimal N-subgroup M is generated by an idempotent e which is a right identity of M.

We now turn to minimal ideals.

<u>3.53 PROPOSITION</u> (Scott (3)). DCCL, I minimal ideal, $M \leq_N N$, M nilpotent, $M \subseteq I$. Then $IM = \{0\}$.

<u>Proof</u>. Let M^k be $= \{0\}$, $k \geq 2$, $M^{k-1} \neq \{0\}$. Then $M^{k-1} \cdot M = \{0\}$, so M^{k-1} is contained in the ideal $(0:M)$. Hence the ideal J generated by M^{k-1} is contained in $(0:M)$. Since $\{0\} \neq J \subseteq I$, $J = I$ and $IM = \{0\}$.

The following theorem is certainly a very nice one.

<u>3.54 THEOREM</u> (Scott (5)). $N \in \mathcal{N}_0$, DCCN, I a minimal ideal. Then I is a finite direct sum of N-isomorphic minimal left ideals of N (and therefore completely reducible when considered as ${}_N I$).

<u>Proof</u>. We need 3 lemmata and keep the assumptions of the theorem.

<u>Lemma 1.</u> Let ${}_N\Gamma$ be faithful and Δ be a minimal ideal of ${}_N\Gamma$. Let $\{0\} \neq L \subseteq (\Delta:\Gamma)$ be a left ideal such that $\forall\, \gamma \in \Gamma$: $N\gamma = \Gamma \vee L\gamma = \{o\}$. Then L is a finite direct sum of N-isomorphic minimal left ideals of N.

<u>Proof</u>. (a) Since ${}_N\Gamma$ is faithful and $L \neq \{0\}$, $\exists\, \gamma_1 \in \Gamma$: $L \not\subseteq (o:\gamma_1)$, so $(o:\gamma_1)_L = (o:\gamma_1) \cap L \subset L$. If $(o:\gamma_1)_L \neq \{0\}$ then $\exists\, \gamma_2 \in \Gamma$: $(o:\gamma_1)_L \supset$ $\supset (o:\gamma_1)_L \cap (o:\gamma_2)_L$.

Proceeding in this manner, by the DCCN we eventually obtain elements $\gamma_1, \gamma_2, \ldots, \gamma_n \in \Gamma$ with $\bigcap_{i=1}^{n} (o:\gamma_i)_L = \{0\}$.

Thus $(o:\{\gamma_1,\dots,\gamma_n\})_L = \bigcap_{i=1}^{n} (o:\gamma_i)_L = \{0\}$.
Anyhow, we get a non-empty subset Σ of $\{\gamma_1,\dots,\gamma_n\}$ of minimal order for the property that $(o:\Sigma)_L = \{0\}$.
Set $\Sigma =: \{\sigma_1,\dots,\sigma_k\}$.

(b) Define $L_1 := L$ if $k = 1$ and $L_i := (o:\Sigma\setminus\{\sigma_i\})_L$ if $k>1$ $(i\in\{1,\dots,k\})$.

We now show that $h_i : L_i \to \Delta$, $\ell \to \ell\sigma_i$ are N-isomorphisms.

$k = 1$: Then $\{0\} \neq L = L_1$ and $(o:\sigma_1)_L=\{0\}$.
Thus $L\sigma_1 = L_1\sigma_1 \neq \{o\}$ and $N\sigma_1 = \Gamma$.
Since $L\subseteq(\Delta:\Gamma)$, $L_1\sigma_1\subseteq\Delta$. By 3.4(a), $L_1\sigma_1 \trianglelefteq_N N\sigma_1 = \Gamma$.
Since Δ is minimal, $L_1\sigma_1 = \Delta$ and h_1 is surjective.
Also, $\operatorname{Ker} h_1 = (o:\sigma_1)_L = \{0\}$; hence h_1 is an N-isomorphism.

$k > 1$: Suppose that $\exists\, j\in\{1,\dots,k\}: L_j\sigma_j = \{o\}$. Then $L_j\subseteq(o:\sigma_j)$, so $L_j\subseteq(o:\Sigma)_L = \{0\}$, a contradiction to the minimality of Σ.
Hence all $L_i\sigma_i \neq \{o\}$ and (as above) $L_i\sigma_i = \Delta$. Also, $\operatorname{Ker} h_i = (o:\sigma_i)\cap L_i = (o:\sigma_i)\cap(o:\Sigma\setminus\{\sigma_i\})\cap L = (o:\Sigma)_L = \{0\}$, and again h_i is an N-isomorphism.

(c) Let i be $\in\{1,\dots,k\}$.
L_i is a minimal left ideal of N:
By 3.4(e), $N/_{(o:\sigma_i)} \cong_N N\sigma_i = \Gamma$. By (b), $L_i\sigma_i = \Delta$ is minimal.
Thus $L_i+(o:\sigma_i)/_{(o:\sigma_i)}$ is a minimal ideal of the N-group $N/_{(o:\sigma_i)}$. Since $L_i \cap (o:\sigma_i) = \{o\}$ (by (b)), 2.8 gives

$L_i \cong_N L_i+(o:\sigma_i)/_{(o:\sigma_i)}$, so L_i is a minimal left ideal of N.

(d) Since all $L_i \cong_N \Delta$, the L_i's are N-isomorphic.

(e) We show that $L = L_1+\ldots+L_k$. We may assume that $k>1$.
If $\ell\varepsilon L$, $\forall\ i\varepsilon\{1,\ldots,k\}$: $\ell\sigma_i\varepsilon\Delta$. By (b), $L_i\sigma_i = \Delta$, so $\exists\ \ell_i\varepsilon L_i$: $\ell\sigma_i = \ell_i\sigma_i$.
Set $\ell' := \ell_1+\ldots+\ell_k$.
If $i \neq j$, $\ell_j\varepsilon L_j\subseteq(o:\sigma_i)$, so $\ell_j\sigma_i = o$.
So for all $i\varepsilon\{1,\ldots,k\}$ $\ell'\sigma_i = \ell_i\sigma_i$.
Therefore $(\ell-\ell')\Sigma = \{o\}$, so $\ell-\ell'\varepsilon(o:\Sigma)\cap L = \{o\}$. Hence $\ell = \ell' = \ell_1+\ldots+\ell_k$.

(f) The sum in (e) is direct:
If $\ell = \ell_1+\ldots+\ell_k = \mu_1+\ldots+\mu_k$ $(\ell_i,\mu_i\varepsilon L_i)$
Then $\forall\ i\varepsilon\{1,\ldots,k\}$: $\ell\sigma_i = \mu_i\sigma_i = \ell_i\sigma_i$.
Thus $\forall\ i\varepsilon\{1,\ldots,k\}$: $\ell_i-\mu_i\varepsilon(o:\Sigma)\cap L$ and the proof is complete.

<u>Lemma 2.</u> If $IN \neq \{0\}$ and $M \leq_N N$ is minimal for $IM \neq \{0\}$ then

(a) M contains a right identity.

(b) $I\cap M$ is minimal amongst all non-zero ideals of M which are also N-subgroups.

(c) $I\cap M$ is the sum of minimal ideals of ${}_NM$.

<u>Proof</u>. (a) We shall show that ${}_MM$ is monogenic.
If $\forall\ m\varepsilon M$: $Mm \neq M$, $IMM = \{0\}$ since $\forall\ m\varepsilon M$: $I(Mm) = \{0\}$ by the minimality of M.
Denote the ideal generated by IM by J.
Since IM is contained in the ideal $(0:M)$, $J\subseteq(0:M)$ and $JM = \{0\}$.
But $IM \neq \{0\}$, $IM\subseteq I$ and so $J = I$ and we arrive at the contradiction $IM = \{0\}$.

So M is monogenic and the result (a) follows from 3.13.

(b) Since $IM \neq \{0\}$ and $IM \subseteq I \cap M$, $I \cap M$ is a non-zero ideal of M.
Let $\{0\} \neq K \trianglelefteq M$ be such that $K \leq_N N$ and $K \subseteq I \cap M$. Since $KM \subseteq K \subseteq I$, $K \subseteq (K:M) \cap I$ and $(K:M) \cap I \neq \{0\}$, so $(K:M) \cap I = I$ and $I \subseteq (K:M)$. It follows that $IM \subseteq (K:M)M \subseteq K$.
By (a), M contains a right identity e, so $(I \cap M)e = I \cap M$. Thus $I \cap M \subseteq (I \cap M)M \subseteq IM \subseteq K$ and (b) follows.

(c) Since $I \cap M \neq \{0\}$, there exists a minimal ideal W of ${}_N M$ in $I \cap M$.
Take some $m \in M$.
If $Mm = M$ then Wm is either $= \{0\}$ or a minimal ideal of ${}_N M$ (since Wm is an N-endomorphic image of W).
If $Mm \neq M$ then $IMm = \{0\}$. But M contains a right identity e, so $We = W$ and thus $Wm = Wem \subseteq IMm = \{0\}$.

Hence $\forall\ m \in M$: $Wm = \{0\}$ or Wm is a minimal ideal of ${}_N M$.

Set $L := \sum_{m \in M} Wm \subseteq I$. $L \leq_N M$.

$W = We \subseteq L$ and $L \neq \{0\}$.
So L is the sum of minimal ideals of ${}_N M$.
Of course, $L \trianglelefteq M$. Also $\forall\ \bar{m} \in M$: $L\bar{m} =$
$= \sum_{m \in M} Wm\bar{m} \subseteq L$, so $L \trianglelefteq M$, and L is a non-zero ideal of M contained in $I \cap M$. By (b), $L = I \cap M$ and (c) is shown.

Although the reader might be gasping for breath, we need a third Lemma, which will be used in the proof of 4.47.

Lemma 3. If $IN \neq \{0\}$ there exists an N-group $_N\Gamma$ and a minimal ideal $\Delta \trianglelefteq_N \Gamma$ such that

(a) $(o:\Gamma) \cap I = \{0\}$.

(b) $I+(o:\Gamma) \subseteq (\Delta:\Gamma)$.

(c) $\forall\ \gamma\varepsilon\Gamma$: $I\gamma = \{o\}$ v $N\gamma = \Gamma$.

Proof. Let M, W, e be as in lemma 2. By part (c) of this lemma, $I \cap M$ is the sum of minimal ideals of $_NM$. By lemma 2 (b) and the fact that $W \subseteq I \cap M$, we conclude from 2.48(g) that W is a direct summand.
So $\exists\ U \trianglelefteq_N I \cap M$: $I \cap M = W \dot{+} U$.
Define $\Gamma := M/U$ and $\Delta := I\cap M/_U = W\dot{+}U/_U \cong_N W$.
So Δ is a minimal ideal of $_N\Gamma$.
We now prove the lemma.

(a) $(o:\Gamma) \cap I = \{0\}$ or $= I$. If $(o:\Gamma)\cap I = I$ then $I \subseteq (o:\Gamma)$. So $I\Gamma = \{o\}$ and $IM \subseteq U$.
But $(I \cap M)e = I \cap M$, so $I \cap M \subseteq (I \cap M)M \subseteq IM \subseteq U$.
Thus $\Delta = \{o\}$, a contradiction.

(b) Since $IM \subseteq I \cap M$, $I \subseteq (\Delta:\Gamma)$ and $I+(o:\Gamma) \subseteq I+(\Delta:\Gamma) = (\Delta:\Gamma)$.

(c) If $\gamma\varepsilon\Gamma$, $\exists\ m\varepsilon M$: $\gamma = m+U$.
If $Nm \subset M$, the minimality of M gives us $I(N\gamma) = \{o\}$. So $IN \subseteq (o:\gamma)$. If $(IN)_\ell$ is the left ideal generated by IN then $(IN)_\ell \subseteq (o:\gamma)$. By 1.52, $(IN)_\ell$ is the ideal generated by IN and therefore equals I.
So $I\gamma = \{o\}$, completing the proof.

Tired, but happy we are ready for the

Proof of the theorem. Suppose that $IN = \{0\}$ and L is a minimal left ideal of N contained in I. So $LN = \{0\}$ and $L \trianglelefteq N$. Thus $L = I$ and the theorem holds.

If $IN \neq \{0\}$, let Γ,Δ be as in Lemma 3. If $N/_{(o:\Gamma)} =: N'$, ${}_{N'}\Gamma$ is faithful and has Δ as a minimal ideal.

Set $I' := I+(o:\Gamma)/_{(o:\Gamma)}$.

By lemma 3(c), $\forall\ \gamma\varepsilon\Gamma$: $(N'\gamma = \Gamma \vee I'\gamma = \{o\})$.
By lemma 3(a), $I \cong_N I/_{\{0\}} \cong_N I'$. Thus by lemma 1, $I \cong_N I'$ is the direct sum of minimal N-isomorphic left ideals.

The proof is now complete.

Note that if $N\varepsilon\mathcal{N}_0$ is simple and has the DCCN then ${}_NN$ is completely reducible and 2.50 is applicable. Cf. 4.46 and 4.47.

3.55 COROLLARY (Scott (3),(5)). $N\varepsilon\mathcal{N}_0$, DCCN, I a non-nilpotent minimal ideal of N; $Q(N)$:= sum of all nilpotent left ideals. Then $Q(N)\cap I = \{0\}$ and $Q(N)$ is nilpotent.

Proof. By 3.54 and 2.48, $Q(N)\cap I$ is a direct summand of ${}_NI$. Let $L \trianglelefteq_\ell I$ be such that $I = Q(N)\cap I \dot{+} L$.

By 2.22, $\forall\ i\varepsilon I\ \forall\ q\varepsilon Q(N)\cap I\ \forall\ \ell\varepsilon L$: $i(q+\ell) \equiv$
$\equiv iq+i\ell \pmod{Q(N)\cap I\cap L}$.
Since $(Q(N)\cap I)\cap L = \{0\}$, $i(q+\ell) = iq+i\ell$.
Hence $I^2 = I(Q(n)\cap I+L) \subseteq I(Q(N)\cap I)+IL$.
But by 3.53, $I(Q(N)\cap I) = \{0\}$, since $Q(N)$ is nilpotent by 2.101(b) and 3.40. So $I^2 \subseteq IL \subseteq I$ and the left ideal generated by I^2, the ideal generated by I^2 (by 1.52) and I (by minimality) coincide.
So I is contained in the left ideal generated by $IL \subseteq L$, $I = L$ and $Q(N)\cap I = \{0\}$.
See Scott (1) or (6) for the proof that $Q(N)$ is nilpotent.

3.56 REMARK There also exist results concerning near-rings with ascending chain conditions. For "Goldie-type" ones, see Oswald (2). For more results, consult Scott (1) and Kaarli (2).

§ 4 PRIMITIVE NEAR - RINGS

This paragraph presents a discussion of the "building stones, near-rings are made of", the so-called "primitive near-rings". Similar to ring theory, the "atoms" are not the simple near-rings as one might expect at a first glance. There is, however, an important connection (4.47). The idea to consider primitive near-rings comes from the bible ("You will recognize them by their fruits"): given a near-ring N, we look at all of its fruits (= N-groups) and ask, whether there are faithful and "enough simple" ones among them. If this is the case, we call N "primitive on this N-group". Since "enough simple" is not precise we fix its meaning in wanting N-groups of type ν. The resulting concept is that of "ν-primitivity".
We get the hierarchy 2-primitivity $\underset{\nLeftarrow}{\Rightarrow}$ 1-primitivity $\underset{\nLeftarrow}{\Rightarrow}$ 0-primitivity, discuss conditions, which force some of these concepts to coincide and make a lot of work towards a density theorem which is comparable to the celebrated one in ring theory due to N. Jacobson. We really get one for 2-primitive near-rings with identity (4.52). Adding a chain condition, we arrive at a Wedderburn-Artin-like structure theorem (4.60). Before that, we get "better and better" density-like structure theorems for 0-, 1- and 2-primitive near-rings. It comes out that many theorems on ν-primitive near-rings can be derived from zero-symmetric ν-primitive near-rings where they are much easier to obtain since these ones behave more like rings. However, many proofs concerning even zero-symmetric near-rings differ totally from the comparable ones in ring theory.
Anyhow, the "building stones" mentioned above (2-primitive near-rings with identity) are shown to be dense in $\mathrm{Hom}_D(\Gamma,\Gamma)$ or $M_{aff}(\Gamma)$ (if N_0 is a ring) or in $M_{G_o\cup\{\bar{o}\}}(\Gamma)$ or $M_{G_o\cup\{\bar{o}\}}(\Gamma)+M_c(\Gamma)$ (if N_0 is a non-ring), where G_o is the fixed-point-free automorphism group $\mathrm{Aut}_{N_0}(\Gamma)$. In particular,

if $G_0 = \{id\}$, the latter two ones are $M_0(\Gamma)$ and $M(\Gamma)$. Finally, the density property is seen to be a kind of an interpolation property and a "purely interpolation-theoretic" result will be obtained. Recall again (p.1) that $\Gamma^* = \Gamma \setminus \{o\}$, and so on.

a) GENERAL

1.) DEFINITIONS AND ELEMENTARY RESULTS

4.1 CONVENTION In all what follows, ν will be any number $\in \{0,1,2\}$ unless otherwise specified.

4.2 DEFINITION

(a) N is called ν-primitive on ${}_N\Gamma$: $\Longleftrightarrow$ ${}_N\Gamma$ is faithful and of type ν.

(b) N is ν-primitive: $\Longleftrightarrow \exists\ {}_N\Gamma \in {}_N\mathcal{G}$: N is ν-primitive on ${}_N\Gamma$.

(c) $I \trianglelefteq N$ is called a ν-primitive ideal of N: $\Longleftrightarrow$ N/I is ν-primitive.

4.3 PROPOSITION Let I be an ideal of N. Then the following conditions are équivalent:

(a) I is ν-primitive.

(b) $\exists\ {}_N\Gamma \in {}_N\mathcal{G}$: $I = (o:\Gamma) \wedge {}_N\Gamma$ is of type ν.

(c) $\exists\ L \trianglelefteq_\ell N$: $I = (L:N) \wedge L$ is ν-modular.

Proof. (a) $\Rightarrow$ (b): I is ν-primitive $\Rightarrow$ N/I is ν-primitive on some ${}_{N/I}\Gamma \Rightarrow {}_N\Gamma$ (as in 3.14(b)) is of type ν and $I = (o:\Gamma)$.

(b) $\Rightarrow$ (c): Let Γ be $= N\gamma \neq \{o\}$. $(o:\gamma) =: L$. Then L is modular. By 3.4(e), $N/_L \cong_N \Gamma$, so L is ν-modular. Finally, $I = (o:\Gamma) = (o:N/L) = (L:N)$.

(c) $\Rightarrow$ (a): Take $N/L =: \Gamma$. Then ${}_N\Gamma$ is of type ν and (as above) $I = (L:N) = (o:\Gamma)$.

4.4 COROLLARY The following conditions are equivalent:

(a) N is ν-primitive.
(b) $\{0\}$ is a ν-primitive ideal.
(c) $\exists\, L \trianglelefteq_\ell N$: L ν-modular $\wedge$ $(L:N) = \{0\}$.

4.5 REMARKS

(a) Observe that (c) in 4.3 and 4.4 give "intrinsic" characterizations of primitivity - that will be extremely helpful, for it enables one to recognize primitivity "within N".

(b) If N is ν-primitive on Γ then $N \neq \{0\}$, $\Gamma \neq \{o\}$ and if $I \trianglelefteq N$ is a ν-primitive ideal then $I \neq N$.

(c) 2-primitivity implies 1-primitivity and this in turn implies 0-primitivity (always on the same group).

(d) The near-rings N_ν of 3.8 are examples of ν-primitive near-rings. (on $\mathbb{Z}_4$).

(e) If N is ν-primitive on Γ then $N \hookrightarrow M(\Gamma)$ (1.48).

(f) See §5 of Betsch (3) for a discussion of the spaces of ν-primitive ideals $(\nu = 1,2)$ of $N \in \mathcal{N}_0$.

4.6 PROPOSITION Let N contain either a left or a right identity e. Then

(a) Every ν-primitive ideal I of N is modular.
(b) If e is a left identity of N then N is 1-primitive iff N is 2-primitive (and in this case e is a two-sided identity).

Proof. (a) If e is a left identity in N then (because N/I is ν-primitive on some ${}_{N/I}\Gamma$) e+I is an identity of N/I by 3.4(c). So $\forall\, n \in N$: $en \equiv ne \equiv n \pmod I$. If e is a right identity, the assertion is trivial.

(b) Let N be 1-primitive on ${}_N\Gamma$. By (a), 4.4 and 3.4(c), e is a two-sided identity for N. By 3.4(b), ${}_N\Gamma$ is unitary. Now apply 3.7(c) and 3.19(a).

4.7 PROPOSITION Let N be simple and $_N\Gamma$ be of type ν. Then N is ν-primitive on Γ.

Proof. $(o:\Gamma) \trianglelefteq N$, so $(o:\Gamma) = \{0\}$ (for $(o:\Gamma) = N$ gives the contradiction $N\Gamma = \{o\}$).

4.8 PROPOSITION (Betsch (3)). Let the ring N be ν-primitive on Γ. Then N is a primitive ring on the N-module Γ ((N. Jacobson), p. 4).

Proof. If $\Gamma = N\gamma$ then $\Gamma \cong_N N/_{(o:\gamma)}$ and $(\Gamma,+)$ is abelian. If $n_1+(o:\gamma), n_2+(o:\gamma)\varepsilon N/_{(o:\gamma)}$ then

$\forall\ n\varepsilon N$: $n(n_1+(o:\gamma)+n_2+(o:\gamma)) = nn_1+(o:\gamma)+nn_2+(o:\gamma) =$

$= n(n_1+(o:\gamma))+n(n_2+(o:\gamma))$.

Hence $\forall\ \gamma_1,\gamma_2\varepsilon\Gamma\ \forall\ n\varepsilon N$: $n(\gamma_1+\gamma_2) = n\gamma_1+n\gamma_2$, and $_N\Gamma$ is a (ring-) module.

Each N-submodule of $_N\Gamma$ is an ideal, so $= \{o\}$ or $= \Gamma$. Finally, $N\Gamma \neq \{o\}$ by assumption, so $_N\Gamma$ is irreducible and N is primitive on Γ.

4.9 COROLLARY (Ramakotaiah (1)). If N is commutative and ν-primitive then N is a field.

Proof. By 4.4(c), $\exists\ L\varepsilon\mathcal{L}_\nu(N)$: $(L:N) = \{0\}$. $L \trianglelefteq N$, since N is commutative. By 3.25, $(L:N)$ is the greatest ideal in L, so $L = \{0\}$ and N contains a right identity. By 1.107(c), N is a ring, hence a primitive ring by 4.8 and by (N. Jacobson), p.7 a field.

2.) THE CENTRALIZER

4.10 DEFINITION

(a) $End_N(\Gamma) = Hom_N(\Gamma,\Gamma) =: C_N(\Gamma) =: C$ is called the <u>centralizer</u> of ${}_N\Gamma$.

(b) $Aut_N(\Gamma) =: G_N(\Gamma) =: G$; $Aut_{N_o}(\Gamma) =: G_o$.

(c) $G^o := \begin{cases} G \cup \{\bar{o}\} & \text{if } \bar{o}\in C; \\ G & \text{otherwise} \end{cases}$; likewise G_o^o.

4.11 REMARKS

(a) $(C,\circ)$ is a monoid, $(G,\circ)$ and $(G_o,\circ)$ are groups, $(G^o,\circ)$ and $(G_o^o,\circ)$ ("groups with zero") are monoids.

(b) $\bar{o}\in C \iff \Omega = \{o\}$.

(c) If ${}_N\Gamma$ is faithful then $N \hookrightarrow M_{C_N(\Gamma)}(\Gamma) \subseteq M_G(\Gamma)$.

4.12 NOTATION If $n\in N$, $f_n: \begin{matrix}\Gamma \to \Gamma \\ \gamma \to n\gamma\end{matrix}$; $F_N(\Gamma) := \{f_n | n\in N\} =: F$.

4.13 PROPOSITION (Mlitz (3)).

(a) If ${}_N\Gamma$ is monogenic then $C_N(\Gamma) =$

$= \{m\in M(\Gamma) | \forall f\in F: m\circ f = f\circ m\} =: M_F(\Gamma)$.

(b) $\forall$ $h\in C_N(\Gamma)$: $h/\Omega = id$.

(c) If $\Omega = \Gamma$ then $C_N(\Gamma) = \{id\}$.

<u>Proof</u>. (a) If $h\in C_N(\Gamma)$ and $f_n\in F_N(\Gamma)$ then $\forall$ $\gamma\in\Gamma$: $(h\circ f_n)(\gamma) = h(n\gamma) = nh(\gamma) = (f_n\circ h)(\gamma)$; so $h\in M_F(\Gamma)$.

Conversely, let f be $\in M_F(\Gamma)$. If $\gamma_1,\gamma_2\in\Gamma = N\gamma$ and $n\in N$ then $\exists$ $n_1,n_2\in N$: $\gamma_1 = n_1\gamma \wedge \gamma_2 = n_2\gamma$. Then $f(n\gamma_1) = (f\circ f_n)(\gamma_1) = (f_n\circ f)(\gamma_1) = nf(\gamma_1)$ and

$f(\gamma_1+\gamma_2) = f(n_1\gamma+n_2\gamma) = (f \circ f_{n_1+n_2})(\gamma) =$
$= (f_{n_1+n_2} \circ f)(\gamma) = (n_1+n_2)f(\gamma) = n_1 f(\gamma)+n_2 f(\gamma) =$
$= f(n_1\gamma)+f(n_2\gamma) = f(\gamma_1)+f(\gamma_2)$.

Hence $f \in C_N(\Gamma)$.

(b) $\forall\ h \in C_N(\Gamma)\ \forall\ no \in \Omega$: $h(no) = nh(o) = no$.

(c) follows from (b).

4.14 NOTATION $\theta_o := \theta_o({}_N\Gamma) := \{\gamma \in \Gamma \mid N\gamma = No = \Omega\}$.

$\theta_1 := \theta_1({}_N\Gamma) := \{\gamma \in \Gamma \mid N\gamma = \Gamma\}$.

4.15 REMARKS (Betsch (5)).

(a) $o \in \theta_o$, so $\theta_o \neq \emptyset$.

(b) $\theta_1 \neq \emptyset \iff {}_N\Gamma$ is monogenic.

(c) $\theta_o \cap \theta_1 = \emptyset \iff \Omega \neq \Gamma$.

(d) ${}_N\Gamma$ is strongly monogenic $\Rightarrow \Gamma = \theta_o \cup \theta_1$.

(e) ${}_N\Gamma$ is unitary $\Rightarrow \theta_o = \Omega$ (for $\gamma \in \theta_o \Rightarrow N\gamma = \Omega \Rightarrow$
$\Rightarrow \gamma = 1\gamma \in \Omega$ and $\omega = no \in \Omega \Rightarrow N\omega = Nno \subseteq No = \Omega \Rightarrow \omega \in \theta_o$).

(f) $G(\theta_o) = \theta_o \wedge G(\theta_1) = \theta_1$, so G induces permutation groups on θ_o and (if $\theta_1 \neq \emptyset$) on θ_1.

The next proposition is a "Schur-type lemma".

4.16 PROPOSITION (Betsch (5), Mlitz (3)).

(a) ${}_N\Gamma$ is simple $\wedge\ \Omega = \{o\} \Rightarrow C = \{\bar{o}\} \cup Mon_N(\Gamma)$ and $(h \in C \wedge \exists\ \gamma \in \theta_1: h(\gamma) \in \theta_1) \Rightarrow h \in G$.

(b) ${}_N\Gamma$ is N-simple $\Rightarrow C = Epi_N(\Gamma,\Omega) \cup Epi_N(\Gamma,\Gamma)$ (if $N \in \mathfrak{N}_0$, $Epi_N(\Gamma,\Omega) = \{\bar{o}\}$!).

(c) ${}_N\Gamma$ is N_0-simple $\Rightarrow C = G^o$.

Proof. (a) follows from the fact that $\forall\ h\in C$: Ker $h \trianglelefteq_N \Gamma$, so either Ker $h = \{o\}$ (then $h\in \mathrm{Mon}_N(\Gamma)$) or Ker $h = \Gamma$ (then $h = \bar{o}$). We may assume that $\Gamma \neq \{o\}$. If $h\in C \wedge \exists\ \gamma\in\theta_1$: $h(\gamma)\in\theta_1$ then $h \neq \bar{o}$, so $h\in \mathrm{Mon}_N(\Gamma)$. Now $h(\Gamma) = h(N\gamma) = Nh(\gamma) = \Gamma$.

(b) $\forall\ h\in C$: Im $h \leq_N \Gamma$, so either Im $h = \Omega$ or Im $h = \Gamma$.

(c) follows from (b).

We are mainly interested in the case that $C = G^o$, in which every non-zero N-endomorphism of Γ is an N-automorphism.

4.17 PROPOSITION (Betsch (5)).

(a) G is fixed-point-free (1.4(b)) on θ_1.

(b) If $_N\Gamma$ is simple then $C = G^o \iff \forall\ \Delta <_N \Gamma$: $\Gamma \not\cong_N \Delta$.

Proof. (a) Assume that for $g\in G$ and $\gamma\in\theta_1$ $g(\gamma) = \gamma$. $\forall\ \delta\in\Gamma\ \exists\ n\in N$: $\delta = n\gamma$. Then $g(\delta) = g(n\gamma) = ng(\gamma) = n\gamma = \delta$. So $g = \mathrm{id}$.

(b) $\Rightarrow$: Assume that h is an N-isomorphism $\Gamma \to \Delta <_N \Gamma$. Then $h\in C = G^o \subseteq \{\bar{o}\}\cup \mathrm{Aut}_N(\Gamma)$, a contradiction.

$\Leftarrow$: If $h\in C$, $h \neq \bar{o}$ then Ker $h \neq \Gamma$, so Ker $h = \{o\}$. Therefore h is a monomorphism and $\Gamma \cong$ Im h. So Im $h = \Gamma$, and $h\in \mathrm{Aut}_N(\Gamma)$.

4.18 COROLLARY (Betsch (5)). If $_N\Gamma$ is of type 1 or if $_N\Gamma$ is simple and finite then $C = G^o$.

Proof. If $_N\Gamma$ is of type 1 then $_N\Gamma$ is simple. Assume that h is an N-isomorphism $\Gamma \to \Delta <_N \Gamma$. Represent Γ as $\Gamma = N\gamma$ and call $h(\gamma) =: \delta$. $N\delta = Nh(\gamma) = h(N\gamma) = h(\Gamma) = \Delta$.
If $\delta\in\theta_1$ then $N\delta = \Gamma$, so $\Gamma = \Delta$, a contradiction.
If $\delta\in\theta_o$ then $N\delta = \Omega = \{o\}$, so $\Delta = \{o\}$ and therefore $\Gamma = \{o\}$, which again is a contradiction.

Now apply 4.15(d) and 4.17(b).

If ${}_N\Gamma$ is simple and finite, apply 4.17(b).

4.19 NOTATION For $\gamma,\delta\in{}_N\Gamma$ we define

$$\gamma \sim \delta :\Longleftrightarrow (o:\gamma)_{N_0} = (o:\delta)_{N_0};$$

$$\gamma \approx \delta :\Longleftrightarrow G_o(\gamma) = G_o(\delta).$$

4.20 REMARKS (Betsch (5)).

(a) $\sim,\approx$ are equivalence relations in Γ.

(b) The equivalence classes of $\approx$ are exactly the orbits of G_o on Γ.

(c) $\forall\ \gamma,\delta\in\Gamma$: $\gamma\approx\delta \Rightarrow \gamma\sim\delta$ (for $\gamma\approx\delta \Rightarrow \exists\ g\in G_o$: $g(\gamma) = \delta \Rightarrow$ $\Rightarrow (o:\gamma) = (o:g(\gamma)) = (o:\delta) \Rightarrow \gamma\sim\delta$).

The reason for defining $\sim,\approx$ via N_0 instead of N stems from 4.13(c): in the frequent case that $\Omega = \Gamma$, $\approx$ would otherwise be the all-relation in any case.

4.21 PROPOSITION (Betsch (5)). If ${}_N\Gamma$ is unitary and $N = N_0$ then $\sim$ and $\approx$ coincide on θ_1.

Proof. If $\gamma\sim\delta$ $(\gamma,\delta\in\theta_1)$ then for all $n_1,n_2\in N$

$n_1\gamma = n_2\gamma \Rightarrow n_1-n_2\in(o:\gamma) = (o:\delta) \Rightarrow n_1\delta = n_2\delta.$

Therefore $h: \begin{array}{l}\Gamma \to \Gamma \\ n\gamma \to n\delta\end{array}$ is well defined. h turns out to be an N-automorphism, so $h\in G$.

Now $h(\gamma) = h(1\gamma) = 1\delta = \delta$, hence $\gamma \approx \delta$.

3.) INDEPENDENCE AND DENSITY

An appropriate frame for our next considerations is given by

4.22 DEFINITION (Mlitz (5)). Let M be an arbitrary set and $\mathfrak{f}(M)$ the set of all finite subsets of M. A map $r: \mathfrak{f}(M) \to \mathbb{N}_0$ is called a rank map if

(a) $r(\emptyset) = 0$

(b) $\forall\ F \in \mathfrak{f}(M)\ \ \forall\ m \in M$: $r(F \cup \{m\}) = r(F) + \sigma$ with $\sigma \in \{0,1\}$

(c) $\forall\ F \in \mathfrak{f}(M)\ \ \forall\ m,n \in M$: $[r(F \cup \{m\}) = r(F \cup \{n\}) = r(F) \Rightarrow$ $\Rightarrow r(F \cup \{m,n\}) = r(F)]$.

F is then called r-independent if $r(F) = |F|$.

4.23 REMARK With respect to r-independence, Steinitz's theorem is fulfilled (see A. Kertész, "On independent sets of elements in algebra", Acta Sci. Math. (Szeged) 21, 1960, 260 - 269).

4.24 EXAMPLES

(a) Define $r(F) := |F|$. Then r is a rank function and every (finite) subset is r-independent.

(b) Take a vector space M over a field F. Set $r(F) := \dim L(F)$ (linear hull). r is a rank function and r-independence is just linear independence.

(c) Take an N-group Γ and define for each $\Phi \in \mathfrak{f}(\Gamma)$ $r(\Phi)$ as the number of non $\sim$-equivalent generators (i.e. $r(\Phi) = |\Phi \cap \theta_1/_\sim|$). Then r is a rank function and $\Phi = \{\gamma_1, \ldots, \gamma_n\}$ is r-independent if $\Phi \subseteq \theta_1$ and $\forall\ i \neq j$: $\gamma_i \not\sim \gamma_j$.

This independence is called $\sim$-independence.
The same can be done for $\approx$.

In the theory of rings each primitive ring R is isomorphic to a "dense" subring $\bar{R}$ of a ring $Hom_D(\Gamma,\Gamma)$ for some irreducible R-module Γ and with $D = Hom_R(\Gamma,\Gamma)$ (the centralizer) making $_D\Gamma$ into a vector space (see (N.Jacobson), p. 26 - 31). Density means here (in our notation) that $\forall\ s\in\mathbb{N}\ \forall\ \{\gamma_1,\ldots,\gamma_s\}$ lin. indep. in Γ $\forall\ \delta_1,\ldots,\delta_s\in\Gamma\ \exists\ \bar{r}\in\bar{R}\ \forall\ i\in\{1,\ldots,s\}: \bar{r}(\gamma_i) = \delta_i$.

(It is clear that only values of independent elements can be arbitrarily prescribed.)

We are going to prove similar theorems for near-rings.

But before doing so we have to take a look at the density concept.

<u>4.25 NOTATION</u> Let M be a subset of some $M(\Gamma)$. We introduce a topology in M as in Betsch (6):

If $m\in M$ and $\gamma\in\Gamma$, define $S(m,\gamma) := \{m'\in M \mid m'(\gamma) = m(\gamma)\}$ and $\mathcal{S} := \{S(m,\gamma) \mid m\in M \wedge \gamma\in\Gamma\}$.

<u>4.26 PROPOSITION</u> (Betsch (6)).

(a) $\mathcal{S}$ is the subbase of some topology $\mathcal{T}$ (the "<u>finite topology</u>") on M.

(b) $N\subseteq M$ is dense in M w.r.t. $\mathcal{T}$ $\Longleftrightarrow$

$\Longleftrightarrow \forall\ s\in\mathbb{N}\ \ \forall\ m\in M\ \ \forall\ \gamma_1,\ldots,\gamma_s\in\Gamma\ \ \exists\ n\in N\ \ \forall\ i\in\{1,\ldots,s\}:$

$: n(\gamma_i) = m(\gamma_i)$.

<u>Proof</u>. straightforward and hence omitted.

In all what follows, "density" means "density with respect to $\mathcal{T}$ of 4.26".

<u>4.27 REMARKS</u>

(a) If M and N are subnear-rings of $M(\Gamma)$ then it is easy to see that N_o is dense in M_o iff $N_o+M_c(\Gamma)$ is dense in $M_o+M_c(\Gamma)$. Note that $N_o+M_c(\Gamma)$ and $M_o+M_c(\Gamma)$ are no near-rings in general (see 4.53(e)), except in some important special cases. (See 4.54 and 4.60.)

(b) If N is dense in M then $N_0 := N \cap M_0(\Gamma)$ is dense in M_0.

(c) Observe that if $H \neq \{id\}$ is a fixed-point-free automorphism group of Γ then $M_H(\Gamma) \subseteq M_0(\Gamma)$ (since $\forall\, m \in M_H(\Gamma)\ \forall\, h \in H: h(m(o)) = m(o))$.

If $H = \{id\}$ then $M_H(\Gamma) = M(\Gamma)$.

(d) We will be mainly interested in near-rings which are dense in $M_{G_o^o}(\Gamma)$ and $\overline{M}_{G_o^o}(\Gamma) := M_{G_o^o}(\Gamma) + M_c(\Gamma)$ (4.52 and others).

<u>4.28 THEOREM</u> (Ramakotaiah (2), Betsch (6)). Let H be a fixed-point-free group of automorphisms of some group Γ.

(a) $\forall\, \gamma \in \Gamma^* \quad \forall\, \delta \in \Gamma \quad \exists!\, m \in M_H(\Gamma): (m(\gamma) = \delta\ \wedge$
$\wedge\ \forall\, \gamma' \in \Gamma \setminus H\gamma: m(\gamma') = o)$.

(b) $\forall\, s \in \mathbb{N}\ \forall\, \gamma_1, \ldots, \gamma_s \in \Gamma^*$, $H\gamma_i \neq H\gamma_j$ for $i \neq j$

$\forall\, \delta_1, \ldots, \delta_s \in \Gamma\ \exists\, m \in M_H(\Gamma)\ \forall\, i \in \{1, \ldots, s\}: m(\gamma_i) = \delta_i$.

(c) If $H \neq \{id\}$, $N \subseteq M_H(\Gamma)$ is dense in $M_H(\Gamma) \iff$

$\iff \forall\, s \in \mathbb{N}\ \forall\, \gamma_1, \ldots, \gamma_s \in \Gamma^*$, $H\gamma_i \neq H\gamma_j$ for $i \neq j$

$\forall\, \delta_1, \ldots, \delta_s \in \Gamma\ \exists\, n \in N\ \forall\, i \in \{1, \ldots, s\}: n(\gamma_i) = \delta_i$.

(d) If $M_c(\Gamma) \subseteq N \subseteq \overline{M}_H(\Gamma)$, N is dense in $\overline{M}_H(\Gamma) \iff$

$\iff \forall\, s \in \mathbb{N}\ \forall\, \gamma_1, \ldots, \gamma_s \in \Gamma$, $H\gamma_i \neq H\gamma_j$ for $i \neq j$

$\forall\, \delta_1, \ldots, \delta_s \in \Gamma\ \exists\, n \in N\ \forall\, i \in \{1, \ldots, s\}: n(\gamma_i) = \delta_i$.

<u>Proof</u>. In any case we may assume that $H \neq \{id\}$, for otherwise $\overline{M}_H(\Gamma) = M_H(\Gamma) = M(\Gamma)$.

(a) $\forall\, \alpha \in H\gamma\ \exists!\, h_\alpha \in H: \alpha = h_\alpha(\gamma)$ (since H is fixed-point-free).

Define $m \in M(\Gamma)$ by $m(\alpha) := \begin{cases} h_\alpha(\delta) & \alpha \in H\gamma \\ o & \alpha \notin H\gamma \end{cases}$ Then clearly $m(\gamma) = \delta$ and $m \in M_H(\Gamma)$; m is uniquely determined by the conditions $m(\gamma) = \delta\ \wedge$ $\wedge\ (\forall\, \gamma' \in \Gamma \setminus H\gamma: m(\gamma') = o)$.

(b) Define maps $m_i \in M_H(\Gamma)$ with $m_i(\gamma_i) = \delta_i$ and $\forall\ \gamma' \notin H\gamma_i$: $m_i(\gamma') = o$ (as in (a)).
Then $m := m_1 + \ldots + m_s$ will do the job.

(c) $\Rightarrow$: By (b) and 4.26(b).
$\Leftarrow$: If $H\gamma_i \neq H\gamma_j$ for $i \neq j$, the result is clear.
If $H\gamma_i = H\gamma_j$, $\forall\ m \in M_H(\Gamma)\ \forall\ n \in N$: $n(\gamma_i) = m(\gamma_i) \Rightarrow$

$\Rightarrow n(\gamma_j) = m(\gamma_j)$ and the result follows again from 4.26(b).

(d) $\Rightarrow$: By 4.27, N_o is dense in $(\overline{M}_H(\Gamma))_o = M_H(\Gamma)$.
If one γ_i (say γ_1) $= o$, take $n_c \in M_c(\Gamma)$ to be the map which is constant $= \delta_1$. Take $n_o \in N_o$ with $n_o(\gamma_i) = \delta_i - \delta_1$ for $i \in \{2, \ldots, s\}$. Then $n := n_o + n_c$ fulfills $\forall\ i \in \{1, \ldots, s\}$: $n(\gamma_i) = \delta_i$.
Two or more γ_i cannot be zero. If all $\gamma_i \neq o$, the result follows from (c).
$\Leftarrow$: If $s \in \mathbb{N}$, $\gamma_1, \ldots, \gamma_s \in \Gamma^*$, $H\gamma_i \neq H\gamma_j$ for $i \neq j$ and $\delta_1, \ldots, \delta_s \in \Gamma$, define $\gamma_{s+1} := o$ and $\delta_{s+1} := o$.
Then $\exists\ n \in N\ \forall\ i \in \{1, \ldots, s+1\}$: $n(\gamma_i) = \delta_i$. Because of $n(o) = o$, $n \in N_o$ and by (c), N_o is dense in $M_H(\Gamma)$, so by 4.27 N is dense in $\overline{M}_H(\Gamma)$.

<u>4.29 THEOREM</u> (Betsch (6)). Let $\Gamma \in \mathcal{G}$, $H \leq Aut(\Gamma)$ and $\mathcal{T}$ as in 4.26. Then the following conditions are equivalent:

(a) $\mathcal{T}$ is discrete in $\overline{M}_H(\Gamma)$.

(b) $\mathcal{T}$ is discrete in $M_H(\Gamma)$.

(c) H has only finitely many orbits on Γ.

<u>Proof</u>. Again the results trivially hold for $H = \{id\}$. (in this case, Γ is finite). So we assume that $H \neq \{id\}$. Then $M_H(\Gamma) \in \mathcal{N}_o$.

(a) $\Rightarrow$ (b): Trivial, since $M_H(\Gamma) \subseteq \overline{M}_H(\Gamma)$.

(b) => (c): Assume that H has infinitely many orbits on Γ. Take $m \in M_H(\Gamma)$ and a neighbourhood U of m.

Then $\exists\, s \in \mathbb{N}\ \exists\, \gamma_1, \ldots, \gamma_s \in \Gamma$: $U \supseteq \bigcap_{i=1}^{s} S(m, \gamma_i)$.

If $H\gamma_i = H\gamma_j$ then $S(m, \gamma_i) = S(m, \gamma_j)$. So we will assume that $H\gamma_i \neq H\gamma_j$ for $i \neq j$.

Since H has infinitely many orbits,
$\exists\, \gamma_{s+1} \in \Gamma \setminus (\{o\} \cup H\gamma_1 \cup \ldots \cup H\gamma_s)$.

Then $\forall\, i \in \{1, \ldots, s+1\}\ \exists\, e_i \in M_H(\Gamma)\ \forall\, j \in \{1, \ldots, s+1\}$:
$: e_i(\gamma_j) = \begin{cases} \gamma_i & i = j \\ o & i \neq j \end{cases}$.

Define $m_1 := m\,(e_1 + \ldots + e_s)$ and $m_2 := m_1 + e_{s+1}$.

Then m_1 and m_2 are $\in \bigcap_{i=1}^{s} S(m, \gamma_i) \subseteq U$.

If $m(\gamma_{s+1}) \neq o$ then $m_1(\gamma_{s+1}) = o \neq m(\gamma_{s+1})$, so $m_1 \neq m$.

If $m(\gamma_{s+1}) = o$ then $m_2(\gamma_{s+1}) = \gamma_{s+1} \neq o = m(\gamma_{s+1})$, so $m_2 \neq m$.

Anyhow, U contains an element $\neq m$ and $\mathcal{T}$ cannot be discrete.

(c) => (a): If H has only finitely many orbits on Γ then each element of $M_H(\Gamma)$ and of $\overline{M}_H(\Gamma)$ is uniquely determined by its effect on finitely many suitable elements of Γ.
So $\mathcal{T}$ is discrete.

b) 0-PRIMITIVE NEAR-RINGS

Now we shall prove a "density-like" structure theorem for 0-primitive near-rings. We start with zero-symmetric ones. We may assume (1.48) that if N is 0-primitive on Γ then $N \subseteq M(\Gamma)$.

4.30 THEOREM (Betsch (5)). Let $N \in \mathfrak{N}_0$ be 0-primitive on Γ.
If N is a ring then N is a primitive ring on the N-module Γ and Jacobson's density theorem is applicable.
If N is a non-ring then we get a kind of a density property:

(D): $\forall\ s \in \mathbb{N}\ \forall\ \gamma_1,\dots,\gamma_s \in \Gamma$, $\sim$-indep. $\forall\ \delta_1,\dots,\delta_s \in \Gamma$ $\exists\ n \in N\ \forall\ i \in \{1,\dots,s\}$: $n\gamma_i = \delta_i$.

Proof. If N is a ring we only have to apply 4.8.
Now let N be a non-ring. In the terminology of (D), let s be > 1 and for $t \in \{1,\dots,s-1\}$ let S(t) be the statement

$$\forall\ k \in \{t+1,\dots,s\}: \bigcap_{i=1}^{t} (o:\gamma_i) \nsubseteq (o:\gamma_k) .$$

Lemma. $\forall\, t \in \{1,\dots,s-1\}$: S(t).

Proof. By induction on t.
Since for $\gamma \in \Theta_1$ $(o:\gamma)$ is a maximal left ideal of N, $\forall\ i,j \in \{1,\dots,s\}$: $(o:\gamma_i) \subseteq (o:\gamma_j) \Rightarrow (o:\gamma_i) = (o:\gamma_j) \Rightarrow \gamma_i \sim \gamma_j \Rightarrow i = j$.
Particularly: S(1): $\forall\ k \in \{2,\dots,s\}$: $(o:\gamma_1) \nsubseteq (o:\gamma_k)$.
Now assume S(t), $s \geq 3$ and $k \in \{t+2,\dots,s\}$.
Then $\bigcap_{i=1}^{t} (o:\gamma_i) \nsubseteq (o:\gamma_k)$ and $(o:\gamma_{t+1}) \nsubseteq (o:\gamma_k)$.
Since $(o:\gamma_k)$ is maximal,

$$\bigcap_{i=1}^{t} (o:\gamma_i) + (o:\gamma_k) = (o:\gamma_{t+1}) + (o:\gamma_k) = N.$$

Since N is not a ring, $\bigcap_{i=1}^{t}(o:\gamma_i) \cap (o:\gamma_{t+1}) \nsubseteq$ $\nsubseteq (o:\gamma_k)$ by 3.4(i), which is nothing else than $S(t+1)$.

Now return to the proof of 4.28 and let $\gamma_1,\ldots,\gamma_s$, $\delta_1,\ldots,\delta_s$ be as in (D). Again we use induction on $t\in\{1,\ldots,s\}$.

If $t = 1$ then $\gamma_1\in\theta_1 \Rightarrow \exists\; n_1\in N$: $n_1\gamma_1 = \delta_1$.

Now assume that $\forall\; t\in\{1,\ldots,s-1\}\;\exists\; n_t\in N$ $\forall\; i\in\{1,\ldots,t\}$: $n_t\gamma_i = \delta_i$.

By the lemma, $L:= \bigcap_{i=1}^{t}(o:\gamma_i)\nsubseteq(o:\gamma_{t+1})$, hence $L\gamma_{t+1} \neq \{o\}$. Since $L\gamma_{t+1} \trianglelefteq_N \Gamma$ (3.4(a)), $L\gamma_{t+1} = \Gamma$.

Therefore $\exists\; \ell\in L$: $\ell\gamma_{t+1} = \delta_{t+1}-n_t\delta_{t+1}$.

Now we take $n_{t+1}:= \ell+n_t$ and get $\forall\; i\in\{1,\ldots,t+1\}$: : $n_{t+1}\gamma_i = \delta_i$, and the proof is complete.

4.31 REMARKS

(a) (D) is no "real" density property since there is no near-ring in sight in which N is dense (w.r.t the finite topology). +)

(b) From (D) it follows (Ramakotaiah (2)) that, if $s\in\mathbb{N}$ and $\gamma_1,\ldots,\gamma_s\in\Gamma$ are $\sim$-independent, $N/\bigcap_{i=1}^{s}(o:\gamma_i) \cong_N \Gamma^{\{\gamma_1,\ldots,\gamma_s\}}$ (where for $f\in\Gamma^{\{\gamma_1,\ldots,\gamma_s\}}$, $(nf)(\gamma_i):= n(f(\gamma_i)))$.

(c) The content of (D) might be very thin: if e.g. $|\theta_1| = 1$, (D) is trivial. So it is not too surprising that the converse of 4.28 does not hold:

+) (Betsch): If one changes $\mathcal{S}$ of 4.25 to $\mathcal{S}':=\{S(m,\gamma)\,|\,m\in M \wedge \gamma\in\theta_1(\Gamma)\}$ then one gets a "real" density theorem w.r.t. the resulting coarser topology.

Let N be the non-ring $\{f \in M_0(\mathbb{Z}_4) \mid f(2) \in \{0,2\} \wedge f(3) = 3f(1)\}$.
In $_N \mathbb{Z}_4$, $\theta_1 = \{1,3\}$, $1 \sim 3$, (D) is fulfilled, but
$\{0,2\} \triangleleft_N \mathbb{Z}_4$, so $_N \mathbb{Z}_4$ is not simple and therefore N
is not 0-primitive on $\mathbb{Z}_4$.

(d) (D) is equivalent to the following property:

(D'): $\forall$ s$\in \mathbb{N}$ $\forall$ $\gamma_1, \ldots, \gamma_s \in \Gamma$, $\sim$-indep. $\forall$ m$\in M(\Gamma)$
$\exists$ n$\in N$ $\forall$ i$\in \{1, \ldots, s\}$: $n\gamma_i = m(\gamma_i)$.

Now we turn to arbitrary near-rings.

4.32 THEOREM

(a) Let N be 0-primitive on Γ.
Case 1: $N_0\Gamma \neq \{o\}$. Then N_0 is 0-primitive on Γ,
so 4.28 is applicable (for $_{N_0}\Gamma$), and
$N_c \subseteq M_c(\Gamma)$.

Case 2: $N_0\Gamma = \{o\}$. Then $N = M_c(\Gamma)$ and Γ is a
non-zero simple group.

(b) Conversely, if either N_0 is 0-primitive on Γ and
$N_c \subseteq M_c(\Gamma)$ or if $N = M_c(\Gamma)$ where $\Gamma \neq \{o\}$ is
simple then N is 0-primitive on Γ.

Proof. (a) Anyhow, $N_c \subseteq M_c(\Gamma)$.
If $N_0\Gamma \neq \{o\}$ then $_{N_0}\Gamma$ is of type 0 by 3.18(a)
and N_0 is 0-primitive on Γ (3.15(a)).

If $N_0\Gamma = \{o\}$, $N_0 = \{0\}$ "by faith", so $\Omega = \Gamma$ and
$N = N_c = M_c(\Gamma)$ by 1.50(b). Since $N_c\Gamma$ is simple
iff Γ is simple, (a) is shown (observe that $N_c\Gamma \neq \{o\}$).

(b) Again by 3.18 (this time by (b)), if $_{N_0}\Gamma$
0-primitive then $_N\Gamma$ is of type 0. Since $N_c \subseteq M_c(\Gamma)$,
N_0 and N_c (and hence N) act faithfully on Γ, so
N is 0-primitive on Γ.
If $N = M_c(\Gamma)$, $\Gamma \neq \{o\}$ and simple, the result is
clear.

4.33 REMARK (D) would not necessarily mean the same in $_N\Gamma$ and in $_{N_0}\Gamma$, if $\sim$ would be defined by $\gamma\sim\delta :\Longleftrightarrow (o:\gamma) = (o:\delta)$ (in N). Cf. 4.19.

4.34 THEOREM (Ramakotaiah (1)). Each 0-primitive ideal is a prime ideal $\neq N$.

Proof. Let I be a 0-primitive ideal of N. Let $_N\Gamma$ be of type 0 with generator γ_0 such that $I = (o:\Gamma)$ (4.3).
Assume that $\exists\, J_1, J_2 \trianglelefteq N: J_1J_2 \subseteq I \wedge J_1 \not\subseteq I \wedge J_2 \not\subseteq I$.
For $i\in\{1,2\}$, $J_i\Gamma = J_iN\gamma_0 \subseteq J_i\gamma_0 \subseteq J_i\Gamma$. Since $J_i \not\subseteq (o:\Gamma)$, $J_i\Gamma = J_i\gamma_0 \neq \{o\}$. By 3.4(a), $J_i\gamma_0 \trianglelefteq_N \Gamma$. So $J_i\gamma_0 = \Gamma$. Now $J_1J_2\Gamma = J_1\Gamma = \Gamma$, so $J_1J_2 \not\subseteq (o:\Gamma) = I$, a contradiction.

4.35 REMARK In 5.40 we will see that the converse of 4.34 holds if $N = N_0$ has the DCCN.

4.36 THEOREM (Ramakotaiah (1)). Every maximal modular ideal I of $N\in\mathfrak{N}_o$ is a 0-primitive one.

Proof. Let I be a modular maximal ideal. By 3.22, I is contained in a modular maximal left ideal L. Since (L:N) is the largest ideal of N contained in L (by 3.25), we get $I\subseteq(L:N)$ and (L:N) is modular by 3.21(a). By the maximality of I, I = (L:N). By 4.3(c), I is 0-primitive, since by 3.29 L is 0-modular.

For the rest of this section, we give a description of a class of 0-primitive near-rings which are not 1-primitive. This discussion is due to Holcombe (5), where the proofs can be found, too.

4.37 NOTATION If ${}_N\Gamma \in {}_N\mathcal{G}$, let $\Delta := \Gamma\setminus\theta_1$ be the set of "non-generators". If $\Delta \leq_N \Gamma$, let $G^\Delta := \mathrm{Aut}_{N/(o:\Delta)}(\Delta)$ (cf. 3.14(a)!).

4.38 DEFINITION If $(\Gamma,+)\in\mathcal{G}$, $B\subseteq\Gamma$ and $H\leq \mathrm{Aut}(\Gamma)$, $H(B)\subseteq B$, we call the triple (Γ,B,H) compatible if at least one of the following conditions is satisfied:

(a) B is no normal subgroup of Γ.

(b) $\exists\ \gamma\in\Gamma\setminus B\ \ \exists\ \beta\in B\ \ \forall\ h\in H : \gamma+\beta \neq h(\gamma)$.

(c) $(\exists\ h'\in H\ \ \exists\ \gamma\in\Gamma\setminus B\ \ \exists\ \beta\in B : \gamma+\beta = h'(\gamma)) \wedge$
$\wedge\ (\exists\ \gamma'\in\Gamma\setminus B : h'(\gamma')-\gamma'\notin B)$.

4.39 THEOREM Let $N\in\mathcal{N}_0$ be 0-primitive on Γ, N a non-ring with identity and DCCL, and let Δ (as in 4.37) be an N-subgroup of Γ such that ${}_N\Delta$ is not faithful, but of type 2. Then N is not 1-primitive on Δ, G (4.10(b)) has finitely many orbits on θ_1, (Γ,Δ,G) is compatible and

if $N/_{(o:\Delta)}$ is a non-ring then $N \cong \{f\in M_{G\cup\{\bar{o}\}}(\Gamma) \mid f/_\Delta \in M_{G^\Delta}(\Delta)\}$,

if $N/_{(o:\Delta)}$ is a ring then $N \cong \{f\in M_{G\cup\{\bar{o}\}}(\Gamma) \mid f/_\Delta \in \mathrm{End}_{G^\Delta\cup\{\bar{o}\}}(\Delta)\}$

(where Δ is a finite dimensional vector space over the division ring $G^\Delta\cup\{\bar{o}\}$).

Conversely:

4.40 THEOREM Let Γ be an additive group and Δ be a non-zero subgroup. Let G^Δ be a group of regular automorphisms of Δ which has only finitely many orbits on Δ. Let H be a subgroup of $\mathrm{Aut}\,(\Gamma,+)$ such that

(a) (Γ,Δ,H) is compatible.

(b) each $h\in H$ is regular on $\Gamma\setminus\Delta$.

(c) H has only finitely many orbits on $\Gamma\setminus\Delta$.

(d) $\forall\ h\in H$: $h/_\Delta \in G^\Delta$.

Then $N = \{f\in M_{H\cup\{\bar{o}\}}(\Gamma) \mid f/_\Delta \in M_{G^\Delta}(\Delta)\}$ is zerosymmetric, 0-primitive, but not 1-primitive on Γ, has an identity and the DCCL.

If moreover $\Gamma \neq \Delta$ and Δ is a finite dimensional vector space over some division ring D and if $\forall\ h \in H$: $h/_{\Delta} \in D$ then $N = \{f \in M_{H \cup \{\bar{o}\}}(\Gamma) \mid f/_{\Delta} \in End_D(\Gamma)\}$ is also 0-primitive, but not 1-primitive on Γ, $N \in \mathcal{N}_o$, $N \in \mathcal{N}_1$, and moreover $N/_{(o:\Delta)}$ is a ring.

4.41 REMARK See also Holcombe (5) for the more general case that Δ is only a finite union of N-subgroups of type 2 with zero intersection.

4.42 REMARK If $G^{\Delta} = \{id\}$ then in the non-ring case of 4.39 we get near-rings of the form $N = \{f \in M_o(\Gamma) \mid f(\Delta) \subseteq \Delta\}$ (see e.g. N_o in 3.8).
Conversely, if $(\Gamma,+)$ is a finite group and Δ a non-trivial subgroup then $N := \{f \in M_o(\Gamma) \mid f(\Delta) \subseteq \Delta\}$ is a finite near-ring with identity, zero-symmetric and 0-, but not 1-primitive on Γ. Δ is just the set of non-generators and is an N-subgroup such that $N/_{(o:\Delta)}$ is a non-ring if $|\Delta| > 2$.

c) 1-PRIMITIVE NEAR-RINGS

Now let N be 1-primitive on Γ. Then $C = G^o$ (by 4.18), Γ is not N-isomorphic to a proper subgroup (4.17(b)), $\Gamma = \theta_o \cup \theta_1$ (by 4.15(d)), $\Omega = \{o\}$ or $\Omega = \Gamma$ (3.2) and $\forall\ L \trianglelefteq_{\ell} N,\ L \neq \{0\}\ \exists\ \gamma \in \Gamma$: $L\gamma = \Gamma$ (by 3.4(a)).
We still assume that $N \subseteq M(\Gamma)$.

4.43 THEOREM

(a) Let N be 1-primitive on Γ. Then

Case 1: $N_o\Gamma \neq \{o\} \wedge \Omega = \Gamma$. Then N_o is 1-primitive on Γ, $N_c = M_c(\Gamma)$ and $\theta_1 = \Gamma$.

If N_o is a ring then N is dense in $M_{aff}(\Gamma)$ where Γ is a vector space over the division ring $D := Hom_{N_o}(\Gamma,\Gamma)$.

If N_0 is not a ring then (D) of 4.30 is applicable.

Case 2: $N_o\Gamma \neq \{o\} \wedge \Omega = \{o\}$. Then $N = N_0$ is 1-primitive on Γ and 4.30 holds.

Case 3: $N_0\Gamma = \{o\}$. Then $N = N_c = M_c(\Gamma)$ and Γ is a simple group $\neq \{o\}$.

(b) Conversely, if a near-ring $N \subseteq M(\Gamma)$ is such that it is 1-primitive on Γ with $N_c \in \{\{0\}, M_c(\Gamma)\}$ or if $N = M_c(\Gamma)$ ($\Gamma \neq \{o\}$ and simple) then N is 1-primitive on Γ.

<u>Proof.</u> (a) If $N_0\Gamma \neq \{o\}$, N_0 is 1-primitive on Γ by 3.18(a). Since each strongly monogenic N-group has either $\Omega = \{o\}$ or $\Omega = \Gamma$, the rest follows from 1.50, 3.9, 3.15(a), 4.27(a) and 4.32.

(b) If N_0 is 1-primitive on Γ and $N_c = \{0\}$ or $N_c \cong M_c(\Gamma)$ then either $\Omega = \{o\}$ or $\Omega = \Gamma$ (1.50), so N is 1-primitive on Γ by 3.18(b) and 3.15(a). If $N \cong M_c(\Gamma)$, Γ simple and $\neq \{o\}$, then clearly N is 1-primitive on Γ.

<u>4.44 REMARK</u> 4.41 is the main reason for defining "strongly monogenic N-groups Γ" as in 3.1(b) and not by the conditions "monogenic" and "$\forall\ \gamma \in \Gamma$: $(N\gamma = \Omega \vee N\gamma = \Gamma)$", for 4.43 would not be true in this case:
Take $\Gamma = \mathbb{Z}_8$, $N_0 := \{f \in M_0(\Gamma) \mid f(2) = f(6) \in \{0,2,4,6\} \wedge \wedge f(4) \in \{0,4\}\}$ and $N_c := \{f \in M_c(\Gamma) \mid f(0) \in \{0,2,4,6\}\}$.

Then one can show that $N := N_0 + N_c$ is a subnear-ring of $M(\Gamma)$ enjoying the following properties:

${}_N\Gamma$ and ${}_{N_0}\Gamma$ are faithful, simple and monogenic. Moreover,

$\forall\ \gamma \in \Gamma$: $(N\gamma = \Omega = \{0,2,4,6\} \vee N\gamma = \Gamma)$. But $\{o\} \neq \Omega \neq \Gamma$, and N_0 is not 1-primitive on Γ (it is not even true that for all $\gamma \in \Gamma$ $N_0\gamma$ is either $= \{o\}$, $= \Omega$ or $= \Gamma$, since $N_0 4 = \{0,4\}$).

From 4.30 and $\Gamma = \theta_o \cup \theta_1$ we get with a straightforward proof

<u>4.45 THEOREM</u> Let the non-ring $N \in \mathcal{N}_o$ be 1-primitive on Γ but without $\sim$-equivalent generators.
Then N is dense in the near-ring $\{f \in M_0(\Gamma) \mid f(\theta_o) = \{o\}\}$.

For 1-primitive near-rings $\in \mathcal{N}_o$ with DCC we get a whole bunch of important results:

<u>4.46 THEOREM</u> (Betsch (3)). Let $N \in \mathcal{N}_o$ be 1-primitive on Γ and endowed with the DCCL. Then

(a) There are only finitely many $\sim$-equivalence classes if N is a non-ring.

(b) $\exists\ s \in \mathbb{N} : {}_N N = \dot{\sum_{i=1}^{s}} L_i$, L_i finitely many pairwise (to Γ) N-isomorphic left ideals and N-groups of type 1 (so 2.50 is applicable!); if N is a non-ring then $s = |\Gamma/\sim| - 1$.

(c) All N-groups of type 0 are N-isomorphic and of type 1.

(d) N contains a right identity (not necessarily two-sided).

(e) N is simple.

(f) N is either 2-primitive on Γ or there is no N-group of type 2.

<u>Proof</u>. If N is a ring, (b) - (f) are either well-known or trivial. So we will assume that N_0 is a non-ring.

(a) Suppose that there are infinitely many $\sim$-equivalence classes with representatives $\gamma_0, \gamma_1, \gamma_2, \ldots$. We may assume that $\gamma_0 \in \theta_o$. Then $(o:\gamma_0) = N \neq (o:\gamma_i)$ for $i \geq 1$, hence $\gamma_1, \gamma_2, \ldots \in \theta_1$. So by (D) of 4.30 $(o:\gamma_0) \supset (o:\gamma_1) \supset (o:\{\gamma_1, \gamma_2\}) \supset \ldots$ which is a contradiction to the DCCL.

(b) Now let $\gamma_0, \gamma_1, \ldots, \gamma_s$ be a complete system of representatives of the $\sim$-equivalence classes with

$\gamma_0 \varepsilon \theta_o$, $\gamma_1,\ldots,\gamma_s \varepsilon \theta_1$. Then $\bigcap_{i=1}^{s} (o:\gamma_i) = \{0\}$, but $L_j := \bigcap_{i \neq j} (o:\gamma_i) \neq \{o\}$. By 3.4(f), $L_1,\ldots,L_s$ are minimal left ideals.

Now apply 2.50(e) to get $N = \dot{\sum_{j=1}^{s}} L_j$. Since $\forall\ j\varepsilon\{1,\ldots,s\}$: $L_j \not\subseteq (o:\gamma_j)$, $L_j \cong_N \Gamma$ by 3.10.

(c) By (b) and 3.11(a).

(d) By (b) and 3.27(d), N contains a right identity e. N_1 of 3.8 shows that e is not necessarily two-sided.

(e) If $I \lhd N$, $\exists\ j\varepsilon\{1,\ldots,s\}$: $L_j \not\subseteq I$. Since L_j is minimal, $L_j \cap I = \{0\}$. But $IL_j \subseteq I \cap L_j = \{0\}$, so $I \subseteq (0:L_j) = 0$ (for $L_j \cong_N \Gamma$), whence $I = \{0\}$.

(f) By 4.7 or by (c).

Note that 4.46(a) is not valid for rings: If Γ is the vector space $\mathbb{R}^2$, considered as an $Hom(\Gamma,\Gamma)$-module, all $(1,x)$ $(x\varepsilon\mathbb{R})$ are pairwise inequivalent w.r.t. $\sim$, $Hom(\Gamma,\Gamma)$ is a near-ring which is primitive on Γ and has the DCCL.

<u>4.47 COROLLARY</u> $N\varepsilon\mathfrak{N}_0$, DCCN, N contains a left identity; $I \trianglelefteq N$, $N \neq \{0\}$. Then

(a) N is 1-primitive <=> N is 2-primitive <=> N is simple.

(b) I is 1-primitive <=> I is 2-primitive <=> I is maximal.

<u>Proof</u>. (a) By 3.4(c) and 3.7(c), 1-primitivity and 2-primitivity coincide. In this case, N is simple. If N is simple then $I = N$ is a minimal ideal and by 3.4(b) and Lemma 3 in the proof of 3.54 (with $I := N$; then $\Delta = \Gamma$) N has a faithful N-group ${}_N\Gamma$ of type 1, so N is 1-primitive.

(b) follows from (a).

See also Kaarli (2).

4.46 and 4.47 are the main reason for studying 1-primitive near-rings.

d) 2-PRIMITIVE NEAR-RINGS

Again we assume that if N is 2-primitive on Γ then $N \subseteq M(\Gamma)$.

1.) 2-PRIMITIVE NEAR-RINGS

The structure of 2-primitive near-rings can be described as follows.

4.48 THEOREM

(a) Let N be 2-primitive on Γ. Then

Case 1: $N_0\Gamma \neq \{o\} \wedge \Omega = \Gamma$. Then N_0 is 2-primitive on Γ, $N = M_c(\Gamma)$ and $\theta_1 = \Gamma$.
If N_0 is a ring then N is dense in $M_{aff}(\Gamma)$ (as in 4.43);
if N_0 is a non-ring then (D) of 4.30 is applicable (for N_0).

Case 2: $N_o\Gamma \neq \{o\} \wedge \Omega = \{o\}$. Then $N = N_0$ is 2-primitive on Γ and 4.30(a) is applicable.

Case 3: $N_0\Gamma = \{o\}$. Then $N = M_c(\Gamma)$ and Γ is a cyclic group of prime order.

(b) Conversely, if N_0 is 2-primitive on Γ with $N_c \in \{\{0\}, M_c(\Gamma)\}$ or if $N = M_c(\Gamma)$ (Γ a cyclic group of prime order) then N is 2-primitive on Γ.

The proof is similar to the one of 4.43 and therefore omitted.

4.49 THEOREM (cf. Fain (1) and Betsch (6)). If N is 2-primitive on Γ and if $I \trianglelefteq N$, $I \neq \{0\}$, then I is 2-primitive on Γ, unless $I = M_c(\Gamma)$ (where Γ is not a cyclic group of prime order).

Proof. (a) We first show this theorem for $N\epsilon\mathcal{N}_0$.

Evidently, ${}_I\Gamma$ is faithful.

Assume that $\Delta \leq_I \Gamma$.

If $I\Delta = \{o\}$ then consider $N\Delta$. If $N\Delta \neq \{o\}$, $\exists\ \delta\epsilon\Delta$: $N\delta \neq \{o\}$. Therefore $N\delta = \Gamma$ and $N\Delta = \Gamma$. But $I\Gamma = IN\Delta \subseteq I\Delta = \{o\}$, so $I = \{0\}$, since ${}_I\Gamma$ is faithful. Hence $N\Delta = \{o\}$, $\Delta \leq_N \Gamma$, whence $\Delta = \{o\}$.

If $I\Delta \neq \{o\}$ then again $\exists\ \delta\epsilon\Delta$: $I\delta \neq \{o\}$. Since $N(I\delta) = (NI)\delta \subseteq I\delta$, $I\delta \leq_N \Gamma$, so $I\delta = \Gamma$. Consequently $\Delta = \Gamma$, for $I\delta \subseteq \Delta$.

Therefore I is 2-primitive on Γ.

(b) Now let N be arbitrary. We may assume that $N \neq N_o$, so case 2 of 4.48 is excluded.

If N falls into case 1, N_o is 2-primitive on Γ. By 2.18 and 1.34(a), $I_o = I \cap N_o \trianglelefteq N_o$. If $I_o \neq \{0\}$ then I_o is 2-primitive on Γ, hence I is 2-primitive on Γ by 3.18(b). If $I_o = \{0\}$ then $I \subseteq N_c = M_c(\Gamma)$. Since $\theta_1 = \Gamma$, $I_o \trianglelefteq_N \Gamma$. $Io = \{o\}$ implies that for all $\gamma = no\epsilon\Gamma$ and for all $i\epsilon I$ $i\gamma = ino = o$, so $I = \{o\}$. Hence $Io \neq \{o\}$ and so $Io = \Gamma$. Take any $m_c\epsilon M_c(\Gamma)$; $m_c o =: \mu$. Because of $Io = \Gamma$, $\exists\ i\epsilon I : io = \mu$. Now $\forall\ \gamma\epsilon\Gamma$: $i\gamma = io = \mu = m_c o = m_c\gamma$, hence $i = m_c$ and we get $I = M_c(\Gamma)$. If $\exists\ p\epsilon\mathbb{P} : \Gamma \cong \mathbb{Z}_p$, I is 2-primitive on Γ; if not, I is not 2-primitive.

If N is in case 3, I is trivially 2-primitive on Γ.

4.50 REMARK 4.49 cannot be transferred to 0- or 1-primitivity, not even for finite, abelian, zerosymmetric near-rings. It is easy to show that if N is e.g. 0-primitive on Γ and $I \trianglelefteq N$ then ${}_I\Gamma$ is faithful and monogenic. But not necessarily simple:

Take $\Gamma := \mathbb{Z}_8$, $\Delta := \{0,2,4,6\}$ and $\mathrm{E} := \{0,4\}$.

$N := \{f\epsilon M_o(\Gamma) \mid f(\Delta) \subseteq \Delta \wedge f(5)=f(1) \wedge f(7)=f(3)\}$, $I := (o:\Delta)$.

N is 0-primitive on Γ, but $E \trianglelefteq_I \Gamma$. Moreover, ${}_I\Gamma$ is strictly monogenic and I has a right identity. I cannot even be 0-primitive on some other group $\Gamma' =: I\gamma_0'$: Assuming that, take (0:1) and (0:3) in ${}_I\Gamma$ and put $L := (o:\gamma_0')$. Then L is a maximal left ideal of I (3.4(f)) and $L \neq (0:1)$, $L \neq (0:3)$ (since (0:1) and (0:3) cannot be maximal). Therefore $(0:1)+L = (0:3)+L = N$, but $(0:1) \cap (0:3) = \{\bar{o}\} \subseteq L$, so I would have to be a ring by 3.4(i), a contradiction.

Seemingly there is no "smaller" counterexample than that above with 4096 elements.

By the way, one can use Zorn's lemma to show that in any case $I \trianglelefteq N$ ($I \neq \{0\}$, N ν-primitive) has some I-groups of type ν.

4.51 COROLLARY (Fain (1)). Let P be a 2-primitive ideal of $N \in \mathcal{N}_0$. Let I be another ideal of N containing P. Then P is a 2-primitive ideal of I.

Proof. $I/_P \trianglelefteq N/_P$, and $N/_P$ is 2-primitive. Since $I \neq P$, $I/_P \neq \{0\}$ and $I/_P$ is 2-primitive. Hence P is 2-primitive in I.

2.) 2-PRIMITIVE NEAR-RINGS WITH IDENTITY

In this case, $\theta_1 = \Gamma^*$ (if $N \in \mathcal{N}_0$) or $\theta_1 = \Gamma$ (if $N \notin \mathcal{N}_0$). Also, if $N = N_0$ then $\sim\, = \,\approx$ (by 4.21).

Recall that a 1-primitive near-ring $\in \mathcal{N}_0$ with a left identity is already a 2-primitive one with identity (4.6(b)).

We are now in a position to get a "real" and fundamental density theorem.

4.52 THEOREM (Wielandt (1), Laxton (2), Ramakotaiah (2), Betsch (5) and (6), Kaarli (2)).

(a) Let N be 2-primitive on Γ with identity. Then N is dense in $(C_o := \mathrm{End}_{N_0}(\Gamma) = G_o^o$ by 4.18, $D := \mathrm{Hom}_N(\Gamma,\Gamma))$:

	$N \neq N_0$ (case 1 of 4.48)	$N = N_0$ (case 2 of 4.48)	
N_0 a ring	$M_{aff}(\Gamma)$	$\mathrm{Hom}_D(\Gamma,\Gamma)$	$_D\Gamma$ a vector space $\neq \{o\}$
N_0 a non-ring	$\overline{M}_{C_o}(\Gamma)$	$M_{C_o}(\Gamma) = M_C(\Gamma)$	G_o fixed-point-free on Γ

(b) Conversely, every near-ring which is dense in $M_{aff}(\Gamma)$ or $\mathrm{Hom}_D(\Gamma,\Gamma)$ (where Γ is a non-zero vector space over some division ring D) or dense in $\overline{M}_{C_o}(\Gamma)$ or $M_{C_o}(\Gamma)$ (G_o fixed-point-free on Γ) is 2-primitive on Γ.

Proof. (a) If N is 2-primitive on Γ and has an identity then case 3 in 4.48 cannot occur. N_0 is therefore 2-primitive on Γ and has an identity. If N_0 is a ring, the statement is clear. If N_0 is not a ring, note that G_o is fixed-point-free on Γ, since $\forall\, g \in G_o: \{\gamma \in \Gamma \mid g(\gamma) = \gamma\} \leq_{N_0} \Gamma$.

If $G_o \neq \{id\}$ then $M_{G_o}(\Gamma) = M_{G_o \cup \{\bar{o}\}}(\Gamma) = M_{G_o^o}(\Gamma) = M_{C_o}(\Gamma)$ and the result follows from (D) of 4.30, 4.28(c) and 4.27(a).

If $G_o = \{id\}$ then (D) of 4.30 implies that N_0 is dense in $M_0(\Gamma)$, which is trivially dense in (since equal to) $M_{C_o}(\Gamma) = M_{\{\bar{o}\}}(\Gamma) = M_0(\Gamma)$.

If $N \neq N_0$, apply again 4.27(a).

(b) Assume now that N is dense in $\mathrm{Hom}_D(\Gamma,\Gamma)$, where Γ is a non-zero vector space over some skew-field D. Then N is a dense subring and therefore a primitive ring on Γ. From this we deduce:
If N is dense in $M_{aff}(\Gamma)$ then N_0 is dense in $\mathrm{Hom}_D(\Gamma,\Gamma)$, so Γ has no non-trivial N_0-subgroups, and N is 2-primitive on Γ. If N is dense in $M_{G_0^o}(\Gamma)$ then $N = N_0$. If $G_0 = \{id\}$, $M_{G_0^o}(\Gamma) = M_0(\Gamma)$ and each dense subnear-ring of that is trivially 2-primitive on Γ. If $G_0 \neq \{id\}$ then $M_{G_0^o}(\Gamma) = M_{G_0}(\Gamma)$ and 4.28(c) shows that ${}_N\Gamma$ cannot contain non-trivial N-subgroups.
Finally if N is dense in $\overline{M}_{C_0}(\Gamma)$ then N_0 is dense in $(\overline{M}_{C_0}(\Gamma))_0 = M_{C_0}(\Gamma)$. As above, Γ cannot contain a non-trivial N_0-subgroup (or one can use 3.18(b)).

4.53 REMARKS

(a) It is not true that each 2-primitive near-ring with identity, $N \neq N_0$ and N_0 a non-ring, is dense in $M_C(\Gamma)$: take Γ finite with $\{id\} \neq G < \mathrm{Aut}(\Gamma)$, G fixed-point-free and $N := \overline{M}_G(\Gamma)$. Then ${}_N\Gamma$ has $\Omega = \Gamma$, so $C_N(\Gamma) = \{id\}$ (4.13(c)) and therefore $M_C(\Gamma) = M(\Gamma)$. But $N \neq M(\Gamma)$, so N cannot be dense in $M(\Gamma)$ by 4.29. This is a late but convincing reason for introducing this crazy $\overline{M}_{C_0}(\Gamma)$, where one first switches down to N_0 (by forming $C_0 = \mathrm{End}_{N_0}(\Gamma)$ and then back up by adding all of the constants : $M_C(\Gamma)$ would be too big in general.

(b) 4.52(a) does neither hold for 0-primitive near-rings with identity nor for 2-primitive near-rings without identity (not even for $N = N_0$ and N finite):

If $\Gamma := \mathbb{Z}_4$ and $\Delta := \{0,2\}$, $N := \{f \in M_0(\Gamma) \mid f(\Delta) \subseteq \Delta\}$ is 0-primitive on Γ with identity, but not dense in $M_{C_0}(\Gamma) = M_0(\Gamma)$ (4.29!). $M := \{f \in M_0(\Gamma) \mid f(3) = 0\}$ is 2-primitive on Γ, without identity and again not dense in $M_{C_0}(\Gamma) = M_0(\Gamma)$.

(c) All 2-primitive near-rings with identity on $\mathbb{Z}_4$, where N_0 is a non-ring, will be classified in 4.63.

(d) 4.32, 4.43, 4.48 and 4.52 reduce the theory of primitive near-rings to those of primitive zero-symmetric near-rings. We will therefore mainly deal with those ones in the sequel.

(e) Recall (4.27(a)) that $\overline{M}_{C_0}(\Gamma)$ is "only a set" in general.
Here is some example: $G = \{id, -id\}$ (with $-id(x) := -x$) is a fixed-point-free automorphism group on $\Gamma = \mathbb{R}$. $C := \{\bar{o}, id, -id\}$. $M_C(\mathbb{R}) = \{f \in M(\mathbb{R}) \mid f(0) = 0 \wedge \forall x \in \mathbb{R}: f(-x) = -f(x)\}$.
If $\overline{M}_C(\mathbb{R}) =: N$, take $n_1 := \sin + \frac{\pi}{2} \in N$ and $n_2 := id + \frac{\pi}{2} \in N$. Consider $n := n_1 \circ n_2 \in M(\mathbb{R})$.
$n_0 = \sin \circ (id + \frac{\pi}{2}) - \sin(\frac{\pi}{2})$ is not an odd function, thus not belonging to $M_C(\mathbb{R})$, whence $n \notin N$ and N is no near-ring.

4.54 COROLLARY If N is 2-primitive on Γ with $Aut_{N_0}(\Gamma) = \{id\}$ then N is dense in either one of the following near-rings (notation as in 4.52): $Hom_D(\Gamma,\Gamma)$, $M_{aff}(\Gamma)$, $M_0(\Gamma)$ or $M(\Gamma)$ (cf. 4.65).

4.55 THEOREM (Ramakotaiah (2)). Let $N \in \mathcal{N}_0$ be a 2-primitive non-ring on Γ with an identity. Then any two equivalence classes w.r.t. $\tilde{\sim}$ (except the zero class) are equipotent.

Proof. Let E be in $\Gamma^*/_{\approx}$ and ε a fixed element of E.

Consider the map $f: \begin{array}{l} G \to E \\ g \to g(\varepsilon) \end{array}$ (with $G = Aut_N(\Gamma)$ again).

Since G is fixed-point-free (4.52), f is injective.

By definition, f is surjective, so f is a bijection.

3.) 2-PRIMITIVE ZERO-SYMMETRIC NEAR-RINGS WITH IDENTITY AND A MINIMAL LEFT IDEAL.

4.56 THEOREM (Betsch (6). Let N be a zero-symmetric near-ring with identity which is 2-primitive on Γ and has a minimal left ideal L. Then

(a) $L \cong_N \Gamma$.

(b) $\exists\, e^2 = e \in L^*$: $L = Ne = Le \wedge (C_N(\Gamma), \circ) \cong (eNe, \cdot)$.

Proof. (a) Since $L\Gamma \neq \{0\}$, $\exists\, \gamma \in \Gamma$: $L\gamma \neq \{o\}$, so $L\gamma = \Gamma$ and $\gamma \in \theta_1$. Now we can apply 3.10.

(b) With γ as above, $\exists\, e \in L^*$: $e\gamma = \gamma$. Therefore $e^2\gamma = e\gamma$ and $e^2 - e \in L \cap (o:\gamma) = \{0\}$ (since L is minimal). Hence $e^2 = e$ and $Le \neq \{0\}$. Since $Le \leq_N L$, $Le = N$. Since $Le \subseteq Ne \subseteq L$, $Ne = Le = L$.

By (a), $C_N(\Gamma) = C_N(L) = C_N(Ne)$ (it can be easily verified that N-isomorphic N-groups have isomorphic centralizer-semigroups).

For $n \in N$, consider $t_n: \begin{array}{l} Ne \to Ne \\ xe \to xene \end{array}$. t_n is well-defined and $\in C_N(Ne)$.

Consider next the map $h: \begin{array}{l} eNe \to C_N(Ne) \\ ene \to t_n \end{array}$. If $ene = eme$ then $t_n = t_m$, so h is well-defined. Clearly, h is a semigroup homomorphism. If $h(ene) = h(eme)$ then $t_n = t_m$ and $\forall\, x \in N$: $xene = t_n(xe) = t_m(xe) = xeme$. Specializing $x = : e$ we get $ene = eme$ and h is shown to be injective.

Finally, $\forall\, c \in C_N(Ne)$ $\exists\, n \in N$: $c(e) = ne$. Therefore $ene = e\, c(e) = c(e^2) = c(e) = ne$ and

$\forall$ $x e \varepsilon Ne$: $c(xe) = xc(e) = xene = t_n(xe)$, so $c = t_n$ and h is surjective, hence a semigroup-isomorphism.

4.57 COROLLARY (Betsch (6)). If $N \varepsilon \mathcal{N}_0$ has an identity and a minimal left ideal L then all faithful N-groups of type 2 (if those exist) are N-isomorphic (to L) and N determines the pair $(\Gamma, C_N(\Gamma))$ uniquely "up to isomorphism". If e is as in 4.56(b), $(eNe,\cdot)$ is a group with zero and the group $(eNe\setminus\{0\},\cdot)$ acts on $L = Ne$ as a fixed-point-free automorphism group (by right multiplication).

4.58 REMARK For more information on these topics (a partial converse of 4.56, the uniqueness of $(\Gamma, C_N(\Gamma))$, etc.) see Betsch (6) and §7a), in particular 7.5.

4.) 2-PRIMITIVE NEAR-RINGS WITH IDENTITY AND MINIMUM CONDITION

4.59 COROLLARY Let $N \varepsilon \mathcal{N}_0$ be a 2-primitive near-ring with DCCL and identity. Then 4.46 is applicable, hence also 2.50 (for $_N N$), and G has finitely many orbits on Γ, (2.50(a) and 4.21), which is the same as to say that τ is discrete on $M_G(\Gamma)$ (4.29).

See § 7a) for the information that if a fixed-point-free automorphism group H of Γ has finitely many orbits on Γ then $M_C(\Gamma)$ has the DCCL.

4.60 THEOREM (Betsch (6)). Let N be 2-primitive on Γ with DCC for the left ideals of N_0 and with an identity. Then N is equal to one of the following near-rings (notation as in 4.52):

	$N \neq N_0$	$N = N_0$	
N_0 a ring	$M_{aff}(\Gamma)$	$Hom_D(\Gamma,\Gamma)$	$\dim_D \Gamma$ finite
N_0 a non-ring	$\overline{M}_{C_0}(\Gamma)$	$M_C(\Gamma)$	G_0 has finitely many orbits on Γ

Proof. follows from 4.52 and 4.59. Note that $\overline{M}_{C_o}(\Gamma)$ is a near-ring in this case (for it equals N).

4.61 COROLLARY If N has an identity, is 2-primitive on Γ and if the non-ring N_o has the DCCL and $\mathrm{Aut}_{N_o}(\Gamma) = \{id\}$ then either $N = M(\Gamma)$ (if $N \neq N_o$) or otherwise $N = M_o(\Gamma)$. In both cases, Γ (and therefore N, too) is finite. So the DCC implies finiteness!

4.62 REMARK These results illustrate some remarks in the preface: while the "elements of ring theory" are rings of linear mappings on Γ, those ones for near-ring theory are near-rings of arbitrary mappings (perhaps with some restrictions) on Γ.

4.63 THEOREM (cf. Ramakotaiah (2)). Let N be 2-primitive on Γ, finite and endowed with an identity. Let $|\Gamma|-1$ ($|\Gamma|$ must be finite, too!) be $\in \mathbb{P}$ and N_o a non-ring.
Then either $N = M(\Gamma)$ or $N = M_o(\Gamma)$ or $N \cong_N \Gamma$ or $|N| = |\Gamma|^2$ (if Γ is abelian, $N \cong_N \Gamma \oplus \Gamma$ in the last case).

Proof. Anyhow, N_o is 2-primitive on Γ and has an identity. Since all equivalence classes (w.r.t. $\sim$) of Γ except the zero class (1 member) are equipotent (4.55) and $|\Gamma|-1 \in \mathbb{P}$, either $G_o = \{id\}$ and we can apply 4.61, or $\exists\ \gamma_o \in \Gamma$: $G_o\gamma_o = \Gamma^*$ (one single non-zero-class). In this case, $N_o = M_{G_o}(\Gamma) \cong_{N_o} \Gamma$ using the map

$$h: \begin{array}{l} M_G(\Gamma) \to \Gamma \\ f \longrightarrow f(\gamma_o) \end{array}$$

. If $N \neq N_o$, $N_c = M_c(\Gamma)$.

$$\overline{h}: \quad N \longrightarrow {}_N\Gamma \oplus {}_N\Gamma$$

$$f = f_o + f_c \to (f_o(\gamma_o), f_c(o))$$

is a bijection and an N-isomorphism if Γ (and hence, N, too) is abelian.

5.) AN APPLICATION TO INTERPOLATION THEORY

4.64 DEFINITION If $\Gamma \in \mathcal{G}$ and $N \subseteq M(\Gamma)$, N is said to fulfill the finite interpolation property if

$\forall\ s \in \mathbb{N}\ \ \forall\ \gamma_1,\dots,\gamma_s \in \Gamma,\ \ \gamma_i \neq \gamma_j$ for $i \neq j\ \ \forall\ \delta_1,\dots,\delta_s \in \Gamma$

$\exists\ n \in N\ \ \forall\ i \in \{1,\dots,s\}: n(\gamma_i) = \delta_i$.

There is an obtrusive similarity to the density concepts. In fact:

4.65 THEOREM Let $N \leq M(\Gamma)$ with $N \neq N_0$ and N_0 not a ring. Then the following conditions are equivalent:

(a) N_0 is 2-fold transitive on Γ^*.

(b) N is 2-primitive on Γ with $G_0 = \{id\}$.

(c) N fulfills the finite interpolation property.

Proof. (a) => (b): ${}_N\Gamma$ is trivially faithful, 2-fold transitivity implies 1-fold transitivity and this in turn that $N\Gamma \neq \{o\}$. If $\{o\} \neq \Delta \leq_{N_0} \Gamma$, take some $\delta \in \Delta^*$ Then $\forall\ \gamma \in \Gamma\ \exists\ n_0 \in N_0: n_0\delta = \gamma$. So $N_0\delta = \Gamma$ and $\Delta = \Gamma$.

If $G_0\gamma = G_0\delta$ but $\gamma \neq \delta$ and (say) $\delta \neq o$, take some $n_0 \in N_0$ with $n_0\gamma = o \wedge n_0\delta \neq o$. Then $(o:\gamma)_{N_0} \neq (o:\delta)_{N_0}$, so $\gamma \not\sim \delta$ in ${}_{N_0}\Gamma$, hence $G_0\gamma \neq G_0\delta$ (4.20(c)), a contradiction. Therefore $G_0 = \{id\}$.

(b) => (c): by 4.54 and 4.28(d).

(c) => (a): trivial.

4.66 REMARKS

(a) So if a near-ring N of mappings on Γ interpolates at 0 and 2 other places then N interpolates already on an arbitrary (finite) number of places. Compare this with the corresponding "linear" result in ring theory ((N. Jacobson), Corollary to theorem 1 on p. 32). This is a "purely interpolation-theoretic" result.

(b) It can be shown that if N fulfills the finite interpolation property and $|\Gamma| > 3$ then N_0 is a non-ring.

4.67 COROLLARY Take $\Gamma = (\mathbb{R},+)$. Then any one of the following near-rings and all near-rings containing one of them have the properties that N is 2-primitive on Γ with $N \neq N_0$, $G_0 = \{id\}$ and N_0 not a ring:

$N_1 := \mathbb{R}[x]$, N_2: the near-ring of all step functions on $\mathbb{R}$, N_3: the subnear-ring of $M(\mathbb{R})$ generated by the trigonometric polynomials.

For all of them fulfill the finite interpolation property which qualifies them for 4.65.

The author hopes that near-rings of interpolating functions become interesting for approximation theory (because these functions can be iterated w.r.t. $\circ$).

After all that complicated stuff the reader will possibly agree with the author that the primitive near-rings have successfully revenged their discriminating name.

One fills the trash into some bags
With these one only calculates.
The rubbish which you still can smell
Is often called the "radical".

Better unknown author, 1975 A.D.

§5 RADICAL THEORY

This paragraph equals on harvest: the strains of previous paragraphs are highly rewarded by the fact that many results of this § 5 are easy consequences of previous ones (cf. e.g. 5.48 or §5 c),d)).
A near-ring N might have no faithful N-group of type ν. The next general case is that all N-groups of type ν work together to get the intersection $\bigcap(o:\Gamma)$ to be zero. N is then called "ν-semisimple". Anyhow, this intersection "measures" how far N is away to be ν-semisimple and is called the ν-radical $\mathfrak{J}_\nu(N)$. It contains all disgusting guys, for factoring out $\mathfrak{J}_\nu(N)$ gives a ν-semisimple near-ring $N/_{\mathfrak{J}_\nu(N)}$.
First we give several characterizations of $\mathfrak{J}_\nu(N)$, using ν-modular left ideals. We get $\mathfrak{J}_0(N)\subseteq\mathfrak{J}_1(N)\subseteq\mathfrak{J}_2(N)$ immediately. Between $\mathfrak{J}_0(N)$ and $\mathfrak{J}_1(N)$ there is another radical-like object $\mathfrak{J}_{1/2}(N)$, the intersection of all 0-modular left ideals. We discuss, when $\mathfrak{J}_\nu$ is hereditary and prove that for all ν, $\mathfrak{J}_\nu(\oplus N_i) = \oplus\mathfrak{J}_\nu(N_i)$. Also for $\nu \neq \frac{1}{2}$, $\mathfrak{J}_\nu(N_o)\subseteq\mathfrak{J}_\nu(N)$.
N is ν-semisimple iff N is a subdirect product of ν-primitive near-rings. With chain conditions this subdirect product becomes a finite direct sum and we get (5.31) in special cases that N is ν-semisimple iff N is a finite direct sum of simple ν-primitive near-rings with DCCL. In 5.32 we get a "Wedderburn-Artin-like" structure theorem for ν-semisimple near-rings.

$\mathcal{J}_2(N)$ contains all nil N-subgroups, $\mathcal{J}_{1/2}(N)$ all nil left ideals and $\mathcal{J}_o(N)$ all nil ideals. However, in contrast to the ring case, $\mathcal{J}_2(N)$ is not necessarily nil if N is finite (cf.5.48).

Finally we consider the nil and the prime radical of a near-ring.

a) JACOBSON-TYPE RADICALS: COMMON THEORY

1.) DEFINITIONS AND CHARACTERIZATIONS OF THE RADICALS

As usual, let N be a near-ring and $\nu \in \{0,1,2\}$. Recall our convention about the intersection of an empty collection of subsets on page 1.

5.1 DEFINITION $\mathcal{J}_\nu(N) := \bigcap_{{}_N\Gamma \text{ of type } \nu} (o:\Gamma)$ is called the *ν-radical* of N.

5.2 THEOREM $\mathcal{J}_\nu(N) = \bigcap_{I\ \nu\text{-pr.id.of } N} I = \bigcap_{\substack{L\ \nu\text{-mod.} \\ \text{left id.of } N}} (L:N)$.

Proof. 4.3.

The relations between the radicals are easily described:

5.3 PROPOSITION

(a) $\mathcal{J}_o(N) \subseteq \mathcal{J}_1(N) \subseteq \mathcal{J}_2(N)$.

(b) If $N \in \mathcal{N}_1$ then $\mathcal{J}_1(N) = \mathcal{J}_2(N)$.

(c) If N is a ring then $\mathcal{J}_o(N) = \mathcal{J}_1(N) = \mathcal{J}_2(N) = \mathcal{J}(N)$ (Jacobson-radical of N).

Proof. (a): by 3.7(a).

(b) : by 3.7(c) and 3.19(a).

(c): If ${}_N\Gamma$ is of type ν and N is a ring then one sees as in 4.8 that ${}_N\Gamma$ is an N-module.

5.4 THEOREM (Betsch (3)). If $\nu \neq 0$ then $\mathcal{J}_\nu(N) = \bigcap_{\substack{L\ \nu\text{-mod.}\\ \text{left id. in } N}} L$.

Proof. By definition, $\mathcal{J}_\nu(N) = \bigcap_{{}_N\Gamma \text{ of type } \nu} (o:\Gamma)$. But ${}_N\Gamma$ is strongly monogenic, so $(o:\Gamma) = \bigcap_{\gamma\in\Gamma} (o:\gamma)$, where each $(o:\gamma)$ is $= N$ or a ν-modular left ideal (3.23).

Conversely, let L be a ν-modular left ideal of N. Then $\exists\ {}_N\Gamma \in {}_N\mathcal{G}\ \exists\ \gamma_0 \in \Gamma$: $\Gamma = N\gamma_0 \wedge L = (o:\gamma_0)$ (3.23). $N/L \cong_N \Gamma$ (by 3.4(e)) is of type ν. Hence the $(o:\gamma)$'s are just all ν-modular left ideals (or $= N$) and the result follows.

This raises the question what happens with the intersection of all 0-modular (= maximal modular) left ideals of N.

5.5 DEFINITION $\mathcal{J}_{1/2}(N) := \bigcap_{\substack{L\ 0\text{-mod. left}\\ \text{ideal of } N}} L$.

5.6 REMARK $\mathcal{J}_{1/2}(N)$ is often denoted by "D(N)" in the literature (see e.g. Betsch (3)). Our notation is motivated by the fact that $\mathcal{J}_{1/2}(N)$ is in general only "half of an ideal" (a left, but not necessarily a two-sided ideal) and by its location:

5.7 PROPOSITION $\mathcal{J}_0(N) \subseteq \mathcal{J}_{1/2}(N) \subseteq \mathcal{J}_1(N)$.

Proof. $\mathcal{J}_0(N) = \bigcap_{{}_N\Gamma \text{ of type } 0} (o:\Gamma) = \bigcap_{{}_N\Gamma \text{ of type } 0} \bigcap_{\gamma\in\Gamma} (o:\gamma) \subseteq$

$\subseteq \bigcap_{{}_N\Gamma \text{ of type } 0} \bigcap_{\gamma\in\theta_1(\Gamma)} (o:\gamma)$.

These $(o:\gamma)$'s are (as in 5.4) exactly all 0-modular left ideals. Hence $\mathcal{J}_0(N) \subseteq \mathcal{J}_{1/2}(N)$. $\mathcal{J}_{1/2}(N) \subseteq \mathcal{J}_1(N)$ is a trivial consequence of 3.7(a) and 5.4.

The following result comes from Fain (1).

5.8 PROPOSITION $L \trianglelefteq_\ell N \wedge \exists\, k \in \mathbb{N}: L^k \subseteq \mathfrak{J}_\nu(N) \wedge \nu \in \{1,2\} \Rightarrow L \subseteq \mathfrak{J}_\nu(N)$. Hence $\mathfrak{J}_1(N)$ and $\mathfrak{J}_2(N)$ are semiprime ideals.

Proof. If $L^k \subseteq \mathfrak{J}_\nu(N)$, but $L \not\subseteq \mathfrak{J}_\nu(N)$, then $\exists\, {}_N\Gamma \in {}_N\mathfrak{G}$: ${}_N\Gamma$ is of type ν and $L\Gamma \neq \{o\}$. So $\exists\, \gamma \in \Gamma$: $L\gamma \neq \{o\}$. Hence $\gamma \notin \theta_0$, so $\gamma \in \theta_1$ and $L\gamma \trianglelefteq_N \Gamma$ by 3.4(a). Thus $L\gamma = \Gamma$ and $L\Gamma = \Gamma$. Therefore $\Gamma = L\Gamma = L^2\Gamma = \ldots = L^k\Gamma = \{o\}$, a contradiction.

5.9 REMARK If $\nu = 2$ in 5.8, the result remains valid if $L \leq_N \Gamma$.

5.10 COROLLARY $\mathfrak{J}_1(N)$ contains all nilpotent left ideals and $\mathfrak{J}_2(N)$ contains moreover all nilpotent N-subgroups.

Cf. 5.37 and 5.45 for more results in this connection.

5.11 EXAMPLES The following examples shall show that no two of $\mathfrak{J}_0, \mathfrak{J}_{1/2}, \mathfrak{J}_1, \mathfrak{J}_2$ generally coincide, not even for finite zero-symmetric near-rings.

(a) $N_1 := \{f \in M_0(\mathbb{Z}_4) \mid f(2) = f(3) = 0\}$. N_1 is 1-primitive on Γ, hence $\mathfrak{J}_1(N_1) = \{0\}$, but not 2-primitive, hence by 4.46(f) $\mathfrak{J}_2(N_1) = N_1$. So $\mathfrak{J}_1(N) \neq \mathfrak{J}_2(N)$ in general.

(b) (Betsch (3)). $N_2 := \{f \in M_0(\mathbb{Z}_4) \mid f(2) \in \{0,2\}\}$. By 3.8, N_2 is 0-primitive, but not 1-primitive on Γ. Since each map $\in N_2$ is determined by its effect on 1,2,3, N_2 is the sum of the left ideals $L_1 := (0:2) \cap (0:3)$, $L_2 := (0:1) \cap (0:3)$ and $L_3 := (0:2) \cap (0:3)$. Since $(0:1) \cap (0:2) \cap (0:3) = \{\bar{o}\}$, $N_2 = L_1 \dot{+} L_2 \dot{+} L_3$.

The map $\begin{matrix}\mathbb{Z}_4 \to L_1\\ \gamma \to f_\gamma\end{matrix}$ with $f_\gamma(x) := \begin{cases} 0 & x \neq 1 \\ \gamma & x = 1 \end{cases}$ is an N_2-isomorphism. Hence $L_1 \cong_{N_2} \mathbb{Z}_4$. Similarly,

$L_3 \cong_{N_2} \mathbb{Z}_4$. $|L_2| = 2$ and is an N_2-group of type 2. Therefore L_2+L_3, L_1+L_3 and L_1+L_2 are 0-modular left ideals. Their intersection is $\mathfrak{J}_{1/2}(N_2) = \{0\}$.

Since N contains an identity, $\mathfrak{J}_1(N) = \mathfrak{J}_2(N)$ (by 5.3(b)). But each N_2-group of type 2 is $\cong_N L_2$ by 3.11(a). Hence $\mathfrak{J}_2(N_2) = (0:L_2) = (0:2) \neq \{0\} =$
$= \mathfrak{J}_{1/2}(N_2)$.

Observe that $\mathfrak{J}_1(N_2) = \mathfrak{J}_2(N_2) = (0:2)$ is not nilpotent - in striking contrast to the situation in ring theory! Compare 5.45!

(c) $N_3 := \{f \in M_0(\mathbb{Z}_4) \mid f(3) = 0 \wedge f(2) \in \{0,2\}\}$. This near-ring of order 8 has (as can be easily seen "by hand") only one non-trivial left ideal, namely $(0:2)$ (= the near-ring N_1 of (a)). $N_3/(0:2)$ is an N_3-group of order 2 and type 2. Hence $\mathfrak{J}_{1/2}(N_3) = \mathfrak{J}_1(N_3) = \mathfrak{J}_2(N_3) =$ $= (0:2)$. But N_3 is 0-primitive on $\mathbb{Z}_4$, hence $\mathfrak{J}_o(N_3) = \{0\}$. So $\mathfrak{J}_o(N_3) \neq \mathfrak{J}_1(N_3)$.

<u>5.12 EXAMPLE</u> If $N = N_c$ then $\mathfrak{J}_o(N) = \mathfrak{J}_{1/2}(N) = \mathfrak{J}_1(N)$ = intersection of all maximal normal subgroups of $(N,+)$ (=Baer-radical" of $(N,+)$), while $\mathfrak{J}_2(N)$ = intersection of all normal maximal subgroups of $(N,+)$.
(Apply 3.21(c), 3.29, the fact that each 0-modular left ideal in $N = N_c$ is 1-modular and 5.2).

2.) RADICALS OF RELATED NEAR-RINGS

To be able to treat $\mathfrak{J}_o$, $\mathfrak{J}_1$ and $\mathfrak{J}_2$ jointly (at least for a while) we introduce the following definition which comes from universal algebra (see (Hoehnke) and Mlitz (5)).

5.13 DEFINITION A map $\mathcal{R}$ which assigns to each near-ring N an ideal $\mathcal{R}(N)$ of N is called a radical (map) if for every $N, N' \in \mathcal{N}$:

(a) $\mathcal{R}(N/_{\mathcal{R}(N)}) = \{0\}$

(b) If $h \in \mathrm{Hom}(N,N')$ then $h(\mathcal{R}(N)) \subseteq \mathcal{R}(h(N))$.

5.14 DEFINITION If $\mathcal{R}$ is some radical map then $N \in \mathcal{N}$ is called

(a) $\mathcal{R}$-semisimple: $\iff \mathcal{R}(N) = \{0\}$.

(b) $\mathcal{R}$-radical: $\iff \mathcal{R}(N) = N$.

If $I \trianglelefteq N$ and $K \subseteq N$ denote $\{k+I \mid k \in K\}$ by $K+I/I$.

5.15 PROPOSITION If $\mathcal{R}$ is a radical map and N, N' are $\in \mathcal{N}$ then

(a) If $h: N \twoheadrightarrow N'$ and N is $\mathcal{R}$-radical then N' is $\mathcal{R}$-radical.

(b) If N is $\mathcal{R}$-radical then $\forall\, I \trianglelefteq N$: $N/_I$ is $\mathcal{R}$-radical.

(c) $\forall\, I \trianglelefteq N$: $\mathcal{R}(N/_I) \supseteq \mathcal{R}(N)+I/_I$.

(d) $\forall\, I \trianglelefteq N\ \forall\, K \subseteq N$: $(\mathcal{R}(N/_I)) = K/_I \Rightarrow K+I \supseteq \mathcal{R}(N)$.

(e) If N is simple then either N is $\mathcal{R}$-radical or $\mathcal{R}$-semisimple.

Proof. (a): by 5.13(b).

(b): by (a).

(c): Consider the canonical epimorphism $\pi: N \to N/I$ and apply 5.13(b).

(d): by (c).

(e): this holds because $\mathcal{R}(N) \trianglelefteq N$.

It would have been silly to introduce 5.13 if the $\mathcal{J}_\nu$'s would not be radicals. In fact, Betsch (3) has shown the following

5.16 THEOREM For $\nu\in\{0,1,2\}$, $N \to \mathcal{J}_\nu(N)$ is a radical map.

Proof. Clearly $\mathcal{J}_\nu(N) \trianglelefteq N$.

Let $n+\mathcal{J}_\nu(N)$ be $\in\mathcal{J}_\nu(N/\mathcal{J}_\nu(N))$ and let Γ be an N-group of type ν. By 3.14(a), Γ is an $N/\mathcal{J}_\nu(N)$-group of type ν since $\mathcal{J}_\nu(N) \subseteq (o:\Gamma)$. Hence $n\Gamma = (n+\mathcal{J}_\nu(N))\Gamma = \{o\}$. Since Γ was arbitrary, $n\in\bigcap(o:\Delta)$, where Δ ranges over all N-groups of type ν. Thus $n\in\mathcal{J}_\nu(N)$ and $n+\mathcal{J}_\nu(N) = \mathcal{J}_\nu(N)$, so 5.13(a) is shown.

To see 5.13(b), let h be $\in\text{Hom}(N,N')$ and $n\in\mathcal{J}_\nu(N)$. Im h =: N". Let Γ be an N"-group of type ν. Since $N'' \cong N/_{\text{Ker } h}$, Γ can be considered as N-group of type ν (see 3.14(b)). Therefore $n\Gamma = \{o\}$. This implies that $h(n)\Gamma = n\Gamma = \{o\}$. Again, Γ is arbitrary, so $h(n)\in\mathcal{J}_\nu(h(N))$.

5.17 REMARK For $\mathcal{J}_\nu(N)$ $(\nu\in\{0,1,2\})$, 5.15(e) can be improved if N is simple then either N is $\mathcal{J}_\nu$-radical or ν-primitive (since all $(o:\Gamma) \trianglelefteq N$). The near-ring N_1 of 5.11(a) is an example of a simple $\mathcal{J}_2$-radical near-ring.

5.18 THEOREM Let ν be $\in\{0,1/2,1,2\}$ and $I \trianglelefteq N=N_0$ be a direct summand of N. Then $\mathcal{J}_\nu(I) \supseteq \mathcal{J}_\nu(N)\cap I$.
Moreover, $\mathcal{J}_2(I) = \mathcal{J}_2(N)\cap I$.

Proof. (a) Let ν be $\in\{1/2,1,2\}$. Then $\mathcal{J}_\nu(I)$ is the intersection of all ν-modular (for $\nu = 1/2$: 0-modular) left ideals L of I. (Again, recall the convention of page 1 if there are no ν-modular left ideals at all.) For each such L there exists some ν-modular (for $\nu = 1/2$: 0-modular) left ideal $\bar{L}$ of N with $\bar{L}\cap I = L$ (3.30). Let $\bar{\mathcal{L}}_\nu$ be the set of all these $\bar{L}$'s and $\mathcal{L}_\nu$ the set of all ν-modular (for $\nu = 1/2$: 0-modular) left ideals of N (cf. 3.28). Then by 3.30

$$\mathcal{J}_\nu(I) = \bigcap_{\bar{L}\in\bar{\mathcal{L}}_\nu}(\bar{L}\cap I) \supseteq \bigcap_{\bar{L}\in\mathcal{L}_\nu}(\bar{L}\cap I) = \Big(\bigcap_{\bar{L}\in\mathcal{L}_\nu}\bar{L}\Big)\cap I = \mathcal{J}_\nu(N)\cap I.$$

(b) $\nu = 0$. Observe that $\forall\, \mathfrak{L}\in\overline{\mathcal{L}}_0$: $((\mathfrak{L}\cap I):I)_I =$
$= ((\mathfrak{L}\cap I):N)_I$: "$\supseteq$" is trivial. So take some
$x\in((\mathfrak{L}\cap I):I)_I$. Then $x\in I$ and $xI\subseteq\mathfrak{L}$. Since $\mathfrak{L}\in\overline{\mathcal{L}}_0$,
$\mathfrak{L}\cap I \neq I$. Hence $I\not\subseteq\mathfrak{L}$, so $\mathfrak{L}+I = N$ since $\mathfrak{L}$ is a maximal left ideal. Take some $n\in N$. Then $n = \bar{\ell}+i$ with $\bar{\ell}\in\mathfrak{L}$ and $i\in I$. Therefore $xn = x(\bar{\ell}+i)-xi+xi\in$ $\in\mathfrak{L}+\mathfrak{L} = \mathfrak{L}$. Moreover, $xn\in IN\subseteq I$, hence $x\in((\mathfrak{L}\cap I):N)_I$.
Making repeated use of 1.44 we now get

$$\mathfrak{J}_0(I) = \bigcap_{\substack{L\ 0\text{-mod.}\\ \text{in } I}} (L:I)_I = \bigcap_{\mathfrak{L}\in\overline{\mathcal{L}}_0} ((\mathfrak{L}\cap I):I)_I =$$

$$= \bigcap_{\mathfrak{L}\in\overline{\mathcal{L}}_0} ((\mathfrak{L}\cap I):N)_I = ((\bigcap_{\mathfrak{L}\in\overline{\mathcal{L}}_0} \mathfrak{L}\cap I):N)\cap I \supseteq$$

$$\supseteq ((\bigcap_{\mathfrak{L}\in\mathcal{L}_0} \mathfrak{L}\cap I):N)\cap I = (\bigcap_{\mathfrak{L}\in\mathcal{L}_0} (\mathfrak{L}:N))\cap(I:N)\cap I =$$

$$= (\bigcap_{\mathfrak{L}\in\mathcal{L}_0} (\mathfrak{L}:N))\cap I = \mathfrak{J}_0(N)\cap I.$$

(c) (Fain (1)). One may assume that $I\not\subseteq\mathfrak{J}_2(N)$. Then there is some 2-primitive ideal P of N with $I\not\subseteq P$. So $P\subset I+P$ and $I+P/_P \trianglelefteq N/_P$. By 4.49, $I+P/_P$ is a 2-primitive near-ring, and the same applies to $I/_{I\cap P}$ by 2.8. Therefore $I\cap P$ is a 2-primitive ideal of I from which one gets that $\mathfrak{J}_2(I)\subseteq\mathfrak{J}_2(N)\cap I$.

5.19 EXAMPLES

(a) Let N be the near-ring N_2 of 5.11(b). Let $I:= (0:2)$ ($= N_1$ of 3.8). Since I is 1-primitive on $\mathbb{Z}_4$, $\mathfrak{J}_1(I)=\{0\}$. But $\mathfrak{J}_1(N) = (0:2) = I$ (5.11(b)), so $\mathfrak{J}_1(I)\subset\mathfrak{J}_1(N)\cap I$.

(b) Let the notation and situation be as in 4.50. Since N is 0-primitive, $\mathfrak{J}_0(N) = \{0\}$. I contains the nilpotent ideal $\{f\in I \mid \forall\, \gamma\in\Gamma: f(\gamma)\in\{0,4\}\}$. By 5.37(d), $\mathfrak{J}_0(I) \neq \{0\}$, whence $\mathfrak{J}_0(I)\supset\mathfrak{J}_0(N)\cap I$.

See also 5.33 for $\nu = 1$ and 6.34 for $\nu = 2$.

5.20 THEOREM Let N_i $(i\in I)$ be a family of near-rings. Then for $\nu\in\{0,1/2,1,2\}$,

$$\mathcal{J}_\nu(\prod_{i\in I} N_i) \subseteq \prod_{i\in I}\mathcal{J}_\nu(N_i) \quad\text{and}\quad \mathcal{J}_\nu(\bigoplus_{i\in I} N_i) \subseteq \bigoplus\mathcal{J}_\nu(N_i)$$

with equality in the latter case if $\nu = 2$ and all $N_i\in\mathcal{N}_0$.

Proof. We will give the proof for "$\subseteq$" and "Π". The argument for "$\subseteq$" and "$\oplus$" is similar. Moreover, we use the notation of 3.33. Also, let $\mathcal{L}_\nu, \mathcal{L}_\nu^{(i)}$ be the sets of all ν-modular left ideals in $N = \prod_{i\in I} N_i$ (in N_i, respectively) (recall 3.28). First we start with $\nu\in\{0,1,2\}$.

$$\mathcal{J}_\nu(N) = \bigcap_{L\in\mathcal{L}_\nu} (L:N) \subseteq \bigcap_{i\in I}\ \bigcap_{L_i\in\mathcal{L}_\nu^{(i)}}(\bar{L}_i:N) \quad \text{(by 3.33)}$$

Since $\bigcap_{L_i\in\mathcal{L}_\nu^{(i)}}(\bar{L}_i:N) = \overline{\bigcap_{L_i\in\mathcal{L}_\nu^{(i)}}(L_i:N_i)}$ (as can be seen), $\mathcal{J}_\nu(N) \subseteq \bigcap_{i\in I}\ \overline{\bigcap_{L_i\in\mathcal{L}_\nu^{(i)}}(L_i:N_i)} = \bigcap_{i\in I}\overline{\mathcal{J}_\nu(N_i)} =$

$= \Pi\mathcal{J}_\nu(N_i)$.

Finally, if $\nu = 1/2$ then $\mathcal{J}_\nu(N) = \bigcap_{L\in\mathcal{L}_0} L \subseteq$

$\subseteq \bigcap_{i\in I}\ \bigcap_{L_i\in\mathcal{L}_0^{(i)}}\bar{L}_i = \bigcap_{i\in I}\ \overline{\bigcap_{L_i\in\mathcal{L}_0^{(i)}}L_i} = \bigcap_{i\in I}\overline{\mathcal{J}_{1/2}(N_i)} =$

$= \prod_{i\in I}\mathcal{J}_{1/2}(N_i)$.

The equality for "$\oplus$" in the situation indicated in the statement follows from 1.57 and 5.18.

5.21 COROLLARY If $N = N_0 = \sum^{\bullet}_{\alpha\in A} I_\alpha$ then $\mathcal{J}_\nu(N) \subseteq \sum^{\bullet}_{\alpha\in A}\mathcal{J}_\nu(I_\alpha)$ for $\nu\in\{0,1/2,1,2\}$ (with equality if $\nu=2$).

5.22 REMARK If only $(N,+) = \sum_{\alpha\in A}^{\bullet}(L_\alpha,+)$ with $\forall\,\alpha\in A$: $L_\alpha \trianglelefteq_\ell N$ then 5.21 does not hold: $N := M_0(\mathbb{Z}_4)$, $L_1 := (0:2)$, $L_2 := (0:1)\cap(0:3)$. Then $N = L_1 \dot{+} L_2$. N is 2-primitive on $\mathbb{Z}_4$ (4.52), hence $\mathfrak{J}_2(N) = \{0\}$. But $\mathfrak{J}_2(L_1) = L_1$, $\mathfrak{J}_2(L_2) = \{0\}$, so $\mathfrak{J}_2(N) \neq \mathfrak{J}_2(L_1)\dot{+}\mathfrak{J}_2(L_2)$.

We conclude our troublesome trip to relatives of N by a consideration of the behaviour of $\mathfrak{J}_\nu(N)$ on the one hand and $\mathfrak{J}_\nu(N_0)$, $\mathfrak{J}_\nu(N_c)$ on the other hand. Our first result is an immediate consequence of 2.18:

5.23 COROLLARY $\forall\,\nu\in\{0,1,2\}$: $\mathfrak{J}_\nu(N) = (\mathfrak{J}_\nu(N))_o + (\mathfrak{J}_\nu(N))_c$.

This is not very much, indeed. It would be fine to be able to compute $\mathfrak{J}_\nu(N)$ via $\mathfrak{J}_\nu(N_0)$ and $\mathfrak{J}_\nu(N_c)$ (5.12!), perhaps as $\mathfrak{J}_\nu(N) = \mathfrak{J}_\nu(N_0)+\mathfrak{J}_\nu(N_c)$, similar to 5.21.
This is not the case (see also 9.77):

5.24 EXAMPLE If $N = \{f\in M(\mathbb{Z}_4) \mid f(0) = f(2) = f(3)\}$. One can show that $\mathfrak{J}_2(N) = N$, $\mathfrak{J}_2(N_0) = N_0$ (5.11(a)!), but $\mathfrak{J}_2(N_c)$ consists only of the maps which are constant $= 0$ or $= 2$.

From that one sees that there is no obvious simple connection between $\mathfrak{J}(N)$, $\mathfrak{J}(N_0)$ and $\mathfrak{J}(N_c)$. But $\mathfrak{J}(N_0)$ is always in $\mathfrak{J}(N)$:

5.25 PROPOSITION $\forall\,\nu\in\{0,1,2\}$: $\mathfrak{J}_\nu(N_0)\subseteq(\mathfrak{J}_\nu(N))_o$; in particular, $\mathfrak{J}_\nu(N_0)\subseteq\mathfrak{J}_\nu(N)$.

Proof. It suffices to prove the "in particular", for $\mathfrak{J}(N_0)$ is trivially contained in N_0.
By 3.18(a), $\mathfrak{J}_\nu(N_0) = \bigcap_{{}_{N_0}\Gamma \text{ of type } \nu} (o:\Gamma) \subseteq$

$$\subseteq \bigcap_{{}_N\Gamma \text{ of type } \nu} (o:\Gamma) = \mathfrak{J}_\nu(N).$$

See 5.67 (u) for $(\mathfrak{J}_2(N))_c$.

3.) SEMISIMPLICITY

Throughout this number, let ν be $\varepsilon\{0,1,2\}$, unless otherwise indicated.

5.26 DEFINITION N is ν-semisimple: <=> N is $\mathcal{J}_\nu$-semisimple.

5.27 EXAMPLE N is ν-primitive => N is ν-semisimple.

5.28 COROLLARY

(a) Each direct sum or direct product of ν-semisimple near-rings is ν-semisimple.

(b) If $\forall\ i\varepsilon I$: $N_i \varepsilon \mathcal{N}_0$ then $\bigoplus_{i\varepsilon I} N_i$ is ν-semisimple <=>

<=> $\forall\ i\varepsilon I$: N_i is ν-semisimple.

Proof. 5.20.

5.29 THEOREM (Betsch (3)). N is ν-semisimple <=> N is isomorphic to a subdirect product of ν-primitive near-rings.

Proof. Consider the set of ν-primitive ideals and apply 5.2, 1.58 and 4.2(c).

5.30 THEOREM Let N have the DCCI (DCCL). Then N is ν-semisimple <=> N is isomorphic to a subdirect product of finitely many ν-primitive near-rings with DCCI (DCCL).

Proof. =>: The family $(I_\alpha)_{\alpha\varepsilon A}$ of all ν-primitive ideals of N has $\bigcap_{\alpha\varepsilon A} I_\alpha = \{0\}$. We claim that it suffices to take finitely many I_α's to get a zero intersection. If not, take some I_{α_0} $(\alpha\varepsilon A)$. Since $\bigcap_{\alpha\varepsilon A} I_\alpha = \{0\}$, there is some $\alpha_1 \varepsilon A$: $I_{\alpha_0} \cap I_{\alpha_1} \subset I_{\alpha_0}$. Continuing in this way we get a chain $I_{\alpha_0} \supset I_{\alpha_0} \cap I_{\alpha_1} \supset \ldots$

which does not terminate and we arrive at a contradiction. Hence N is isomorphic to a subdirect product of finitely many ν-primitive near-rings. The rest follows by 2.35.

<=: follows from 5.29.

<u>5.31 THEOREM</u> (Betsch (3), Blackett (2)). Let $N = N_0$ and $\nu\in\{1,2\}$. Then N is ν-semisimple with DCCL <=> N is a finite distributive sum of ideals which are ν-primitive simple near-rings with a right identity and DCCL.

<u>Proof</u>. =>: by 5.30, 4.46(e) and (d) and 2.52(b).

<=: By 5.29, N is ν-semisimple. Using 2.35(b) repeatedly (by induction) one sees that N has the DCCL.

5.31 has many interesting corollaries (which mostly stem from Betsch (3)), which represent the "non-linear" version of the celebrated Wedderburn-Artin-structure theorem.

<u>5.32 THEOREM</u> If $N = N_0$ is ν-semisimple ($\nu\in\{1,2\}$) with DCCL then

(a) N has a right identity (so 3.43 is applicable).

(b) N is completely reducible, so all of 2.50 is at hand (for N).

(c) $N = (L_{11}\dot{+}\dots\dot{+}L_{1n_1})\dot{+}(L_{21}\dot{+}\dots\dot{+}L_{2n_2})\dot{+}\dots\dot{+}(L_{k1}\dot{+}\dots\dot{+}L_{kn_k})$,

where for $i\in\{1,\dots,k\}$, $L_{i1},\dots,L_{in_i}$ are pairwise $N_i := \sum_{j=1}^{n_i} L_{ij}$-isomorphic left ideals of N_i and N_i-groups of type ν. All L_{ij} are also simple left ideals of N and N-groups of type ν. Each N-group of type ν is N-isomorphic to one of them; 2.50 can be applied for $_N N$.

(d) N has only finitely many classes of non-N-isomorphic N-groups of type ν.

(e) Every ideal of N is again ν-semisimple.

Proof. (a) and (b) follow directly from 5.31.

(c): by 5.31, 4.46(b), 2.49, 2.48(e), 2.12, 3.14(b) and 3.11(a).

(d): by (c) and 3.11(b).

(e): by 5.31 and 2.55(a).

For the following result cf. 5.18.

5.33 THEOREM Let $N \in \mathfrak{N}_0$ have the DCCL. If $I \trianglelefteq N$ then $\mathcal{J}_1(I) \subseteq \mathcal{J}_1(N) \cap I$.

Proof. $N/\mathcal{J}_1(N)$ is 1-semisimple with DCCL. Since

$I+\mathcal{J}_1(N)/\mathcal{J}_1(N) \trianglelefteq N/\mathcal{J}_1(N)$, $I+\mathcal{J}_1(N)/\mathcal{J}_1(N)$ is 1-semisimple by 5.32(e); the same applies to $I/\mathcal{J}_1(N) \cap I$ by 2.8. 5.15(c) and 5.16 tell us that $\mathcal{J}_1(I)+(\mathcal{J}_1(N) \cap I)/\mathcal{J}_1(N) \cap I = \{0\}$, whence $\mathcal{J}_1(I) \subseteq$ $\subseteq \mathcal{J}_1(N) \cap I$.

Some decompositions of N induce decompositions of ${}_N\Gamma$:

5.34 THEOREM (Betsch (3)). If $\nu \in \{1,2\}$ and $N = N_0$ has DCCL and is ν-semisimple and if ${}_N\Gamma$ is monogenic then

(a) ${}_N\Gamma = \dot{\sum}_{i=1}^{k} \Delta_i$ with $\Delta_i \trianglelefteq_N \Gamma$ and ${}_N\Delta_i$ of type ν.

(b) Each $(o:\gamma)$ is either $= N$ or a finite intersection of ν-modular left ideals.

(c) If ${}_N\Gamma$ is of type 0 then it is of type ν.

Proof. (a) By 5.32(c), $N = \dot{\sum}_{i=1}^{s} L_i$ where each $L_i \trianglelefteq_\ell N$ is an N-group of type ν. Let Γ be $= N\gamma_0$. Then $\Gamma = N\gamma_0 = (\sum_{i=1}^{s} L_i)\gamma_0 = \sum_{i=1}^{s} (L_i\gamma_0)$. Each $L_i\gamma_0 \trianglelefteq_N \Gamma$

by 3.4(a).
If $L_i\gamma_0 \neq \{o\}$ then $L_i \cong_N L_i\gamma_0$ and $L_i\gamma_0$ is of type ν, hence simple. As in (b) => (c) of 2.48 one can choose some subset S of $\{1,\ldots,s\}$ with $\Gamma = \sum^{\bullet}_{i\in S}(L_i\gamma_0)$.

(b) If $n\in(o:\gamma)$ $(\gamma\in\Gamma)$ and $\gamma = \sum_{i=1}^{k}\delta_i$ then

$o = n\gamma = n(\sum_{i=1}^{k}\delta_i) = \sum_{i=1}^{k} n\delta_i$ by 2.30. Since the sum of the Δ_i's is direct, $\forall\ i\in\{1,\ldots,k\}$: $n\delta_i = o$.

Hence $n\in\bigcap_{i=1}^{k}(o:\delta_i)$.

If $\delta_i\in\Theta_o(\Delta_i)$ then $(o:\delta_i) = N$.
If $\delta_i\in\Theta_1(\Delta_i)$ then $\Delta_i \cong_N N/_{(o:\delta_i)}$, hence $(o:\delta_i)$ is ν-modular.

(c) Use 5.32(a) and the fact that ${}_N\Gamma$ is simple (so $k = 1$).

See Choudhari (1) (no. 3.34) for more characterizations of $\mathfrak{J}_\nu(N)$ (especially via quasi-regularity). If $N\in\mathfrak{N}_o\cap\mathfrak{N}_1$ is 2-semisimple with DCCL and if $N = \oplus I_k$ (I_k finitely many simple ideals - 5.31) then the center of $(N,\cdot)$ is isomorphic to the direct product of the centers of $(I_k,\cdot)$ (as semigroups). In this case, if $I_k = M_{G_o^o}(\Gamma)$ (4.52), the center of $(I_k,\cdot)$ is a group, isomorphic to the center of G_o (see Holcombe (2)).

b) JACOBSON-TYPE RADICALS: SPECIAL THEORY

1.) $\mathfrak{J}_0$ and $\mathfrak{J}_{1/2}$.

5.35 THEOREM (Ramakotaiah (1)). $\mathfrak{J}_0(N) = (\mathfrak{J}_{1/2}(N): N)$.

Proof. $(\mathfrak{J}_{1/2}(N): N) = (\bigcap_{L\in\mathcal{L}_0} L:N) = \bigcap_{L\in\mathcal{L}_0} (L:N) = \mathfrak{J}_0(N)$.

5.36 THEOREM (Ramakotaiah (1)). $\mathfrak{J}_0(N)$ is the greatest ideal of N contained in $\mathfrak{J}_{1/2}(N)$.

Proof. Let I be another ideal of N which is in $\mathfrak{J}_{1/2}(N)$.
If $\mathcal{L}$ is the set of all 0-modular left ideals of N then $\forall\ L\in\mathcal{L}$: $I\subseteq L$.
Hence $\forall\ L\in\mathcal{L}$: $I\subseteq(L_i:N)$, so $I \subseteq \bigcap_{L\in\mathcal{L}}(L:N) = \mathfrak{J}_0$.

Observe that in the finite case this also follows from 3.27(b) and 3.25.

5.37 THEOREM (Ramakotaiah (1)). Let N be $\in\mathcal{N}_0$.

(a) $\mathfrak{J}_{1/2}(N)$ is the greatest quasiregular left ideal of N.

(b) $\mathfrak{J}_{1/2}(N)$ contains all nil left ideals.

(c) $\mathfrak{J}_0(N)$ is the greatest quasiregular ideal of N.

(d) $\mathfrak{J}_0(N)$ contains all nil ideals.

Proof. (a)1) We show that $\mathfrak{J}_{1/2}(N)$ is quasiregular.
Let z be $\in\mathfrak{J}_{1/2}(N)$. If $z\notin L_z$ (3.35) then Zorn's lemma (3.27(b)!) guarantees the existence of a (by z) modular left ideal L, maximal for having $z\notin L$.
If L' is another left ideal containing L then $z\in L'$; since $\forall\ n\in N$: $n-nz\in L'$, $L' = N$. Hence L is a maximal left ideal, so $\mathfrak{J}_{1/2}(N)\subseteq L$ and we arrive at a contradiction.
Therefore $\mathfrak{J}_{1/2}(N)$ is quasiregular.

2) Now let Q be a quasiregular left ideal. If $\mathfrak{J}_{1/2}(N) = N$ then clearly $Q \subseteq \mathfrak{J}_{1/2}(N)$. So assume that $\mathfrak{J}_{1/2}(N) \neq N$. Let L be a modular (by e, say) maximal left ideal.

If $Q \not\subseteq L$ then $Q+L = N$. So $\exists\, q \in Q\ \exists\, \ell \in L$: $e = q+\ell$. $\forall\, n \in N$: $ne-nq = n(q+\ell)-nq \in L$. Hence $\forall\, n \in N$: $n-nq = n-ne+ne-nq \in L+L = L$. This shows that L $(\neq N)$ is modular via q. By 3.38(c), q cannot be quasiregular, a contradiction. So $Q \subseteq L$.

Hence $Q \subseteq \bigcap_{\substack{L\ 0\text{-mod.}\\ \text{in } N}} L = \mathfrak{J}_{1/2}(N)$.

(b): by (a) and 3.38(b).

(c): by (a) and 5.36.

(d): by (c) and 3.38(b).

Another intersection of big things is contained in $\mathfrak{J}_2(N)$:

<u>5.38 THEOREM</u> If $N = N_0$, the intersection M of all maximal N-subgroups is quasiregular and contained in $\mathfrak{J}_2(N)$.

<u>Proof</u>. Take $z \in M$. Assume that $z \notin L_z$. Then there is some $\overline{M} <_N N$ containing L_z which is maximal w.r.t. not containing z. As in part 2) of the proof of 5.37(a) one shows that $\overline{M}$ is a maximal N-subgroup, so $z \in \overline{M}$, a contradiction.
The rest will be obvious from 5.44.

<u>5.39 THEOREM</u> (Betsch (3)). Let $N \in \mathfrak{N}_0$ have the DCCL. Then

(a) $\mathfrak{J}_{1/2}(N) = \{0\} \iff N = \sum_{i=1}^{s} {}^{\bullet} L_i$ where the L_i's are left ideals and N-groups of type 0.

(b) In this case, N contains a right identity and 2.50 is applicable for ${}_N N$.

Proof. (a) =>: If $\mathfrak{J}_{1/2}(N) = \{0\}$, the intersection of the 0-modular left ideals = $\{0\}$. As in the proof of 5.30 it suffices to take finitely many of them, say $L_1,\dots,L_k$, minimal for having intersection = $\{0\}$. Apply 2.50(g) to see that $N = \dot{\sum}_{i=1}^{k} K_i$, where $K_i = \sum_{j \neq i} L_j \cong_N N/L_i$ are N-groups of type 0.

<=: If $N = \dot{\sum}_{i=1}^{k} L_i$ as indicated, $\{K_1,\dots,K_k\}$ (as above) are 0-modular left ideals with $\bigcap_{i=1}^{k} K_i = \{0\}$. Hence $\mathfrak{J}_{1/2}(N) = \{0\}$.

(b): by 3.27(d) and (a).

5.40 THEOREM (Ramakotaiah (1)). Let $N \in \mathfrak{N}_0$ have the DCCN. Then

(a) $\mathfrak{J}_{1/2}(N)$ is nilpotent.

(b) N is 0-semisimple <=> N has no non-zero quasiregular ideal <=> N has no non-zero nil ideal <=> N has no non-zero nilpotent ideal.

(c) Each prime ideal $\neq N$ is 0-primitive.

Proof. (a) and (b) are immediate consequences of 5.37 and 3.40.

(c): Let $P \triangleleft N$ be prime. By 2.104, $\bar{N} := N/P$ has no non-zero nilpotent ideals and is therefore 0-semisimple by (b). Since $N \neq P$, $\bar{N} \neq \{\bar{0}\}$. So $\bar{N}$ has DCCN (2.35) and 0-primitive ideals $\bar{P}_1,\dots,\bar{P}_k$ with $\bigcap_{i=1}^{k} \bar{P}_i = \{\bar{0}\}$. Hence $\bar{P}_1\bar{P}_2 \dots \bar{P}_k = \{\bar{0}\}$, so (since $\{\bar{0}\}$ is prime) some $\bar{P}_i = \{\bar{0}\}$. So $\{\bar{0}\}$ is 0-primitive and the result follows.

2.) $\mathfrak{J}_1$

5.41 REMARKS If $N \in \mathfrak{N}_0$, $\mathfrak{J}_1(N)$ contains of course all nil, nilpotent and all quasiregular left ideals of N (5.37), but not necessarily all nil N-subgroups of N (but $\mathfrak{J}_2(N)$ has this property - see 5.45):
In $N = \{f \in M_0(Z_4) \mid f(2) = 0\}$ is $(\{0:2\}:Z_4) \leq_N N$ nilpotent, but $\mathfrak{J}_1(N) = \{0\}$. Also, $\mathfrak{J}_1(N)$ is not nil in general (see 5.11(b)). On the other hand, $\mathfrak{J}_1(N)$ contains one more item:

5.42 THEOREM $N = N_0$, DCCL. Let M denote the intersection of all maximal ideals of N (cf. Mlitz (1), (2)). Then

(a) $\mathfrak{J}_1(N) \supseteq M(N)$.

(b) If $N \in \mathfrak{N}_1$ has the DCCN then $\mathfrak{J}_1(N) = \mathfrak{J}_2(N) = M(N)$.

Proof. (a) Let $P \triangleleft N$ be a 1-primitive ideal. Then N/P is 1-primitive with DCCL, hence simple by 4.46(e). Hence P is maximal.

(b): by 4.47(b).

Recall also 5.33.

3.) $\mathfrak{J}_2$

5.43 PROPOSITION If $N \in \mathfrak{N}_1$ has a minimal N_0-subgroup then N is not 2-radical.

Proof. Let $M \leq_N N$ be a minimal N_0-subgroup. Then ${}_NM$ is of type 2. If $NM = \{0\}$ then $M = \{0\}$ since $N \in \mathfrak{N}_1$, a contradiction.
Hence $\mathfrak{J}_2(N) \neq N$.

Now we look who is contained in $\mathfrak{J}_2(N)$. See also 5.67 (u).

5.44 THEOREM (Ramakotaiah (1)). If $N = N_0$, $\mathfrak{J}_2(N)$ contains all quasiregular N-subgroups.

Proof. We proceed similar to 5.37(a). We may assume that $\mathfrak{J}_2(N) \neq N$. Let $Q \leq_N N$ be quasiregular. If $Q \not\subseteq \mathfrak{J}_2(N)$ then there is some 2-modular left ideal L with $Q \not\subseteq L$. Let e be a right identity modulo L. By 2.15, we see that $L+Q \leq_N N$ and $L \subset L+Q$. L is a maximal N-subgroup (3.29), so $L+Q = N$. Let $e = :\ell+q$ ($\ell \in L$, $q \in Q$). If $n \in N$ then $ne-nq = n(\ell+q)-nq \in L$. Hence $\forall\ n \in N$: $n-nq = n-ne+ne-nq \in L+L = L$ and L is modular by q. By 3.38(c), q is not quasiregular and we arrive at a contradiction.

From 3.38 we deduce

5.45 COROLLARY If N is zero-symmetric then $\mathfrak{J}_2(N)$ contains all nil(potent) N-subgroups.

Recall from 5.11(b) that $\mathfrak{J}_2(N)$ is not necessarily nil (not even for finite zero-symmetric near-rings with identity (not even under the assumption of distributive generation - see Laxton (4))). 5.48 will characterize the case that $\mathfrak{J}_2(N)$ is nilpotent.

$\mathfrak{J}_2(N)$ is the intersection of all 2-primitive left ideals, while $\mathfrak{J}_{1/2}(N)$ is $\mathfrak{J}_2(N)$ intersected with all 0-, but not 2-modular left ideals. $\mathfrak{J}_2(N)/\mathfrak{J}_{1/2}(N)$ swallows up all 0-modular ones (Laxton (6)):

5.46 PROPOSITION Let $N = N_0$ have the DCCL. Then $\mathfrak{J}_2(N)/\mathfrak{J}_{1/2}(N)$ is zero or a finite direct sum of N-groups of type 0, which are not of type 2.

Proof. $\bar{N} := N/\mathfrak{J}_2(N)$ is a finite direct sum $\sum\limits_{i=1}^{k}{}^{\bullet}\Gamma_i$ of $\bar{N}$-groups of type 2 by 5.31. $\bar{N} \Rightarrow \sum\limits_{i=2}^{k}\Gamma_i \Rightarrow \ldots \Rightarrow \Gamma_k \Rightarrow \{0\}$ is a principal sequence. Assume that $\mathfrak{J}_2(N)$ is contained in some 0-, but not 2-modular left ideal L. of N. Then $N/\mathfrak{J}_2(N) \Rightarrow L/\mathfrak{J}_2(N) \Rightarrow \{0\}$ can be refined to another principal sequence (2.40) and the first factor $N/\mathfrak{J}_2(N) \big/ L/\mathfrak{J}_2(N) \cong_N N/L$ is simple, so $\bar{N}$-isomorphic to some Γ_i. But Γ_i is of type 2 and $N/\mathfrak{J}_2(N) \big/ L/\mathfrak{J}_2(N)$ is not, so we arrive at a contradiction. Hence for each 0-, but not 2-modular left ideal L we get $L+\mathfrak{J}_2(N) = N$ and consequently

$N/L \cong_N L+\mathfrak{J}_2(N)/L \cong_N \mathfrak{J}_2(N)/L\cap\mathfrak{J}_2(N)$.

This shows that $L\cap\mathfrak{J}_2(N) =: L'$ is a 0-modular left ideal of $\mathfrak{J}_2(N)$. In $\mathfrak{J}_2(N)/\mathfrak{J}_{1/2}(N)$, all these $L'/\mathfrak{J}_{1/2}(N)$'s are 0-modular left ideals with "trivial" $(=\mathfrak{J}_{1/2}(N))$ intersection, so by 2.50 $\mathfrak{J}_2(N)/\mathfrak{J}_{1/2}(N)$ is a finite direct sum of N-groups of type 0, but not type 2.

From this we deduce

<u>5.47 THEOREM</u> (Laxton (6)). Let $N\varepsilon\mathfrak{N}_0$ have the DCCN and let L be a left ideal. Then

$L\subseteq\mathfrak{J}_2(N) \iff \exists\ K \trianglelefteq_\ell N$: $K\subseteq L$, K nilpotent and L/K is zero or a finite direct sum of N-groups of type 0, but not of type 2.

Proof. =>: $L \cap \mathfrak{J}_{1/2}(N) =: K$ is nilpotent by 5.40(a).

$L/K = L/_{L\cap\mathfrak{J}_{1/2}(N)} \cong_N L+\mathfrak{J}_{1/2}(N)/_{\mathfrak{J}_{1/2}(N)} \trianglelefteq_\ell$

$\trianglelefteq_\ell \mathfrak{J}_2(N)/_{\mathfrak{J}_{1/2}(N)}$. By 5.46 and 2.55, L/K is a finite direct sum of N-groups of type 0, but not of type 2.

<=: $L+\mathfrak{J}_2(N)/_{\mathfrak{J}_2(N)} \cong_N L/_{\mathfrak{J}_2(N)\cap L}$. By 5.45, $K \subseteq \mathfrak{J}_2(N) \cap L$.

Hence $L/_K / _{(\mathfrak{J}_2(N)\cap L/_K)} \cong L/_{\mathfrak{J}_2(N)\cap L}$. L/K is a finite direct sum of N-groups of type 0, but not of type 2, so the same applies to $L/_{\mathfrak{J}_2(N)\cap L}$ by 2.55, and

hence also to $L+\mathfrak{J}_2(N)/_{\mathfrak{J}_2(N)} \trianglelefteq_\ell N/_{\mathfrak{J}_2(N)}$. But

$N/_{\mathfrak{J}_2(N)}$ is a direct sum of N-groups of type 2, so $L+\mathfrak{J}_2(N) = \mathfrak{J}_2(N)$ and $L \subseteq \mathfrak{J}_2(N)$.

From this result one can construct a decomposition of $\mathfrak{J}_2(N)$ into a nilpotent and a "totally nilpotent" part (Laxton (6)) in special cases. For more general cases, see Laxton-Machin (1) and Scott (1).

Now we characterize the case that $\mathfrak{J}_2(N)$ is nilpotent.

5.48 THEOREM (Ramakotaiah (1)). Let $N \in \mathfrak{N}_0$ have the DCCN. Then the following conditions are equivalent:

(a) $\mathfrak{J}_2(N)$ is nil.

(b) $\mathfrak{J}_2(N)$ is nilpotent.

(c) $\mathfrak{J}_2(N)$ is quasiregular.

(d) $\mathfrak{J}_2(N) = \mathfrak{J}_1(N) = \mathfrak{J}_{1/2}(N) = \mathfrak{J}_0(N)$.

(e) Each N-group of type 0 is of type 2.

(f) $\forall$ $I \trianglelefteq N$: I is 0-primitive <=> I is 1-primitive <=> I is 2-primitive.

(g) Each 0-modular left ideal of N is 2-modular.

(h) $\forall$ $\{0\} \neq I \triangleleft N$: I is prime => I is 2-primitive.

Proof. (a) <=> (b) <=> (c) holds by 3.40.

(c) <=> (d): by 5.37(c).

(d) => (e): If ${}_N\Gamma$ is of type 0 then $(N/\mathcal{J}_2(N))^\Gamma$ is of type 0 by 3.14(a) (since $\mathcal{J}_2(N) = \mathcal{J}_0(N) \subseteq (o:\Gamma)$). Since $N/J_2(N)$ is 2-semisimple (5.16), $N/\mathcal{J}_2(N)^\Gamma$ is of type 2 by 5.34, so ${}_N\Gamma$ is of type 2 by 3.14(b).

(e) => (f): is immediate.

(f) => (d): trivial.

(f) <=> (g): by 4.3(c).

(f) <=> (h): by 5.40(c) and 4.34.

Cf. also 5.61(b).

Finally, we are going to describe 2-semisimple near-rings with chain condition more closely.

5.49 THEOREM (Choudhari-Tewari (1), Oswald (3)). Let $N \in \mathcal{N}_0$ have a right identity. Then the following statements are equivalent:

(a) N is 2-semisimple with DCCL.

(b) N is a direct sum of finitely many N-simple left ideals.

(c) If ${}_N\Gamma$ is monogenic then $\forall\ \Delta \leq_N \Gamma\ \exists\ E \trianglelefteq_N \Gamma: \Gamma = \Delta + E\ \wedge$ $\wedge\ \Delta \cap E = \{o\}$.

(d) $\forall\ M \leq_N N\ \exists\ L \trianglelefteq_\ell N: N = M+L\ \wedge\ M \cap L = \{0\}$.

(e) Each exact sequence $\{0\} \to M_1 \to M_2$ $(M_1, M_2 \leq_N N)$ splits (definitions as usual).

(f) $\forall\ M_1, M_2 \leq_N N,\ M_1 \subseteq M_2\ \forall\ {}_N\Gamma \in {}_N\mathcal{G}\ \forall\ h \in Hom_N(M_1,\Gamma)$ $\exists\ \bar{h} \in Hom_N(M_2,\Gamma): \bar{h}/_{M_1} = h$.

(g) N satisfies the DCCN and has no non-zero nilpotent N-subgroup (cf. Blackett (2), where near-rings with this condition are called "semisimple").

Proof. (a) => (b): by 5.32.

(b) => (a): By 2.50, N has the DCCL. If $N = \sum_{i=1}^{k}{}^{\bullet} L_i$, L_i N-simple left ideals, then $K_i := \sum_{j \neq i} L_j$ are 2-modular left ideals and (as in 2.50(g)) $\bigcap_{i=1}^{k} K_i = \{0\}$, hence $\mathcal{J}_2(N) = \{0\}$.

(b) => (c): If $N = \sum_{i=1}^{k}{}^{\bullet} L_i$ as above, then by (a) and 5.34(a) there is a subset S of $\{1,\dots,k\}$ with $\Gamma = \sum_{i \in S}{}^{\bullet} L_i\gamma_0$ $(\gamma_0 \in \theta_1(\Gamma))$, all $L_i\gamma_0 \neq \{o\}$ and of type 2.

If $\Delta \leq_N \Gamma$, take some maximal element $\bar{E}$ (Zorn!) in $\mathcal{L} := \{E \trianglelefteq_N \Gamma \mid \Delta \cap E = \{o\}\}$. If $\Delta + \bar{E} \neq \Gamma$, $\exists\, i \in S$: $L_i\gamma_0 \nsubseteq \Delta + \bar{E}$. But $L_i\gamma_0$ is of type 2, hence $L_i\gamma_0 \cap (\Delta+\bar{E}) = \{o\}$. Therefore $\Delta \cap (L_i\gamma_0 + \bar{E}) = \{o\}$ which contradicts the maximality of $\bar{E}$. Hence $\Delta + \bar{E} = \Gamma$ and (c) is shown.

(c) => (d) is trivial since ${}_N N$ is monogenic (by e).

(d) => (e): Let $\{0\} \to M_1 \xrightarrow{f} M_2$ be exact. By (d) $\exists\, L \trianglelefteq_\ell N$: $f(M_1)+L = N \wedge f(M_1) \cap L = \{0\}$. Then the "projection" $p: N \to f(M_1)$ defined by $p(n) = p(f(m_1)+\ell) := f(m_1)$ is an N-homomorphism. $p/_{M_2} =: \bar{p}$. Then $f^{-1} \circ \bar{p}: M_2 \to M_1$ is an N-homomorphism with $f^{-1} \circ \bar{p} \circ f = id_{M_1}$. Hence $\{0\} \to M_1 \to M_2$ splits.

(e) => (f): Let M_1, M_2, Γ, h be as in the statement, and let $\iota: M_1 \to M_2$ be the injection map. Then $\{0\} \to M_1 \xrightarrow{\iota} M_2$ splits (say by $g: M_2 \to M_1$). Then $\bar{h} := h \circ g: M_2 \to \Gamma$ does the required job.

(f) => (e): Let $\{0\} \to M_1 \xrightarrow{f} M_2$ be exact. The identity map $id_{f(M_1)}$ can be extended to an N-homomorphism $h: M_2 \to f(M_1)$. Then clearly $f^{-1} \circ h$ is a splitting N-homomorphism $M_2 \to M_1$.

(e) => (g): First we show that N has the DCCN.
Let $N = M_0 \supset M_1 \supset M_2 \supset, ..$ be a chain of N-subgroups of N.

$\forall\, i \in \mathbb{N}$: $\{0\} \to M_i \xrightarrow{\iota_i} M_{i-1}$ (ι_i the injection maps) splits.

Let $g_i: M_{i-1} \to M_i$ be corresponding splitting N-homomorphisms. Then $h_1 := g_1$ is a splitting N-homomorphism for $\{0\} \to M_1 \to N$, $h_2 := g_2 \circ g_1$ one for $\{0\} \to M_2 \to N$, et cetera.

If $L_i := \text{Ker } h_i$ then $L_i \trianglelefteq_\ell N$ and (as easily seen) $N = M_i + L_i$ with $M_i \cap L_i = \{o\}$. Furthermore, $L_1 \subseteq L_2 \subseteq \ldots$.

But ${}_N N$ is completely reducible (2.48(e)), finitely generable (since $e \in N$), so endowed with the ACCL (2.50(e)) which causes $L_1 \subseteq L_2 \subseteq \ldots$ to stop after finitely many steps. Hence the same applies to $M_0 \supset M_1 \supset \ldots$.

Now we show that N has no non-zero nilpotent N-subgroups. Let M be such one. As before, $\exists\, L \trianglelefteq_\ell N$: $N = L+M$, $L \cap M = \{0\}$.

Let e be $= \ell_0 + m_0$. ($\ell \in L, m_0 \in M$). As in 3.43, ℓ_0 turns out to be a right identity for M, hence M cannot be nilpotent and the proof is accomplished.

(g) => (b): (Blackett (2)). Let M_0 be a minimal N-subgroup. By 3.52, $\exists\, e_0 = e_0^2 \in N$: $Ne_0 = M_0 e_0 = M_0$.
By 1.13, $N = N_0 e_0 + (0:e_0) = M_0 + (0:e_0)$ with $M_0 \cap (0:e_0) = \{0\}$.

If $M_0 = N$ there is nothing to prove. So let $M_0 \neq N$. Hence $N \supset (o:e_0) \neq \{0\}$.

Either $(o:e_0)$ is a minimal N-subgroup or it contains (by applying the above considerations to $(o:e_0)$ instead of N) another smaller N-subgroup of the form $(o:e_0) \cap L$ where L is some left ideal of N. The DCCN assures that after finitely many steps we arrive at a minimal N-group M_1 which is the intersection of $(o:e_0)$ with a left ideal of N, hence a normal subgroup. Take some idempotent $e_1 \in N$

with $Ne_1 = M_1e_1 = M_1$. Hence $N = M_1+(o:e_1)$ (as groups). Repeating this procedure with $(o:e_1)$ (if necessary) yields $N = M_1 \dot{+} \dots \dot{+} M_k$ where M_i are minimal (hence N-simple) N-subgroups of N.

Now by 5,40 $\mathcal{J}_{1/2}(N) = \{0\}$, so N is the direct sum $\dot{\sum}_{i=1}^{s} L_i$ of left ideals of type 0 (5.39). The M_i's of above are N-groups of type 2, hence N-isomorphic to some L_j's.

Therefore N is the finite direct sum of left ideals which are N-groups of type 2.

5.50 REMARKS

(a) As Choudhari-Tewari (1) have shown one can add to 5.49 the following condition if $N \in \mathfrak{n}_1$:

(i) Each N-subgroup of N is monogenic, projective (definition again as usual) and generated by an idempotent.

See there for the proof (c) => (i) => (e) .

(b) Oswald (3) proved that a near-ring N fulfilling the conditions of 5.49 has the ACCN (we know from 2.50 that N has the ACCL). This follows from 5.49 (d).

(c) Observe that 4.46 and 5.49 show that a 2-primitive near-ring with DCCL has DCCN (Scott (1)), so N has all chain conditions.

(d) As Mason (3) pointed out, there exists no non-trivial injective $N = N_0$-group. See more on that in his papers (3) and (4), in Prehn (1) and in Banaschewski-Nelson (1).

c) THE NIL RADICAL

5.51 DEFINITION The sum of all nil ideals of N is called the nil radical of N and denoted by $\mathfrak{N}(N)$ (by $\mathfrak{J}_{-1}(N)$ in Ramakotaiah (1) and Polin (2)).

5.52 THEOREM

(a) $\mathfrak{N}(N)$ is the greatest nil ideal of N.
(b) $\mathfrak{N}(N)$ is the smallest ideal I of N such that N/I has no non-zero nil ideals.

Proof. (a): by 2.101(b).

(b): Let $\pi: N \to N/\mathfrak{N}(N) =: \bar{N}$ be the canonical projection.
If $\bar{I} \trianglelefteq \bar{N}$, $\bar{I}$ nil, look at $I := \pi^{-1}(\bar{I}) \trianglelefteq N$.
If $i \in I$ then $\exists\, k \in \mathbb{N}: \pi(i^k) = \pi(i)^k = \bar{0}$ (zero in $\bar{N}$), hence $i^k \in \operatorname{Ker} \pi = \mathfrak{N}(N)$. But $\mathfrak{N}(N)$ is nil, so $\exists\, \ell \in \mathbb{N}: (i^k)^\ell = 0$ and i is nilpotent. Hence I is nil, therefore $I \subseteq \mathfrak{N}(N)$ and we get $\bar{I} = \{\bar{0}\}$.
Now assume that N/I is without non-zero nil ideals. By 2.103, $I+\mathfrak{N}(N)/_I$ is nil in N/I, so $I+\mathfrak{N}(N) \subseteq I$ and we arrive at $\mathfrak{N}(N) \subseteq I$.

5.53 COROLLARIES

(a) $\mathfrak{N}$ is a radical map (in the sense of 5.13).
(b) $\mathfrak{N}(N) \subseteq \mathfrak{N}(N_0) \subseteq \mathfrak{J}_0(N_0) \subseteq \mathfrak{J}_0(N)$.
(c) N is $\mathfrak{N}$-semisimple iff N has no non-zero nil ideals.
(d) Each constant near-ring is $\mathfrak{N}$-semisimple.

Proof. (a): by 2.100 and 5.52(b).

(b): by 2.99, 5.52(a), 5.37(d) and 5.25.

(c): by 5.52(b).

(d): by (b).

It is clear that for rings $\mathcal{N}(N)$ coincides with the usual nil radical of rings. $\mathcal{N}$ is subhereditary on direct summands:

5.54 THEOREM (cf. Maxson (1)). If $I \trianglelefteq N$ is a direct summand then $\mathcal{N}(I) \subseteq \mathcal{N}(N) \cap I$.

The proof follows from 2.12.

5.55 REMARK $\mathcal{N}(N)$ is also identical with the "upper nil radical" $U = s(0)$ of Van der Walt (1). See this paper for a characterization of $\mathcal{N}(N)$ via "s-systems".

d) THE PRIME RADICAL

5.56 DEFINITION The intersection of all prime ideals of N is called the prime radical of N and denoted by $\mathcal{P}(N)$ (other notations: $\mathcal{J}_{-2}(N)$, $L_r(N)$, $m(0)$).

Again, this is just the usual prime radical in the case of rings.

5.57 PROPOSITION $\mathcal{P}$ is a radical map.

Proof. (a) $\mathcal{P}(N) \trianglelefteq N$.

(b) If $h: N \twoheadrightarrow \bar{N}$ and $\bar{P} \trianglelefteq \bar{N}$ is prime then $P := h^{-1}(\bar{P})$ is prime in N by 2.64 and 2.17(a), showing that $h(\mathcal{P}(N)) \subseteq \mathcal{P}(\bar{N})$.

(c) If $\bar{P}$ is a prime ideal of $\bar{N} := N/\mathcal{P}(N)$ then, as in (b), $\pi^{-1}(\bar{P}) =: P$ is prime in N.
Conversely, if $P \trianglelefteq N$ is prime then $\pi(P)$ is prime in $\bar{N}$.
Hence if $\bar{x}$ is in each prime ideal of $\bar{N}$ then each $x \in h^{-1}(\{\bar{x}\})$ is in each prime ideal of N, so $x \in \mathcal{P}(N)$ and $\bar{x}$ is zero.
Therefore $\mathcal{P}(N/\mathcal{P}(N))$ is zero.

The connection to 2.93 is given by

5.58 REMARK $\mathcal{P}(N) = \mathcal{P}(\{0\})$ and this is a semiprime ideal.

5.59 PROPOSITION $\mathcal{P}(N)$ is a nil ideal and contains the sum of all nilpotent ideals.

Proof: by 2.105.

From this we can locate $\mathcal{P}(N)$:

5.60 COROLLARY $\mathcal{P}(N) \subseteq \mathcal{N}(N) \subseteq \mathcal{J}_0(N) \subseteq \mathcal{J}_{1/2}(N) \subseteq \mathcal{J}_1(N) \subseteq \mathcal{J}_2(N)$ (and all inclusions can be strict).

The first two inclusions can even be strict in the case of rings.

5.61 THEOREM Let $N \in \mathcal{N}_0$ have the DCCN. Then

(a) $\mathcal{P}(N) = \mathcal{N}(N) = \mathcal{J}_0(N)$.

(b) $\mathcal{J}_2(N)$ is nilpotent (cf. 5.48) $\Longleftrightarrow \mathcal{P}(N) = \mathcal{N}(N) = \ldots = \mathcal{J}_2(N)$.

Proof. (a) follows from 4.34 and 5.40(c).

(b): "$\Leftarrow$" is trivial.
"$\Rightarrow$": follows from (a) and 5.48(d).

5.62 THEOREM (Maxson (1)). If $I \trianglelefteq N$ is a direct summand then

$$\mathcal{P}(I) \subseteq \mathcal{P}(N) \cap I.$$

This result follows from 2.63.

2.69 yields

5.63 EXAMPLE Each prime (e.g. each constant) near-ring is $\mathcal{P}$-semisimple.

More generally:

5.64 PROPOSITION N is $\mathcal{P}$-semisimple iff N is isomorphic to a subdirect product of prime near-rings.

This is a direct consequence of 1.58 and 2.67.

5.65 PROPOSITION Each $\mathcal{P}$-semisimple near-ring has no non-zero nilpotent ideals.

This follows from 2.104.

See more on that in Scott (1) and Holcombe (2).

e) CONCLUSIVE REMARKS

5.66 SUMMARY We summarize some properties of our radicals (we include $\mathcal{J}_{1/2}$ although it is not a radical map).

Radical $\mathcal{R}$	$\mathcal{J}_2$ $\supseteq$	$\mathcal{J}_1$ $\supseteq$	$\mathcal{J}_{1/2}$ $\supseteq$	$\mathcal{J}_0$ $\supseteq$	$\mathcal{N}$ $\supseteq$	$\mathcal{P}$
$\mathcal{R}(N)$ quasiregular	-	-	(+)	(+)	(+)	(+)
$\mathcal{R}(N) \supseteq$ all quasiregular N-subgroups	(+)	-	-	-	-	-
$\mathcal{R}(N) \supseteq$ all quasiregular left ideals	(+)	(+)	(+)	-	-	-
$\mathcal{R}(N)$ nil	-	-	-	-	+	+
$\mathcal{R}(N) \supseteq$ all nil N-subgroups	(+)	-	-	-	-	-
$\mathcal{R}(N) \supseteq$ all nil left ideals	(+)	(+)	(+)	-	-	-
$\mathcal{R}(N) \supseteq$ all nil ideals	+	+	+	+	+	-
$\mathcal{R}(N) \supseteq$ all nilpotent ideals	+	+	+	+	+	+
$\mathcal{R}(N)$ semiprime	+	+	?	?	?	+
$\mathcal{R}(N)$ is the greatest quasiregular left ideal	-	-	(+)	-	-	-
$\mathcal{R}(N)$ is the greatest quasiregular ideal	-	-	-	(+)	-	-
$\mathcal{R}(N)$ is the greatest nil ideal	-	-	-	-	+	-

"+" means "yes"
"-" means "no"
"?" means "unknown (to the author)"

"(+)" stands for "yes, if $N \in \mathcal{N}_0$)" (otherwise unknown to the author)

If $N \in \mathcal{N}_1$ has a minimal N_0-subgroup then all radicals are $\neq N$.
If $N \in \mathcal{N}_0$ has the DCCN then $\mathcal{J}_2(N)$ is nil iff all radicals are equal.

5.67 SOME MORE REMARKS

(a) See Beidleman (1), (3) and (8) about the connection between $\mathcal{J}_2(N)$ and "(strictly) small" left ideals. Similar considerations can be found in Riedl (1) and Mlitz (1), (2). They adopt a lattice-theoretic point of view (the intersection of all maximal ideals (...) = sum of all "small" ideals (...)).

(b) Ramakotaiah (1) showed that each biregular near-ring (3.49) is 0-semisimple.

(c) The author suggests not to use the notations $\mathcal{J}_{-1}, \mathcal{J}_{-2}$ for $\mathcal{N}$ and $\mathcal{P}$, respectively, because these are not Jacobson-type radicals.

(d) Ramakotaiah (1) also defines a radical "$\mathcal{J}_{-3}(N)$" contained in $\mathcal{P}(N)$, as the intersection of all ideals I such that N/I has no nilpotent ideals. Polin (2) claims that $\mathcal{J}_{-3} = \mathcal{J}_{-2}$, but the author cannot follow this proof. See the paper by Choudhari(1) for another proof of this statement and for other characterizations of $\mathcal{P}(N)$ (such as the intersection of all semiprime ideals (cf. Maxson (1)) and of $\mathcal{N}(N)$.

(e) See Freidman (1) and Plotkin (2) for a "Levitzky-type" radical.

(f) Ramakotaiah (4) also defined a radical corresponding to the Brown-McCoy-radical (<u>$\mathcal{G}$-radical</u>) in ring theory as the intersection $\mathcal{G}(N)$ of all maximal modular ideals.

If $z \in N$ is called "G-regular" if the ideal generated by $\{n-nz \mid n \in N\}$ equals N then $\mathcal{G}(N)$ turns out to be the intersection of all ideals I of N, such that N/I has no G-regular ideals. $N/\mathcal{G}(N)$ has no G-regular ideals and is a subdirect product of simple near-rings with a right identity. See also Choudhari-Tewari (2).

(g) Laxton (3) defined one more "radical-like" ideal S(N) of N as the intersection of all "s-primitive ideals". He showed that $\mathfrak{J}_{1/2}(N) \subseteq S(N) \subseteq \mathfrak{J}_1(N)$ if $N \in \mathfrak{N}_0$ and gives an example of a dg. near-ring with $\mathfrak{J}_{1/2}(N) \subset S(N) \subset \mathfrak{J}_1(N)$.

(h) Another radical was defined by Deskins (1) (see also Williams (1)). If $N = N_0$ has the DCCN then semisimplicity w.r.t. this radical is equivalent to 2-semisimplicity, and this in turn to semisimplicity in the sense of Blackett (1), (2) (see 5.49).

(i) Beidleman considered in (3) the "radical subgroup" $R_s(N)$ as the intersection of all maximal N-subgroups in near-rings $\in \mathfrak{N}_0$.
By 5.38 we know that in this case $R_s(N) \subseteq \mathfrak{J}_2(N)$. Beidleman proved e.g. that $R_s(N) = \mathfrak{J}_2(N) \Leftrightarrow \mathfrak{J}_2(N)$ is quasiregular (in his sense - see 3.37(c)).
Cf. 5.48(d).

(j) The "quasi-radical" $Q(N) = \bigcap L$, where L ranges over all maximal left ideals, was also considered (by various authors). If N has a right identity then $Q(N) = \mathfrak{J}_{1/2}(N)$ (3.29). This and more radicals can be found in Choudhari (1).

(k) Gorton (1) called an N-group $_N\Gamma$ to be of class λ (λ a non-zero cardinal number) if
$\forall\ \Delta \subseteq \Gamma,\ |\Delta| \leq \lambda\ \ \forall f \in M(\Gamma)\ \ \exists\ n \in N\ \ \forall\ \gamma \in \Gamma: f(\gamma) = n\gamma$.
N is called λ-complete if N has a faithful N-group of class λ. A radical $C_\lambda(N)$ is defined as the intersection of all $(o:\Gamma)$, where Γ is an N-group of class λ.

He showed that if N is λ-complete on Γ then $N_c \cong_N \Gamma$, and that N is 1-complete iff N_c is faithful (a base of equality - 1.91). If $N \in \mathcal{N}_0$ then $\mathcal{J}_2(N) \subseteq C_1(N)$. Defining C_λ-modular left ideals as those modular left ideals L such that N/L is an N-group of class λ (cf. 3.28) yields a result similar to 5.2.
Also, he gave several examples.

(l) Maxson (1) proved that there is not such a fine connection between injectivity of N-groups (defined as usual) and semisimplicity as in the ring-case. He showed that if each N-group is injective then $\mathcal{J}_2(N) = \{0\}$, but gave an example that the converse does not hold.

(m) Ferrero developed a radical theory for "p-singular near-rings" in (18).

(n) A radical (corresponding to $\mathcal{J}_2$) for N-groups was considered by Beidleman in (1), (3) and (4) and by Choudhari in (1).

(o) Van der Walt (1) called an ideal I of N a nil radical if I is nil, but N/I has no nil ideals any more. He proved that the sum of all nil radicals of N equals $\mathcal{N}(N)$, which is the greatest nil radical of N, while the intersection of all nil radicals (the smallest nil radical) coincides with $\mathcal{P}(N)$. Therefore he called $\mathcal{N}$ ($\mathcal{P}$) the upper (lower) nil radical of N.

(p) Mlitz (2), (3) and Polin (2) generalized this radical theory to what they called "m-Ω-near-rings".

(q) See also other papers of Mlitz for a radical theory in universal algebras. However, these radical concepts turn out to be "too less near-ring-specific".

(r) Another attempt to get a radical theory for zero-symmetric near-rings was made by Scott in (3). He used a method similar to that of (Divinsky) for rings. As an example he studies the Baer-lower-radical, which turns out to be $= \mathcal{P}(N) = \mathcal{J}_{-3}(N)$ for near-rings with DCC on N-subgroups.

(s) The answer to the question whether or not
$\forall I \trianglelefteq N: \mathcal{J}_2(I) = \mathcal{J}_2(N) \cap I$
(the "heredity property of $\mathcal{J}_2$") is true for general near-rings, is still unknown (The proof in Fain (1) seems to have a lack).
The following conditions on the near-ring N and the ideal I assure that the equation above holds for N and for I:
(i) $\forall i \in I$: $Ni = Ii$ (see 3.30 and 5.18).
(ii) (S.D. Scott, private communication) $N = N_0$ has the DCCN.
(iii)(M.L.W. Holcombe, private communication). Call an N-group $_N\Gamma$ to be of <u>type 3</u> if $_N\Gamma$ is of type 2 , if $N \in \mathcal{N}_0$ and if
$\forall \gamma, \gamma' \in \Gamma$: $(\forall n \in N: n\gamma = n\gamma' \Rightarrow \gamma = \gamma')$.
Moreover, let $\mathcal{J}_3(N) := \bigcap_{_N\Gamma \text{ of type } 3} (o:\Gamma)$.

Holcombe could show that $\mathcal{J}_3$ is hereditary.
If N is a zero-symmetric near-ring with identity then $\mathcal{J}_3(N) = \mathcal{J}_2(N)$.
Hence one might add:
(iv) N is zero-symmetric and has an identity.
(v) N is a dgnr. (see 6.34).
(vi) N is a tame near-ring (see 6.35 (d)). The proof is similar to the one of 6.34.
Anyhow, the crucial point is to extend $_M M/L$ to $_N M/L$ in the proof of 3.30.

(t) It is easy to see that if N is a zero-symmetric near-ring with identity and the DCCN and if e is some idempotent in N then $\mathcal{J}_\nu(N)e = Ne \cap \mathcal{J}_\nu(N)$ for all $\nu \in \{0,1,2\}$.
If Ne is a minimal non-nilpotent N-subgroup of N and if $\mathcal{J}_2(N)$ is nilpotent then $\mathcal{J}_2(N)e$ is the greatest proper N-subgroup of Ne, so $Ne/\mathcal{J}_2(N)e$ is an N-group of type 2 (it is harder to see that all N-groups of type 2 arise in this way).

See Lausch-Nöbauer (1), where these results are formulated and proved for dgnr.'s - but they are valid in the general case.

(u) Let, for the moment, $\mathcal{M}$ denote the set of all "strictly maximal" (cf. 3.29) ideals of ${}_{N_0}N_c$ (i.e. those ideals of ${}_{N_0}N_c$ which are at the same time maximal N_0-subgroups of N_c). Routine arguments give the following information on $(\mathcal{J}_2(N))_c = \mathcal{J}_2(N) \cap N_c$:

If $L \in \mathcal{L}_2(N)$ then $L \cap N_c = N_c$ or $L \cap N_c \in \mathcal{M}$

(for if $M \leq_{N_0} N_c$ contains $L \cap N_c$ then $L+M \leq_{N_0} N$, whence $L+M = N$ and $M = N_c$).

Conversely, if $M \in \mathcal{M}$ then $N_0 + M \in \mathcal{L}_2(N)$.

Hence $\mathcal{J}_2(N) \cap N_c = \bigcap_{M \in \mathcal{M}} M$.

See also 5.12, 5.23, 5.24 and 9.77.

PART III

SPECIAL CLASSES OF NEAR-RINGS

To keep this monograph within a reasonable size we will only cite, but not give proofs of some statements which lie a little bit away from the main stream of discussion (but might be equally important).

§ 6 DISTRIBUTIVELY GENERATED NEAR-RINGS

In this paragraph we discuss these types of near-rings which are still more "ring-like" than zero-symmetric near-rings. In fact, every dgnr. is $\in \mathcal{N}_0$. If N is a dgnr. then the ideals of ${}_N\Gamma$ are exactly the normal N-subgroups, but this nice feature does not seem to help a lot. For instance, all near-ring radicals can still be different (even for finite dgnr.'s). Abelian dgnr.'s are rings.

We also discuss the open problem of embedding a zero-symmetric near-ring into a dgnr. .

In the case of near-ring homomorphisms those ones deserve particular interest which carry the distributive generators into the ones of the image. These "(N,D)-(N',D')-homomorphisms" are characterized. Although the dgnr.'s form no variety, it is possible to speak about "free near-rings distributively generated by a given semigroup". N-groups Γ are studied which have the property that the distributive generators of N act "distributive" (= as endomorphisms) over Γ.

Finally we study the structure of dgnr.'s: 2-primitive finite dg. non-rings with identity are just the $M_0(\Gamma)$'s for a finite, non-abelian invariantly simple group Γ. In the finite case, $M_0(\Gamma) = E(\Gamma)$ iff Γ is of this kind.

a) ELEMENTARY

N is distributively generated (<u>dg.</u>, better: distributively generable) if there is a subsemigroup D of $(N_d,\cdot)$ generating $(N,+)$.

<u>6.1 NOTATION</u> If D generates N we denote this by (N,D).

<u>6.2 EXAMPLES</u>

(a) If $(\Gamma,+)\in\mathcal{G}$, define $E(\Gamma)$ by the set of all finite sums $\Sigma\sigma_i e_i$, where $\sigma_i\in\{-1,+1\}$ and $e_i\in\text{End }\Gamma$. $E(\Gamma)$ is a subnear-ring of $M(\Gamma)$, distributively generated by $(\text{End }\Gamma,\circ)$ and called the "<u>endomorphism near-ring on Γ</u>" (see 1.15).

(b) (H. Neumann (1), (2); Fröhlich (1), (2)).
Let $(\Gamma_n,+)$ be a reduced free group with generators $\{e_1,\ldots,e_n\} =: E$ (i.e. each map $E\to\Gamma_n$ can uniquely be extended to an endomorphism on Γ_n; Γ_n is then the free group in some variety of groups).
Define for the set $\text{End }\Gamma_n$ two binary operations $\oplus,\cdot$ by

$$(\phi_1 \oplus \phi_2)(e_i) := \phi_1(e_i)+\phi_2(e_i)$$

$$(\phi_1 \cdot \phi_2)(e_i) := \phi_1(\phi_2(e_i)) \quad \text{(and extend from E}$$

to Γ_n).
Then $(\text{End }\Gamma_n,\oplus,\cdot)$ turns out to be a dgnr., generated by

$$D := \{\phi_{ij} \mid \phi_{ij}(e_k) = \begin{cases} e_j & i = k \\ o & i \neq k \end{cases}\}.$$

Remark that $+$ and $\oplus$ are different if Γ_n is not abelian, for e.g.

$(\phi_1+\phi_2)(e_1+e_2) = \phi_1(e_1+e_2)+\phi_2(e_1+e_2) = \phi_1(e_1)+\phi_1(e_2)+$ $+\phi_2(e_1)+\phi_2(e_2)$, while $(\phi_1\oplus\phi_2)(e_1+e_2) = (\phi_1\oplus\phi_2)(e_1)+$ $+(\phi_1\oplus\phi_2)(e_2) = \phi_1(e_1)+\phi_2(e_1)+\phi_1(e_2)+\phi_2(e_2)$.

In Fröhlich's papers, + is referred to as the "addition of the first type" and ⊕ as the "addition of the second type".

(c) Similar to (a), the near-rings $A(\Gamma)$ and $I(\Gamma)$, defined as the subnear-rings of $M(\Gamma)$ generated by the automorphisms (inner automorphisms) of $(\Gamma,+)$, are dgnr.'s.

6.3 REMARKS

(a) $E(\Gamma)$, $A(\Gamma)$ and $I(\Gamma)$ will be studied in §7c).

(b) The End Γ_n's were introduced and studied by H. Neumann in (1) and (2). Her results on these types of near-rings include:
End Γ_n contains no identity, but all ϕ fixing some e_i and sending the other e_j's into zero are distributive and can be viewed as "relative units".
There is a 1-1-correspondence ψ between the set $\mathfrak{F}_n$ of all fully invariant subgroups of Γ_n and the set $\mathfrak{I}_n$ of all ideals of End Γ_n by
$\psi: \mathfrak{F}_n \to \mathfrak{I}_n$

$$A \to \{\phi \mid \forall i \in \{1,\dots,n\}: \phi(e_i) \in A\}$$

All homomorphic images of End Γ_n are also some End Γ_m's. Each End Γ_n is the homomorphic image of End Φ_n, where Φ_n is the free group on n generators.
Similar results hold for the near-rings of the kind $\bigoplus_{n\in\mathbb{N}}$ End Γ_n, which are also dg. (see 6.9(d)).

(c) See Fitting (1) for the problem, which automorphisms of a (non-abelian) group have the property that their sum is an automorphism again. Cf. also Heerema (1) and Robinson (1) for similar questions.

(d) See Plotkin (2) for the connection between the representations of a group Γ and those of $E(\Gamma)$.

Now we study some elementary properties of dgnr.'s. Note, that if N is dg. by D then each $n\in N$ is a finite (ordered) sum $n = \Sigma\sigma_i d_i$ with $\sigma_i = \pm 1$, $d_i \in D$.

6.4 PROPOSITION Let N be dg. by D.

(a) $\forall\ n\in N\ \forall\ d\in D$: $d(-n) = (-d)n = -(dn)$.

(b) $N\in\mathfrak{N}_0$.

(c) $\forall\ n,n'\in N\ \forall\ d\in D$: $d(n+n') = dn+dn' \wedge (-d)(n+n') =$
$= (-d)n'+(-d)n = -dn' - dn$.

(d) If $n = \sum_i \sigma_i d_i$ and $n' = \sum_j \sigma_j' d_j'$ then

$$nn' = \sum_i \sigma_i (\sum_j \sigma_j' d_i d_j').$$

The proof is accomplished by easy computations and therefore omitted.

6.5 PROPOSITION (Seth-Tewari (1)). Let N be dg. by D, let ${}_N\Gamma$ be $\varepsilon_N\mathfrak{G}$ and $\Delta\subseteq\Gamma$. Then the ideal of ${}_N\Gamma$ generated by Δ is given by the set $\bar{\Delta}$ of all finite sums of the form $\Sigma(\gamma_i+d_i\delta_i-\gamma_i)$, where $\gamma_i\in\Gamma$, $d_i\in D$ and $\delta_i\in\Delta$.

Proof. The set of all finite sums of the form $\Sigma d_i\delta_i$ is a subgroup $\underline{\Delta}$ of $(\Gamma,+)$. $\bar{\Delta}$ is then just the usual normal closure of $\underline{\Delta}$ in $(\Gamma,+)$.
To see that $\bar{\Delta} \trianglelefteq_N \Gamma$, consider $n(\bar{\delta}+\gamma)-n\gamma$, decompose n as $n = \Sigma\sigma_i d_i$ and $\bar{\delta}$ as $\bar{\delta} = \Sigma(\gamma_i+d_i'\delta_i-\gamma_i)$ and proceed as usual. (The next result shows that it suffices to show that $N\bar{\Delta}\subseteq\bar{\Delta}$.)

b) SOME AXIOMATICS

6.6 PROPOSITION Let N be dg. by D and Γ be an N-group.

(a) If Δ is a normal subgroup of $(\Gamma,+)$ then
$\Delta \trianglelefteq_N \Gamma \iff \Delta \leq_N \Gamma$.
(This is one step towards the situation in rings, since the ideals of ${}_N\Gamma$ are just the normal N-subgroups.)

(b) N^2 is abelian $\iff$ N is distributive. +)

(c) N is abelian $\iff$ N is a ring.

(d) If $N \in \mathcal{N}_1$ then N is distributive $\iff$ N is abelian $\iff$ N is a ring.

Proof. (a) If $n = \sum_{i=1}^{k} \sigma_i d_i \in N$, $\gamma \in \Gamma$ and $\delta \in \Delta \leq_N \Gamma$ then

$$n(\delta+\gamma)-n\gamma = \sum_{i=1}^{k} \sigma_i d_i(\delta+\gamma) - \sum_{i=1}^{k} \sigma_i d_i \gamma = \sigma_1 d_1(\delta+\gamma)+$$
$$+\ldots+\sigma_k d_k(\delta+\gamma)-\sigma_k d_k \gamma-\ldots-\sigma_1 d_1 \gamma.$$

Since $d_k(\delta+\gamma)-d_k\gamma = d_k\delta+d_k\gamma-d_k\gamma = d_k\delta \in \Delta$ and (using 6.4) $(-d_k)(\delta+\gamma)-(-d_k)\gamma = (-d_k)\gamma+(-d_k)\delta-(-d_k)\gamma =$
$= -d_k\gamma-d_k\delta+d_k\gamma \in \Delta$, we see that in any case
$\sigma_k d_k(\delta+\gamma)-\sigma_k d_k \gamma \in \Delta$.
Proceeding in this way yields $\Delta \trianglelefteq_N \Gamma$.
The converse follows from 1.34(b) and 6.4(b).

(b) $\Rightarrow$: If N^2 is abelian then for $n,n',n'' \in N$, $n = \Sigma\sigma_i d_i$ we get $n(n'+n'') = \Sigma\sigma_i d_i(n'+n'') =$
$= \Sigma\sigma_i(d_i n'+d_i n'') = \Sigma\sigma_i d_i n'+\Sigma\sigma_i d_i n'' = nn'+nn''$.
$\Leftarrow$: Conversely, the proof of 1.107(c) shows that $N = N_d$ implies N^2 to be abelian.

(c) $\Rightarrow$: If N is abelian, the same applies to N^2. So N is distributive and therefore a ring.

$\Leftarrow$: trivial.

+) "N^2 abelian" stands for "$\forall\, a,b,c,d \in N$: $ab + cd = cd + ab$".

(d) follows from (b) and (c).

The next result examines the rôle of identities in dgnr.'s.

<u>6.7 THEOREM</u> (Ligh (1)). Consider the dgnr. (N,D).

(a) If D contains a left (right, two-sided) identity e then e serves as the same for N.

(b) If N contains exactly one left (or right) identity e then e is two-sided.

<u>Proof</u>. (a) If e is a left identity of D then

$\forall\ n = \Sigma\sigma_i d_i \varepsilon N$: $en = e\Sigma\sigma_i d_i = \Sigma e(\sigma_i d_i) = \Sigma\sigma_i(ed_i) =$
$= \Sigma\sigma_i d_i = n$, and similar for right identities.

(b) Assume that e is the only left identity of N. Then

$\forall\ n\varepsilon N\ \forall\ x\varepsilon N$: $(ne-n+e)x = nex-nx+ex = nx-nx+x = x$,

hence $\forall\ n\varepsilon N$: $ne-n+e = e$. Therefore $\forall\ n\varepsilon N$: $ne = n$ and e is two-sided.

Suppose now that e is the unique right identity of N. Again, take n and $x = \Sigma\sigma_i d_i$ arbitrary εN.
$x(en-n+e) = \Sigma\sigma_i d_i(en-n+e) = \Sigma\sigma_i d_i e = (\Sigma\sigma_i d_i)e = xe$.
As above, $en-n+e = e$, so e is again two-sided.

<u>6.8 REMARKS</u>

(a) Observe that 6.7(b) holds for general near-rings in the "left-case". But see Ligh (1) for examples that in all other cases 6.7 does not hold for general near-rings.

(b) See Ligh (12) for a proof of "A finite dgnr. N is commutative <=> all zero divisors are central". This does not hold in the infinite case.

(c) Ligh (10) gives characterizations of all dgnr.'s N with $(N,+) = S_n$ $(n \geq 5)$ or $(N,+)$ a dihedral group D_{2p} $(p\varepsilon\mathbb{P}\setminus\{2\})$.

c) CONSTRUCTIONS OF DISTRIBUTIVELY GENERATED NEAR-RINGS

Dgnr.'s have no particular stench, so it is not quite easy to recognize them among other nr.'s. The next result might help in some cases.

6.9 THEOREM

(a) If $M \leq N$ and M is dg. then N is not dg. in general.
(b) If $M \leq N$ and N is dg. then M is not dg. in general.
(c) Every homomorphic image of a dgnr. is dg. .
(d) Every direct sum of dgnr.'s is a dgnr. .
(e) Every direct summand of a dgnr. is itself dg..

Proof. (a) Take some $N \in \mathfrak{N}$ with $N \neq N_0$. Then N is not dg. by 6.4(b), but $M := \{0\}$ is

(b) See Lyons-Malone (1) for an example of a subnear-ring of $E(S_3)$ which is not dg. .

(c) If $h: N \twoheadrightarrow N'$ is an epimorphism and if N is dg. by D then a routine check shows that $h(D) \subseteq N'_d$ and $h(D)$ generates N'.

(d) Let the near-rings N_i $(i \in I)$ be dg. by $D_i =: \{d_{ij} \mid j \in J_i\}$. Define $\overline{d}_{ij} := (\ldots,0,d_{ij},0,\ldots) \in$ $\in \bigoplus_{i \in I} N_i =: N$, where d_{ij} stands in the i-th component. Put $\{\overline{d}_{ij} \mid i \in I \wedge j \in J_i\} =: D$.
It is easy to show that $D \subseteq N_d$.
If $n = (\ldots,n_i,\ldots) \in N$, decompose each n_i as $n_i = \sum_{j_i} \sigma_{ij_i} d_{ij_i}$, where each $j_i \in J_i$ and the sum is a finite one.
Then $n = \sum_{i \in I} \sum_{j_i} \sigma_{ij_i} \overline{d}_{ij_i}$ (this is again a finite sum). Consequently, D generates $(N,+)$.

(e) follows from (c).

From 6.9(b) we see that the class of all dgnr.'s is no variety. Hence "distributive generation" cannot be defined by "equations" (see (Grätzer)). This brings up the question about the smallest variety which contains all dgnr.'s (the variety "generated" by the dgnr.'s). We state without proof (it uses a lot of universal algebra)

<u>6.10 THEOREM</u> (Meldrum (1)). The following varieties coincide:

(a) The variety generated by all nr.'s of the type $I(\Gamma)$.

(b) The variety generated by all dgnr.'s.

(c) $\mathfrak{N}_0$.

So every zero-symmetric near-ring can be embedded into some "descendant" of a dgnr. (see (Grätzer), pp. 152/153). For finite nr.'s $N \in \mathfrak{N}_0$ we get more:

<u>6.11 THEOREM</u> (Malone (6)). Every finite zero-symmetric near-ring can be embedded into $I(\Gamma) = E(\Gamma) = M_0(\Gamma)$, where Γ is a certain finite non-abelian simple group.

<u>Proof</u>. Embed N into some $M_0(\Gamma)$ (1.88(c)) with $n \geq 3$; embed Γ into some S_n (e.g. via Cayley's theorem). Next, embed S_n into the alternating group A_{n+2} in the following way:

(1) If $\pi \in \mathfrak{P}_n$ is even, let

$$\bar{\pi} := \begin{pmatrix} 1 & 2 & & n & n+1 & n+2 \\ \pi(1) & \pi(2) & \dots & \pi(n) & n+1 & n+2 \end{pmatrix}.$$

(2) If $\pi \in \mathfrak{P}_n$ is odd, let

$$\bar{\pi} := \begin{pmatrix} 1 & 2 & & n & n+1 & n+2 \\ \pi(1) & \pi(2) & \dots & \pi(n) & n+2 & n+1 \end{pmatrix}.$$

The map $h: \begin{matrix} \mathfrak{P}_n \to A_{n+2} \\ \pi \to \bar{\pi} \end{matrix}$ is a group monomorphism, hence an embedding map. A_{n+2} is a finite simple non-abelian group containing properly a homomorphic image of S_n. By 1.99, $N \hookrightarrow M_o(\Gamma) \hookrightarrow M_o(A_{n+2})$. We will see in 7.46 that $M_o(A_{n+2}) = E(A_{n+2}) = I(A_{n+2})$.

<u>6.12 COROLLARY</u> (Malone (6)). Every finite $N \in \mathcal{N}_0$ can be embedded into a finite dg. non-ring with identity.

Several questions remain open:

Is every zerosymmetric near-ring embeddable into some dgnr.?

Is every dgnr. embeddable into some dgnr. with identity ?

Is every $E(\Gamma)$ embeddable into some $I(\Gamma)$?

Is every dgnr. embeddable into some $E(\Gamma)$?

And so on. Cf. also Heatherly-Malone (2) and 1.90.

<u>6.13 REMARK</u> In (1) and (3), H. Lausch developed an extension theory (via homological algebra) for dgnr.'s.

d) DISTRIBUTIVELY GENERATED NEAR-RINGS WITH FINITENESS CONDITIONS

<u>6.14 THEOREM</u> (Ligh (3)). Let $N \neq \{0\}$ be a dgnr. with DCC on monogenic N-subgroups. Then

(a) $N \in \mathcal{N}_1 \iff N^*$ contains an element which is no divisor of zero.

(b) $\forall\, n \in N^*: (N \in \mathcal{N}_1$ and $n \in N$ is invertible$) \iff n$ is no zero divisor.

<u>Proof</u>. (a) "$\Rightarrow$" is clear. So assume that $x\ (\neq 0)$ is not a divisor of zero. Now $Nx \supseteq Nx^2 \supseteq \ldots$. Therefore $\exists\, k \in \mathbb{N}: Nx^k = Nx^{k+1} = \ldots$. This implies that $\exists\, e \in N: x \cdot x^k = e \cdot x^{k+1}$. So $(x-ex)x^k = 0$. Hence $x-ex = 0$ and we get $ex = x$.

Also, $(xe-x)x = 0$ and thus $xe = x$.
So $\forall$ $m \in N$: $(me-m)x = 0$ whence $me = m$.
Now take some arbitrary $n \in N$. Decompose x as
$x = \Sigma\sigma_i d_i$. Then $x(en-n) = \Sigma\sigma_i d_i(en-n) =$
$= \Sigma\sigma_i(d_i n - d_i n) = 0$, implying that $en = n$.

(b) "$\Rightarrow$" is clear again.
Let $n \neq 0$ be no zero divisor. Then $N \in \mathfrak{n}_1$ by (a).
As in (a), $\exists$ $k \in \mathbb{N}$: $Nn^k = Nn^{k+1} = \ldots$.
So $\exists$ $m \in N$: $n^k = 1n^k = m \cdot n^{k+1}$. This implies that
$(1-mn)n^k = 0$, so $1 = mn$. Also, $(nm-1)n = 0$,
so $nm = 1$ and n is invertible.

6.15 REMARK (Ligh (13)). If N is a finite dgnr. then $(N,+)$ is a perfect group (i.e. N coincides with its commutator subgroup).

There are several connections between chain conditions, solvability of $(N,+)$ and "weak distributivity" (see Fröhlich (1), Def. 4.3.1). We state without proof the following collection of results (see also Beidleman (11)).

6.16 THEOREM Let N be a dgnr.

(a) (Fröhlich (1)). If $(N,+)$ is solvable then N is weakly distributive. If $N^2 = N$, the converse also holds.

(b) (Beidleman (4)). If N is finite and if ${}_N\Gamma \in {}_N\mathfrak{g}$ then $(\Gamma,+)$ is solvable iff ${}_N\Gamma$ is solvable (i.e. Γ has a normal sequence (2.37) with abelian quotients).

(c) (Beidleman (4)). If $(N,+)$ is solvable and N has the DCCN then every maximal left ideal is modular and contains the commutator subgroup of $(N,+)$. $\mathfrak{J}_2(N)$ is nilpotent and $N/\mathfrak{J}_2(N)$ is a ring. Also, N has a certain kind of ACC.

(d) (Ligh (3)). If $(N,+)$ is solvable such that not all elements are divisors of zero. Then the DCCL implies the ACCL.

e) "FREE" DISTRIBUTIVELY GENERATED NEAR-RINGS

Since the dgnr.'s do not form a variety, there is no guarantee for the existence of "free dgnr.'s". Moreover, this concept does not seem to be appropriate for this class of near-rings. We are now going to define a similar concept. First of all we need a "refined" version of homomorphisms between dgnr.'s.

<u>6.17 DEFINITION</u> Let (N,D), (N',D') be dgnr.'s. A homomorphism $h: N \to N'$ is called an <u>(N,D)-(N',D')-homomorphism</u> if $h(D) \subseteq D'$.

<u>6.18 EXAMPLE</u> Each dgnr.-homomorphism $N \to N'$ is an (N,N_d)-(N',N'_d)-homomorphism.

<u>6.19 PROPOSITION</u> (Fröhlich (2)). Let (N,D), (N',D') be dg. and let $h: (N,+) \to (N',+)$ be a group homomorphism and a semigroup homomorphism $(D,\cdot) \to (D',\cdot)$. Then h is an (N,D)-(N',D')-homomorphism.

<u>Proof</u>. It only remains to show that $\forall\, n,n' \in N$: $h(nn') = h(n)h(n')$.

Let $n = \Sigma\sigma_i d_i$, $n' = \Sigma\sigma'_j d'_j$. Then, using 6.4(d),

$$h(nn') = h(\sum_i \sigma_i(\sum_j \sigma'_j d_i d'_j)) = \sum_i \sigma_i(\sum_j \sigma'_j h(d_i d'_j)) =$$

$$= \sum_i \sigma_i(\sum_j \sigma'_j h(d_i)h(d'_j)) = (\sum_i \sigma_i h(d_i))\cdot(\sum_j \sigma'_j h(d'_j)) =$$

$$= h(n)h(n').$$

We are now going to define something like a "free near-ring dg. by a given semigroup $(D,\cdot)$".
We use a slight modification of a method due to Fröhlich (4) and Meldrum (2). Cf. also Zeamer (1).

<u>6.20 DEFINITION</u> Let $(D,\cdot)$ be a semigroup and $\mathfrak{V}$ a variety of groups. Denote by $(F_{D,\mathfrak{V}},+)$ the free group in $\mathfrak{V}$ on (the set) D.

$F_{D,\mathcal{V}}$ consists of all finite sums $\Sigma\sigma_i d_i$, where equality is determined by $\mathcal{V}$. If e.g. $\mathcal{V} = \mathcal{G}$, then "equality" is "formal equality".

Defining $(\Sigma\sigma_i d_i)\cdot(\Sigma\sigma'_j d'_j) := \sum_i \sigma_i(\sum_j \sigma'_j d_i d'_j)$ yields

<u>6.21 THEOREM</u>

(a) $\cdot$ is well-defined.

(b) $(F_{D,\mathcal{V}},+,\cdot) =: F$ is a nr., dg. by D, whose additive group belongs to $\mathcal{V}$.

(c) For every dgnr. (N',D'), every semigroup homomorphism $D \to D'$ can uniquely be extended to an (F,D)-(N',D')-homomorphism.

(d) Every dgnr. (N,D) is a (F,D)-(N,D)-homomorphic image of (F,D).

<u>Proof</u>. (a): holds by the definition of equality via laws in $\mathcal{V}$.

(b): By a routine but somewhat nasty calculation one sees that $(F_{D,\mathcal{V}},+,\cdot)$ is a near-ring. By construction, F is free over D in $\mathcal{V}$, so $(F,+)\in\mathcal{V}$ and D generates $(F,+)$.

(c): By definition, every map $f:D \to D'$ can uniquely be extended to a homomorphism $h:(F,+) \to (N,+)$. If f is moreover a semigroup homomorphism, h is an (F,D)-(N',D')-homomorphism by 6.19.

(d): Considering the diagram (ι is the inclusion map)

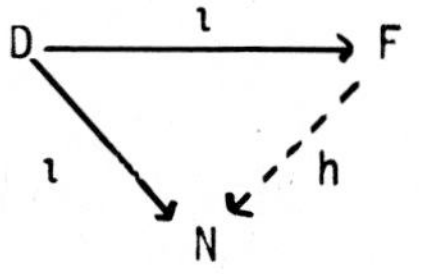

and remembering group theory (or making a routine diagram argument) gives the information that h is a group-epimorphism. Now $h/D = id_D$, whence h is a (F,D)-(N,D)-epimorphism by 6.19.

f) D-GROUPS AND (N,D)-GROUPS

Like nr. homomorphisms of dgnr.'s, the concept of N-groups can be "refined" for a dgnr. (N,D): we want the elements of D to "distribute over Γ".

6.22 DEFINITION Let (N,D) be a dgnr. . ${}_N\Gamma \in {}_N\mathcal{G}$ is called an (N,D)-group if $\forall\ \gamma_1,\gamma_2 \in \Gamma\ \ \forall\ d \in D$: $d(\gamma_1+\gamma_2) = d\gamma_1 + d\gamma_2$.

6.23 DEFINITION Let $(D,\cdot)$ be a semigroup and $(\Gamma,+)$ a group. Γ is called a D-group if a multiplication $\cdot : D \times \Gamma \longrightarrow \Gamma$, $(d,\gamma) \to d\gamma$ is defined with $\forall\ \gamma_1,\gamma_2 \in \Gamma\ \ \forall\ d \in D$: $d(\gamma_1+\gamma_2) = d\gamma_1 + d\gamma_2$.

6.24 REMARK So if (N,D) is dg. then ${}_N\Gamma$ is an (N,D)-group iff Γ is a D-group (w.r.t. the restricted multiplication of ${}_N\Gamma$).

Now let $(D,\cdot)$ be a semigroup and $F_{D,\mathfrak{V}}$ be the "free nr. dg. by D in $\mathfrak{V}$ " as in 6.21, where $\mathfrak{V}$ is some variety in $\mathcal{G}$.

6.25 THEOREM Every D-group $\Gamma \in \mathfrak{V}$ is an $(F_{D,\mathfrak{V}},D)$-group.

Proof. If $\Sigma\sigma_i d_i \in F_{D,\mathfrak{V}}$ and $\gamma \in \Gamma$, define $(\Sigma\sigma_i d_i)\gamma :=$
$= \Sigma\sigma_i(d_i\gamma)$.
Again this is well defined and checking the $(F_{D,\mathfrak{V}},D)$-group axioms creates no problem.

Again, let $\mathfrak{V}$ be a variety of groups and (N,D) a dgnr. We consider the class ${}_{(N,D),\mathfrak{V}}\mathcal{G}$ of all $\Gamma \in \mathfrak{V}$ which are (N,D)-groups.
Let Ω be the family of operations

$$(+,0,-) \cup (\omega_n)_{n \in N} \quad \text{of type}$$
$$(2,0,1) \cup (1)_{n \in N} .$$

Let $\mathcal{K}$ be the class of (universal) algebras of this type.

Let $\overline{\mathcal{G}}$ be the variety determined by all laws which define, for $\Gamma \varepsilon K$, $(\Gamma,+,0,-)$ to be a group $\varepsilon \mathcal{V}$ and by all laws

$$(\omega_{n+n'}(x),\ \omega_n(x)+\omega_{n'}(x)) \qquad (n,n' \varepsilon N)$$

$$(\omega_n(\omega_{n'}(x)),\ \omega_{nn'}(x)) \qquad (n,n' \varepsilon N)$$

$$(\omega_d(x_i+x_j),\ \omega_d(x_i)+\omega_d(x_j)) \qquad (d \varepsilon D).$$

Then clearly

<u>6.26 THEOREM</u> $\overline{\mathcal{G}} = {}_{(N,D),\mathcal{V}}\mathcal{G}$; so the latter class is a variety.

From universal algebra we now get

<u>6.27 COROLLARIES</u> There exist all free (N,D)-groups; they are unique up to (N,D)-isomorphisms; each (N,D)-group is the (N,D)-(N,D)-homomorphic image of a free (N,D)-group.

<u>6.28 REMARKS</u>

(a) Meldrum (2) used these "free nr.'s dg. by D in $\mathcal{V}$" and the free (N,D)-groups in a suitable non-abelian variety $\mathcal{V}$ to show that not every dgnr. (N,D) has a faithful (N,D)-group (not even in the finite case). Therefore not every dgnr. (even not every finite dgnr.) can be embedded into some $E(\Gamma)$ in such a way that all $d \varepsilon D$ remain distributive on Γ (= become endomorphisms on Γ).
Observe that we know from 6.11 that every finite dgnr. can be embedded into some $E(\Gamma)$, if one does not insist that all $d \varepsilon D$ remain distributive.
Meldrum also constructed in (2) "nearest" dgnr.'s $(\overline{N},D)$, $(\underline{N},D)$ with faithful $(\overline{N},D)$-($(\underline{N},D)$-)groups such that (N,D) is a $(\overline{N},D)$-(N,D)-homomorphic image of $(\overline{N},D)$ and $(\underline{N},D)$ is a (N,D)-$(\underline{N},D)$-homomorphic image of (N,D).
Moreover he considered the "Dorroh-type" adjunction of an identity 1 to a dgnr. (N,D) (one has to adjoin 1 to D) (cf. (Kertész), Th. 3.13).

(b) For more information on (N,D)-groups see Fröhlich (2), (4). In (4), Fröhlich described free sums and products, orthogonal sums, free bases and projectivity in the case of (N,D)-groups. It turns out that the situation is similar to the ring (-module) case.

(c) Fröhlich also studied categories of N- and (N,D)-groups in (5) and developed a "non-abelian homological algebra" via these groups in (6) - (8).

g) STRUCTURE THEORY

We start with a result on generators in N and N/I.

6.29 THEOREM ((Gaschütz), Lausch (4)). Let N ($_N\Gamma$) be fg., N a dgnr. and I (Δ) be a finite ideal. Moreover, let N/I (Γ/Δ) be the N-subgroup generated by $\{\bar{e}_1,\dots,\bar{e}_k\}$. Then $\forall\ i\in\{1,\dots,k\}\ \exists e_i\in\bar{e}_i$: $\{e_1,\dots,e_k\}$ generates the N-subgroup N (Γ).

Proof. As in (Gaschütz) (where it is proved for groups; this proof carries over to groups with operators - see Lausch (3)).

This result can be used to prove 6.31:

6.30 DEFINITION If $N\in\mathfrak{N}_1$ then $I(N) := \{n\in N \mid n$ is invertible in $(N,\cdot)\}$ denotes the "group kernel" of $(N,\cdot)$.+)

6.31 THEOREM (Lausch (3), Lausch-Nöbauer (1), Scott (1)). Let $N\in\mathfrak{N}_1$ be a finite dgnr. and let $h: N \to \bar{N}$ be a nr.-homomorphism. Then

$$h(I(N)) = I(h(N)).$$

+) The elements of I(N) are also called the "units" of N.

Proof. (a) If $i \in I(N)$, then $\exists\ j \in N$: $ij = ji = 1$. Hence $h(i)h(j) = h(j)h(i) = h(1)$, so $h(i) \in I(h(N))$.

(b) Conversely, if $\bar{i} \in I(h(N))$ then $\{\bar{i}\}$ generates the N-subgroup $h(N)$: take $h(n) = \sum_i \sigma_i h(d_i) \in h(N)$

and $\bar{j} = \sum_\ell \sigma'_\ell h(d'_\ell)$ with $\bar{j} \cdot \bar{i} = h(1)$. Then $(\sum_i \sigma_i \sum_\ell \sigma'_\ell h(d_i) h(d'_\ell))\bar{i} = h(n) \cdot \bar{j} \cdot \bar{i} = h(n)$.

So by 6.29 there is some $i \in I$ with $h(i) = \bar{i}$ and such that the N-subgroup generated by i equals N. So there is some $j \in N$ with $ji = 1$. Hence by 1.113, i is invertible.

6.32 REMARK See Lausch (4) for some more general versions of 6.31.

Next, we visit primitive dgnr.'s. with DCCN and get

6.33 THEOREM (Laxton (2)). Let $N \subseteq M(\Gamma)$ be a finite dg. near-ring with a left identity. Then the following conditions are equivalent:

(a) N is 1-primitive on Γ.
(b) N is 2-primitive on Γ.
(c) N is simple.
(d) $N = M_0(\Gamma)$ and moreover Γ is a finite, non-abelian, invariantly simple group.

Proof. (a) <=> (b) <=> (c) follows from 4.47(a).

(b) => (d): Assume that N is 2-primitive on Γ. By 4.6(b), N has an identity. By 4.60, $N = M_{G0}(\Gamma)$. If Γ is abelian, N is abelian by 1.49, so by 6.6(c) N is a ring. Hence Γ is non-abelian. Since N is finite, the same applies to Γ. Γ is monogenic, so $\exists\ \gamma_0 \in \Gamma$: $N/_{(0:\gamma_0)} \cong_N \Gamma$ by 3.4(e). So every $d \in N_d$ is an endomorphism of Γ. Since ${}_N\Gamma$ is N-simple, it cannot contain a non-trivial subgroup invariant under all $d \in N_d$, whence Γ shows up to be invariantly simple.

Now $Aut_N(\Gamma)$ is finite and fixed-point-free, so it consists either of $\{id\}$ alone or contains a fixed-point-free automorphism of prime order. The paper (Thompson) tells us that $(\Gamma,+)$ is nilpotent. But Γ is invariantly simple and therefore abelian, a contradiction. So $Aut_N(\Gamma) = \{id\}$ and $M_{G^0}(\Gamma) = M_0(\Gamma)$.

(d) => (b): If $N = M_0(\Gamma)$ then N is 2-primitive on Γ by 4.52(b).

This theorem has some interesting conclusions (see 7.46).

<u>6.34 THEOREM</u> If $I \trianglelefteq N$, N a dgnr., then $\mathfrak{J}_2(I) = \mathfrak{J}_2(N) \cap I$.

<u>Proof</u>. Recall the situation and terminology of 3.30. If $NL \nsubseteq L$ then $\exists\ \ell_0 \in L$: $N\ell_0 \nsubseteq L$. $N\ell_0 \leq_I I$. L is a maximal I-subgroup of I, hence $L + N\ell_0 = I$. Let $e \in I$ be a right identity modulo L in I. $\exists\ \ell \in L\ \exists\ n \in N$: $e = \ell + n\ell_0$. Since $\forall\ i \in N$: $i - ie \in L$ we get from $\forall\ i \in I\ \exists\ \ell_i \in L$: $ie = i(\ell + n\ell_0) = in\ell_0 + \ell_i \in L + L = L$ the contradiction $L = I$. Hence $NL \subseteq L$. Proceed as in 6.6(a) to see that $\forall n \in N\ \forall i \in I\ \forall \ell \in L$: $n(i+\ell) - ni \in L$. Hence $n(i+L) := ni+L$ is well-defined. Now go on as in 3.30 and 5.18.

<u>6.35 REMARKS</u>

(a) Surprisingly (or unfortunately), 6.6(a) does not force the various radicals of a dgnr. to coincide (not even for finite dgnr.'s). See several papers of Laxton and Beidleman. Also, $\mathfrak{J}_2(N)$ is not necessary nil in this case.

(b) For dgnr.'s $N \in \mathfrak{N}_1$, Beidleman (8) defined "<u>strictly primitive</u>" ideals as 2-primitive maximal ideals. The intersection of these ones contains $\mathfrak{J}_2(N)$ and equals $\mathfrak{J}_2(N)$ in the case of DCCN (this follows from 4.47(b)).

(c) Laxton (3) contains an example of a finite dgnr. N with the property that $\mathfrak{J}_{1/2}(N)$ is no ideal, while N has nilpotent left ideals (but of course no nilpotent ideal).

(d) Call ${}_N\Gamma$ <u>tame</u> if $N \in \mathcal{N}_0$ and every N-subgroup of Γ is an ideal (cf. 6.6(a)). N is <u>tame</u> if N has a faithful tame N-group. $E(\Gamma)$, $A(\Gamma)$, $I(\Gamma)$ and all 2-primitive near-rings are tame. Tame nr.'s have the nice property that $\mathcal{J}_2(N)$ is nilpotent (cf. 5.48).
The sum of all minimal left ideals (the "<u>socle</u>" of N) is an ideal if N has the DCCL.
If N is tame and $\in \mathcal{N}_1$ then the DCCL implies the DCCN and ACCL. In this case and if ${}_N\Gamma$ is tame then every $\Delta \trianglelefteq_N \Gamma$ is f.g. .
All of that and more results on tame near-rings are due to Scott (1). See also 9.74.

(e) See Meldrum (4) for a generalization of group rings to dgnr.'s.

(f) See Tharmaratnam (1), (2) and (3) for "topological dgnr.'s": a topological nr. N (def. as usual - see Beidleman-Cox (1)) is called a <u>topological dgnr.</u> if N_d generates N topologically.
If the topological nr. N is a dgnr. then N is a topological dgnr., but the converse does not hold in general.
Tharmaratnam also described topological (N,D)-groups and the structure of topological dgnr.'s, especially that of a 2-primitive complete topological dgnr..

(g) See Laxton (4) for the behaviour of prime ideals in dgnr.'s.

(h) Plotkin (1), (2) transferres the concept of a dgnr. to universal algebra.

(i) See also §7 c).

§ 7 TRANSFORMATION NEAR-RINGS

This chapter contains results on near-rings of group mappings (the "elements of near-ring-theory" of 4.62) and of near-rings which are related to these (§7 d)). We will mainly be concerned with the ideal structure of these classes of near-rings.

We start with $M_H^o(\Gamma) := M_{H\cup\{\bar{o}\}}(\Gamma) = (M_H(\Gamma))_o$, where H is some fixed-point-free group of automorphisms of the (additively written) group Γ. $M_H^o(\Gamma)$ is shown to fulfill all conditions of 2.50 iff H has finitely many orbits on Γ. In this case, $M_H^o(\Gamma)$ is simple and the finite topology on $M_H^o(\Gamma)$ is discrete. We also answer the question, under which conditions $M_{H_1}^o(\Gamma_1)$ and $M_{H_2}^o(\Gamma_2)$ are isomorphic, using semi-linear transformations as in ring theory. The automorphism group of the $M_H^o(\Gamma)$-group Γ is just H itself.

Turning to $M_o(\Gamma)$ in b) we show that for $M_o(\Gamma)$ the following are equivalent: all conditions of 2.50, ACCL, DCCL, finite generation and finiteness of Γ. All minimal left ideals of $M_o(\Gamma)$ are shown to be the $(o:\Gamma\setminus\{\gamma\})$ for $\gamma\in\Gamma^*$. The maximal ones are all $(o:\gamma)$ $(\gamma\in\Gamma^*)$ and some others, which are less easy to characterize. Concerning ideals we show that $M_o(\Gamma)$ and (if $|\Gamma| \neq 2$) $M(\Gamma)$ are simple near-rings. There are no subnear-rings strictly between $M_o(\Gamma)$ and $M(\Gamma)$.

In c) we study mainly $E(\Gamma)$. $E(\Gamma)$ is 2-primitive on Γ iff Γ is invariantly simple. In this case, $E(\Gamma) = M_o(\Gamma)$. Similar results are obtained for $A(\Gamma)$ and $I(\Gamma)$. $E(\Gamma)$ has all conditions of 2.50 if Γ is the direct sum of finitely many minimal fully invariant subgroups. $E(\Gamma)$ is simple iff Γ is invariantly simple. Aut $I(\Gamma) \cong \Gamma$.

Finally we study near-rings of polynomials $R[x]$ or $\Gamma[x]$ over a commutative ring R with unity or a group Γ and their associated near-rings $\overline{R}[x]$, $\overline{\Gamma}[x]$ of polynomial functions. We show that

$\overline{R}[x] = M(R)$ iff R is a finite field. If F is a field, $F[x]$ is simple iff F is infinite. If F is finite, but $\neq \mathbb{Z}_2$, $F[x]$ contains exactly one maximal ideal: $\{p \in F[x] \mid p$ induces the zero function on F$\}$. If $F = \mathbb{Z}_2$, there are exactly 2 maximal ideals. If F is finite but char $F \neq 2$, each ideal of $F[x]$ is a ring-ideal of $F[x]$ and hence quite well-known, provided that char $F \neq 2$.
$\overline{R}[x]$ is simple iff R is a field $\neq \mathbb{Z}_2$ and $\Gamma[x]$ is simple iff $|\Gamma|>1$. $\overline{\Gamma}[x] = M(\Gamma)$ holds iff $\Gamma = \mathbb{Z}_2$ or Γ is a finite, non-abelian simple group.

a) $M_H^o(\Gamma)$

Now we are going to decompose $M_H(\Gamma)$, where H is some fixed-point-free automorphism group of $(\Gamma,+)$.
In contrast to 3.43, we first decompose the identity and then get a decomposition of $M_H(\Gamma)$. Before doing so, we have to fix some notation.

7.1 NOTATION Throughout this section 7a) let Γ be a non-zero group and H a fixed-point-free automorphism group of Γ. Let $\Gamma = \{o\} \cup \bigcup_{i \in I} B_i$ be a partition of Γ into a disjoint union of orbits of Γ under H.
Denote (for $i \in I$) by e_i the uniquely determined map (4.28(a)) of $M_{H \cup \{\bar{o}\}}(\Gamma)$ with

$$e_i(\gamma) = \begin{cases} \gamma & \text{for } \gamma \in B_i \\ o & \text{for } \gamma \notin B_i \end{cases}$$

(so e_i is like the identity in B_i and o elsewhere). Moreover, we abbreviate sometimes $M_{H \cup \{\bar{o}\}}(\Gamma)$ by $M_H^o(\Gamma)$ or simply by M.

<u>7.2 THEOREM</u> (Betsch (6)). As promised, let $M_H^o(\Gamma) =: M$.

(a) $\{e_i \mid i \in I\}$ is a set of orthogonal idempotents.

(b) All $Me_i = \bigcap_{j \neq i} (o:B_j) =: L_i$ are left ideals of type 2 which are M-isomorphic to Γ and fulfill

$$L_i\gamma = \begin{cases} \Gamma & \text{if } \gamma \in B_i \\ \{o\} & \text{if } \gamma \notin B_i \end{cases}$$

; e_i is a right identity for L_i.

(c) $M \cong_M \prod_{i \in I} L_i$.

(d) If $L := \sum_{i \in I} L_i = \sum_{i \in I}^{\bullet} L_i$ then $L = M_H^o(\Gamma)$ iff H has finitely many orbits on Γ; in this case, $1 = \sum_{i \in I} e_i$.

(e) Every non-zero invariant subnear-ring S of M contains L.

<u>Proof</u>. (a) is established by **an** easy computation.

(b) $Me_i\gamma = \begin{cases} \Gamma & \text{if } \gamma \in B_i \\ \{o\} & \text{if } \gamma \notin B_i \end{cases}$ holds because of 4.28(a), whence $Me_i \subseteq \bigcap_{j \neq i} (o:B_j)$.

Conversely if $m \in \bigcap_{j \neq i} (o:B_j)$ then $m = me_i \in Me_i$, so $Me_i = \bigcap_{j \neq i} (o:B_j)$. By 1.43(a), $L_i \trianglelefteq_\ell N$.

The map $f_\gamma : Me_i \to \Gamma$, $me_i \to me_i\gamma$ is an N-epimorphism for $\gamma \in B_i$.

$\operatorname{Ker} f_\gamma = Me_i \cap (o:\gamma)$. But $(o:\gamma) = (o:B_i)$, so $\operatorname{Ker} f_\gamma = \bigcap_{j \in I} (o:B_j) = \{0\}$, and f_γ is unmasked to be an N-isomorphism from L_i to Γ. Since $M_H^o(\Gamma)$ is 2-primitive on Γ by 4.52(b), the same applies to $L_i \cong_M \Gamma$.

(c) is settled by the M-isomorphism $f: M_H^o(\Gamma) \to \prod_{i \in I} L_i$ sending m into $(\ldots, me_i, \ldots)$.

(d) $\prod_{i\in I} L_i = \bigoplus_{i\in I} L_i$ holds iff I is finite. Now apply 2.30.

(e) If $i\in I$ and S as described, $(o:L_i) = (o:\Gamma)$ by 1.45(b). Suppose that $L_i \cap S = \{0\}$. Then $SL_i \subseteq L_i \cap S = \{0\}$, so $S \subseteq (o:L_i) = (o:\Gamma) = \{0\}$, a contradiction. So $L_i \cap S \neq \{0\}$, whence $L_i \cap S = L_i$ by (b), so all $L_i \subseteq S$ and therefore $L \subseteq S$.

Ramakotaiah (7) showed that the L_i's are exactly all minimal left ideals of $M_G^o(\Gamma)$. He also characterized in this paper all maximal left ideals (also in terms of the finite topology in 4.26).
The following result generalizes Theorem 5.7 of Betsch (6) (notation as above).

7.3 COROLLARY The following statements are equivalent:

(a) $M = L$.
(b) $_M M$ fulfills all conditions of 2.50.
(c) H has finitely many orbits on Γ.

Proof: apply 7.2.

7.4 COROLLARY If M fulfills the conditions of 7.3 then M has no non-trivial two-sided invariant subnear-rings. In particular, M is simple.

This follows from 7.2(e) (simplicity can also be derived from 4.46).

7.5 COROLLARY (Betsch (6)). Let the non-ring $N \in \mathcal{N}_0 \cap \mathcal{N}_1$ be 2-primitive on Γ. Then the following conditions are equivalent:

(a) $_N N$ is "finitely completely reducible" (all conditions of 2.50 are valid).
(b) $N \cong M_G^o(\Gamma)$ and G has finitely many orbits on Γ.
(c) $N \cong M_G^o(\Gamma)$ and the finite topology on $M_G^o(\Gamma)$ is discrete.

Proof. 2.50, 4.60(a), 7.3 and 4.29.

There is an intimate connection between the lattices $\mathcal{G}_H(\Gamma)$ of all H-invariant subgroups of Γ and $\mathcal{S}(M) = \{S \leq M \mid SM \subseteq S\}$ (the "right-invariant subnear-rings" of M).
We mention without proof:

7.6 THEOREM (Laxton (2), Betsch (6), §8). If H has $s \in \mathbb{N}$ orbits on Γ^* then the map $f: \mathcal{G}_H(\Gamma) \to \mathcal{S}(M)$, $\Delta \to (\Delta:\Gamma)$ is a lattice isomorphism with $f^{-1}: \mathcal{S}(M) \to \mathcal{G}_H(\Gamma)$ given by $S \to S\Gamma$.
Moreover, $|\mathcal{G}_H(\Gamma)| = |\mathcal{S}(M)| \leq 2^s$.

Holcombe (6) suggested the following

7.7 DEFINITION Every choice $B := \{b_i \mid i \in I\}$ of representatives $b_i \in B_i$ is called a H-base. $\dim_H(\Gamma) := |B|$ is called the H-dimension of Γ.

This comes from the easy-to-prove (cf. 4.28)

7.8 PROPOSITION (Holcombe (6)).

(a) $\forall\, \gamma \in \Gamma^*\ \exists\, i \in I\ \exists\, h \in H: \gamma = h(b_i)$.

(b) Each map $B \to \Gamma$ can be uniquely extended to a map $\in M_H(\Gamma)$.

Holcombe formulated 7.8(b) more generally: "Every map $B \to \Gamma'$, where Γ' is another group on which H operates (Γ' is an "H-group") can be extended to a unique map $\Gamma \to \Gamma'$ which commutes with H".
So Γ is in a kind "free" on B.
H operates on $M_H^0(\Gamma)$ in a natural way. From 7.8(b) we get

7.9 THEOREM (Holcombe (6)). If $\dim_H(\Gamma) = s \in \mathbb{N}$ then

$$\dim_H(M_H(\Gamma)) = (s+1)^s - 1.$$

<u>7.10 REMARK</u> In (6), pp. 92-97, Betsch studied the distributive elements $D := (M_H^o(\Gamma))_d$ of $M_H^o(\Gamma)$ and "monomial matrices" over D (i.e. matrices over D which contain in each column at most one non-zero entry - cf. also Fröhlich (3)). D is shown to be embeddable into the semigroup $(\mathrm{End}\,(N,+),\circ)$ if H has finitely many orbits on Γ. $(D,\cdot)$ is anti-isomorphic to the monoid of all $f\in\mathrm{End}(\Gamma)$ which commute with all $h\in H$.

Now we consider the following problem: when are $M_{H_1}^o(\Gamma_1)$ and $M_{H_2}^o(\Gamma_2)$ isomorphic ?
For rings, this problem is solved in the following way (see (Jacobson) p. 45 and p. 79): If $h: \mathrm{Hom}_{D_1}(V_1,V_1) \to \mathrm{Hom}_{D_2}(V_2,V_2)$ is an homomorphism (V_1,V_2 vector spaces over the division ring rings D_1,D_2, respectively) then h is an isomorphism iff there is some 1-1-semi-linear transformation $t: V_1 \to V_2$ such that $\forall\ \phi\in\mathrm{Hom}_{D_1}(V_1,V_1)$: $h(\phi) = t\phi t^{-1}$.
We follow in some way Jacobson's discussion and start with

<u>7.11 THEOREM</u> (Holcombe (5), Ramakotaiah (6)). If $H_1 \subseteq H_2$ are (as usual) fixed-point-free groups of automorphisms on Γ then $M_{H_1}^o(\Gamma) = M_{H_2}^o(\Gamma) \Longleftrightarrow H_1 = H_2$.

<u>Proof</u>. We only have to show "$\Longrightarrow$".
Suppose that $H_1 \subset H_2$, and take some $h_2 \in H_1 \setminus H_2$. Take $\gamma\in\Gamma^*$ and consider the orbits B_1,B_2 containing γ with respect to H_1,H_2. Then $B_1 = H_1\gamma$ and $B_2 = H_2\gamma$. Clearly $h_2(\gamma)\in B_2$, but $h_2(\gamma)\notin B_1$ (since otherwise $\exists\ h_1\in H_1 \subseteq H_2$: $h_1(\gamma) = h_2(\gamma)$, so $h_1 = h_2$ since H_2 is fixed-point-free, a contradiction).
4.28(a) guarantees the existence of some $m_1\in M_{H_1}^o(\Gamma)$ with $m_1(\gamma) = h_2(\gamma)$ and $\forall\ \delta\notin B_1$: $m_1(\delta) = o$. Hence $m_1(h_2(\gamma)) = o$.
But $o = m_1(h_2(\gamma)) = h_2(m_1(\gamma)) = h_2(h_2(\gamma))$ since $m_1\in M_{H_1}^o(\Gamma) = M_{H_2}^o(\Gamma)$. Thus $h_2(\gamma) = o$, whence $\gamma = o$, a contradiction. Hence $H_1 = H_2$.

7.12 REMARK Observe that $M_{H_1}(\Gamma) = M_{H_2}(\Gamma) \iff M^o_{H_1}(\Gamma) = M^o_{H_2}(\Gamma)$.

7.13 COROLLARY (Ramakotaiah (5)). $\mathrm{Aut}_{M^o_H(\Gamma)}(\Gamma) = H$.

Proof. If $H' := \mathrm{Aut}_{M_H(\Gamma)}(\Gamma)$ then $H \subseteq H'$. H' is by 4.52 shown to be fixed-point-free with $M_H(\Gamma) = M_{H'}(\Gamma)$. So $H = H'$ by 7.11.

Next we consider $M^o_{H_1}(\Gamma_1)$ and $M^o_{H_2}(\Gamma_2)$, where H_i are fixed-point-free on Γ_i $(i = 1,2)$.

7.14 DEFINITION $S \in \mathrm{Hom}(\Gamma_1, \Gamma_2)$ is called a semi-linear homomorphism if $\exists\, s: H_1 \twoheadrightarrow H_2\ \forall \gamma_1 \in \Gamma_1\ \forall h_1 \in H_1$: $S(h_1(\gamma_1)) = s(h_1)(S(\gamma_1))$.

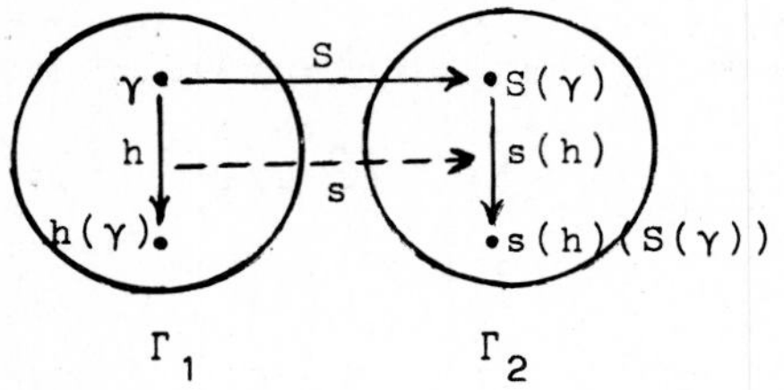

If $S \neq \bar{o}$, s is uniquely determined and called the isomorphism associated with S. We will also speak about the semi-linear Monomorphism (S,s).

7.15 THEOREM (Ramakotaiah (5), for the finite-dimensional (7.7) case also Betsch (6) and Holcombe (6)). A near-ring homomorphism $f: M^o_{H_1}(\Gamma_1) \to M^o_{H_2}(\Gamma_2)$ is an isomorphism $\iff$

$\iff$ there is some semi-linear isomorphism $S: \Gamma_1 \to \Gamma_2$ with $\forall\, m \in M^o_{H_1}(\Gamma)$: $f(m) = S \circ m \circ S^{-1}$.

Proof. We abbreviate $M^o_{H_i}(\Gamma_i)$ by M_i $(i \in \{1,2\})$, and keep this notation for i.

$\Leftarrow$: If (S,s) is a semilinear isomorphism $\Gamma_1 \to \Gamma_2$ then the map
$$f: M_1 \to M_2,\quad m \to SmS^{-1}$$
is an isomorphism:

To see this we first show that f maps M_1 into M_2. Clearly $f(m) \in M_o(\Gamma_2)$. So take $\gamma_2 \in \Gamma_2$ and $h_2 \in H_2$.

$\exists\ \gamma_1\varepsilon\Gamma_1\ \exists\ h_1\varepsilon H_1$: $S(\gamma_1) = \gamma_2 \wedge s(h_1) = h_2$. Also, $h_1 om = moh_1$. So $S(h_1(m(\gamma_1))) = \sigma(h_1)(S(m(\gamma_1))) =$ $= h_2(S(m(\gamma_1))) = h_2(SomoS^{-1}(S(\gamma_1))) = h_2(SomoS^{-1}(\gamma_2))$.

On the other hand, we can compute $S(h_1(m(\gamma_1)))$ in a second way:

$S(h_1(m(\gamma_1))) = S(m(h_1(\gamma_1))) = (SomoS^{-1})(S(h_1(\gamma_1))) =$
$= (SomoS^{-1})(\sigma(h_1)(S(\gamma_1))) = (SomoS^{-1})(h_2(\gamma_2))$.

This holds for each $\gamma_2\varepsilon\Gamma_2$, so we get
$h_2o(SomoS^{-1}) = (SomoS^{-1})oh_2$; hence $SomoS^{-1}\varepsilon M_2$.
It is easy to show that f is an isomorphism.

=>: Assume now that $f: M_1 \to M_2$ is an isomorphism.

(a) M_1 can be considered to be 2-primitive on Γ_2, since
$$M_1\times\Gamma_2 \to\to \Gamma_2 \quad (m_1,\gamma_2) \to f(m_1)\gamma_2$$
does the required job.

(b) We show that there is an isomorphism $S: \Gamma_1 \to \Gamma_2$ with $\forall\ m_1\varepsilon M_1$: $f(m_1) = Som_1oS^{-1}$.
By 7.2(b), M_1 contains a minimal left ideal. By (a), M_1 is 2-primitive on Γ_1 and on Γ_2. 4.56(a) assures that Γ_1 and Γ_2 are M_1-isomorphic (by S, say). So by (a) $\forall\ \gamma_1\varepsilon\Gamma_1\ \forall\ m_1\varepsilon M_1$: $S(m_1(\gamma_1)) =$

$= m_1(S(\gamma_1)) = f(m_1)(S(\gamma_1))$, whence $Som_1 = f(m_1)oS$ or $f(m_1) = Som_1oS^{-1}$.

(c) Now we claim that $\forall\ h_1\varepsilon H_1$: $Soh_1oS^{-1}\varepsilon H_2$.

Clearly $Soh_1oS^{-1}\varepsilon End(\Gamma_2)$.
h_1 commutes with all $m_1\varepsilon M_1$ and by (b) therefore with all $m_2\varepsilon M_2$, so $Soh_1oS^{-1}\varepsilon Aut_{M_2}(\Gamma) = H_2$ (7.13).

(d) Next we observe that $s: H_1 \to H_2$, $h_1 \to Soh_1oS^{-1}$ is an isomorphism, a fact which can be seen by the usual procedures.

(e) Finally, we have to check the semi-linearity

condition 7.14 for (S,s): take some $\gamma_1 \varepsilon \Gamma_1$ and some $h_1 \varepsilon H_1 = Aut_{M_1}(\Gamma)$. Then $S(h_1(\gamma_1)) =$
$= (S \circ h_1 \circ S^{-1})(S(\gamma_1)) = s(h_1)(S(\gamma_1))$.

The proof is now complete.

From that we can deduce interesting results about the automorphism of near-rings of the type $M_H^o(\Gamma)$:

<u>7.16 COROLLARY</u> (Ramakotaiah (5)). If $M = M_H^o(\Gamma)$ and $f \varepsilon End(M)$ then $f \varepsilon Aut\ M \iff$ there exists a semi-linear automorphism S on Γ with $f(m) = S \circ m \circ S^{-1}$ for all $m \varepsilon M$.

This follows from 7.15 by specializing $M_1 = M_2 = M$.

<u>7.17 THEOREM</u> (Ramakotaiah (5)). Let G be the group of semi-linear automorphisms on Γ and $G' := Aut\ M_H^o(\Gamma)$. Then $G/_{G \cap H} \cong G'$.

<u>Proof</u>. Define $\alpha: G \to G'$ as follows: if $S \varepsilon G$, there is some $f \varepsilon Aut\ M_H^o(\Gamma)$ with $\forall\ m \varepsilon M_H^o(\Gamma)$: $f(m) = S \circ m \circ S^{-1}$ (by 7.16). Observe that this f is unique. Put $\alpha(S) := f$.

First we prove that α is a homomorphism. To do this, take $S, T \varepsilon G$. $\alpha(ST) =: g$, $\alpha(S) =: f_1$, $\alpha(T) =: f_2$. Then for all $m \varepsilon M_H^o(\Gamma)$ $g(m) = (ST)m(ST)^{-1} =$
$= STmT^{-1}S^{-1} = Sf_2(m)S^{-1} = f_1(f_2(m))$.

Hence $g = f_1 f_2$, implying that $\alpha(ST) = \alpha(S)\alpha(T)$.

Now α is an epimorphism: if $f \varepsilon G'$ then there is some $S \varepsilon G$ with $\forall\ m \varepsilon M_H^o(\Gamma)$: $S \circ m \circ S^{-1} = f(m)$. Thus $\alpha(S) = f$.

Finally we compute Ker α.

If $S \varepsilon Ker\ \alpha$ then $\forall\ m \varepsilon M_H^o(\Gamma)$: $m = id(m) = SmS^{-1}$.

So $S \varepsilon Aut_{M_H^o(\Gamma)}(\Gamma) = H$ (7.13). Hence $S \varepsilon G \cap H$.

Conversely, each element of $G \cap H$ is in Ker α.

This shows that $G/_{G \cap H} \cong G'$.

Clay (14) determined the group U of units of $M^o_G(\Gamma)$: U is isomorphic to the wreath product of G with the symmetric group on the index set I of the orbits of Γ under G. He also pointed out the intimate connection between U and the general linear groups in linear algebra. He also defined a "determinant function" on U.
See Betsch (8) for a detailed discussion of $M^o_G(\Gamma)$, where G is an arbitrary (not necessarily fixed-point-free) automorphism group on Γ.

b) $M(\Gamma)$ AND $M_0(\Gamma)$

There are a lot of things which we can get by specializing $H = \{id\}$ in the previous section. By 1.13 it is justified to consider primarily $M_0(\Gamma)$. We start by considering left ideals in $M_0(\Gamma)$.

7.18 COROLLARY (Heatherly (1), (4), cf. also Fröhlich (3), Ramakotaiah (7)).

(a) If for $\delta\in\Gamma$, $e_\delta:\Gamma\to\Gamma$ with $e_\delta(\gamma) = \begin{cases}\delta & \text{if } \gamma=\delta\\ o & \text{if } \gamma\neq\delta\end{cases}$

then $\{e_\delta \mid \delta\in\Gamma^*\}$ is a set of orthogonal idempotents.

(b) All $M_0(\Gamma)e_\delta =: L_\delta = (o:\Gamma\setminus\delta)$ are left ideals and $M_0(\Gamma)$-groups of type 2 (hence minimal $M_0(\Gamma)$-subgroups) generated by e_δ and $M_0(\Gamma)$-isomorphic to Γ.

(c) $M_0(\Gamma) \cong \prod_{\delta\in\Gamma^*} L_\delta$.

(d) If $L := \sum^{\bullet}_{\delta\in\Gamma^*} L_\delta$ then $L = M_0(\Gamma)$ iff Γ is finite.

(e) Every non-zero invariant subnear-ring S of $M_0(\Gamma)$ contains L.

Proof: 7.2.

Hence $M_0(\Gamma)$ is a 2-primitive nr. on Γ with identity and a minimal left ideal (see §4d3)). 7.3 and 7.18 give

<u>7.19 COROLLARY</u> (Heatherly (4), M. Johnson (5)). The following are equivalent:

(a) $M_0(\Gamma) = \sum^{\bullet}_{\delta\in\Gamma^*} L_\delta = L$.

(b) $M_0(\Gamma)$ has DCCL.

(c) $M_0(\Gamma)$ has ACCL.

(d) $M_0(\Gamma)$ is completely decomposable into finitely many minimal left ideals.

(e) $M_0(\Gamma)$ has only f.g. left ideals.

(f) $M_0(\Gamma)$ is finite.

(g) Γ is finite.

Clearly (by 7.4) $M_0(\Gamma)$ is simple in this case. However, we will extend this result to the arbitrary case (7.30). But first we examine the left ideals more closely.

<u>7.20 THEOREM</u> (Heatherly (4)) Let L be a left ideal of $M_0(\Gamma)$.

(a) $\forall\ \gamma\in\Gamma$: $L\gamma = \{o\}$ or $L\gamma = \Gamma$.

(b) If $\Delta := \{\delta\in\Gamma \mid L\delta = \Gamma\} \neq \emptyset$ then $\sum_{\delta\in\Delta} L_\delta \subseteq L$.

<u>Proof</u>. (a) If $L\gamma \neq \{o\}$ then $\exists\ \ell\in L$: $\ell\gamma \neq o$. But $\forall\ m\in M_0(\Gamma)$: $m\ell\in L$, whence $\{m\ell\gamma \mid m\in M_0(\Gamma)\} = \Gamma$ and a fortiori $L\gamma = \Gamma$.

(b) It suffices to show that if $L\delta \neq \{o\}$ then $L_\delta \subseteq L$. This trivially holds for $|\Gamma| \leq 2$ since then $|M_0(\Gamma)| \leq 2$. So assume that $|\Gamma| \geq 3$.

Suppose that $\ell_\delta \in L_\delta$ $(\delta\in\Delta)$. Denote $\ell_\delta(\delta) =: \theta$. Choose some $\ell\in L$ with $\ell(\delta) = \theta = \ell_\delta(\delta)$. This is possible by (a).

Take $m,n\in M_0(\Gamma)$ with $m(\gamma) = \begin{cases} \theta & \gamma = \theta \\ o & \gamma \neq \theta \end{cases}$

(so $m = e_\theta$ of 7.18(a)), and $n(\delta) = o$, but for $\gamma \neq \delta$ $n(\gamma) \notin \{\theta, \theta - \ell(\gamma)\}$.

Then $\bar{\ell} := m(n+\ell) - mn \in L$ and for $\gamma \neq \delta$ we get $\bar{\ell}(\gamma) = m(n(\gamma)+\ell(\gamma)) - m(n(\gamma)) = o - o = o = \ell_\delta(\gamma)$, while $\bar{\ell}(\delta) = m(n(\delta)+\ell(\delta)) - m(n(\delta)) = m(\theta) - m(o) = \theta = \ell_\delta(\delta)$.

So $\ell_\delta = \bar{\ell} \in L$ and we are through.

7.21 CORROLLARIES (Heatherly (4), M. Johnson (1), (3)).

(a) The L_δ's ($\delta \in \Gamma^*$) are exactly all minimal left ideals of $M_0(\Gamma)$.

(b) Every left ideal of $M_0(\Gamma)$ which is contained in $L = \sum_{\delta \in \Gamma^*} L_\delta$ is isomorphic to a direct sum of suitable L_δ's.

(c) L cannot be a non-trivial direct summand of $M_0(\Gamma)$.

Proof. (a) is a consequence of 7.20.

(b) follows from (b) and 2.55.

(c) If $L' \trianglelefteq_\ell M_0(\Gamma)$ is such that $L \dot{+} L' = M_0(\Gamma)$ and $L' \neq \{0\}$ then (by 7.20) $\exists\, \delta \in \Gamma^*$: $L_\delta \subseteq L \cap L'$, a contradiction.

Since we have been very successful in determining all minimal left ideals of $M_0(\Gamma)$, we turn to maximal left ideals. We readily get some of them:

7.22 EXAMPLE (Heatherly (1), (4)). For every $\gamma \in \Gamma^*$, $(o:\gamma)$ is a maximal $M_0(\Gamma)$-subgroup (hence also a maximal left ideal) of $M_0(\Gamma)$.

Proof. This holds since by 3.4(a), $N/(o:\gamma) \cong_N L_\gamma \cong_N \Gamma$, where $N = M_0(\Gamma)$. Now apply 3.4(h).

But woe:

<u>7.23 PROPOSITION</u> (Heatherly (4)). If Γ is infinite then L (as in 7.18(d)) is contained in a maximal left ideal of $M_o(\Gamma)$, but not in any $(o:\gamma)$ $(\gamma\varepsilon\Gamma^*)$.

<u>Proof</u>. The first statement is settled by 1.53(a) since $M_o(\Gamma)\varepsilon\mathfrak{N}_1$ and by 7.18(d). Now if $L\subseteq(o:\gamma)$ then $L_\gamma\subseteq(o:\gamma)$ and $e_\gamma(\gamma) = o$, whence $\gamma = o$.

So there are other maximal left ideals beside the $(o:\gamma)$'s, which implies more trouble for us. But, fortunately, M. Johnson has solved this problem. See also Ramakotaiah (7) and (8).

<u>7.24 NOTATION</u> For $m\varepsilon M_o(\Gamma)$ call $\{\gamma\varepsilon\Gamma \mid m(\gamma) = o\} =: Z_m$ (the "<u>zero set of m</u>").

We state without proof

<u>7.25 PROPOSITION</u> (M. Johnson (3), (5)). Let L be a left ideal in $M_o(\Gamma)$. Then

(a) $\ell\varepsilon L \iff (o:Z_\ell)\subseteq L$.

(b) $\forall\ \ell_1,\ell_2\varepsilon L\ \exists\ \ell\varepsilon L$: $Z_{\ell_1}\cap Z_{\ell_2} = Z_\ell$.

<u>7.26 NOTATION</u> Let $\mathcal{L}(\Gamma)$ be for the moment $= \mathcal{L} =$ $= \{L \trianglelefteq_\ell M_o(\Gamma) \mid \forall\ell\varepsilon L$: Z_ℓ is infinite$\}$.

Now we are in a position to characterize the maximal left ideals of $M_o(\Gamma)$.

<u>7.27 THEOREM</u> (M. Johnson (5)). Let L be a left ideal of $M_o(\Gamma)$. L is maximal $\iff$ $(\exists\gamma\varepsilon\Gamma^*$: $L = (o:\gamma))$ v (L is maximal in $\mathcal{L}(\Gamma)$).

<u>Proof</u>. $\Rightarrow$: Suppose that $\forall\ \gamma\varepsilon\Gamma$: $L \neq (o:\gamma)$. Assume moreover that $\exists\ \ell\varepsilon L$: Z_ℓ is finite. For $\gamma\varepsilon\Gamma$ consider e_γ of 7.18(a). Since L is no $(o:\gamma)$, all $e_\gamma\varepsilon L$ by 7.20(b).

Assume that $|Z_\ell| = n \in \mathbb{N}$. We claim that $\exists\, k \in L$: $Z_k = \{o\}$ and prove this by induction on n.

(a) This is trivial for $n = 1$.

(b) Suppose that $n>1$ and the statement holds for $n-1$.

Now $\ell+e_\gamma \in L$ and $Z_{\ell+e_\gamma} = Z_\ell \setminus \{\gamma\}$, so $|Z_{\ell+e_\gamma}| = n-1$.

So $L = M_o(\Gamma)$ by 7.25(a) since $(o:Z_k) = (o:o) = M_o(\Gamma) \subseteq L$. This is a contradiction. Hence $L \in \mathcal{L}(\Gamma)$ and since L is a maximal left ideal, L is maximal in $\mathcal{L}(\Gamma)$.

$\Leftarrow$: In view of 7.22 it suffices to consider maximal elements of $\mathcal{L}$. Let $K \trianglelefteq_\ell M_o(\Gamma)$ properly contain L. Then $\exists\, k \in K$: $|Z_k| = n \in \mathbb{N}$. Again we use induction to show the existence of some $k_1 \in K$ with $Z_{k_1} = \{o\}$, $n = 1$ is trivial again.

If $n>1$, take some $\gamma \in Z_k$, $\gamma \neq o$. Then $e_\gamma \in L_\gamma = (o:\Gamma \setminus \{\gamma\})$ by 7.18(b). $L+(o:\Gamma\setminus\{\gamma\}) \in \mathcal{L}$, so $L+(o:\Gamma\setminus\{\gamma\}) = L$ since L is maximal in $\mathcal{L}$. Hence $e_\gamma \in L$ and so $k+e_\gamma \in K$ with $|Z_{k+e_\gamma}| = n-1$ as before. Hence $K = M_o(\Gamma)$ and L is shown to be a maximal left ideal of $M_o(\Gamma)$.

Since $\mathcal{L}(\Gamma)$ is empty if Γ is finite (see 7.19), we get

7.28 COROLLARY Let Γ be finite. Then

(a) The minimal left ideals of $M_o(\Gamma)$ are exactly the L_γ's. $(\gamma \in \Gamma^*)$.

(b) The maximal left ideals of $M_o(\Gamma)$ are exactly the $(o:\gamma)$'s $(\gamma \in \Gamma^*)$.

(c) The left ideals of $M_o(\Gamma)$ are exactly the $(o:\Delta)$'s $(\Delta \subseteq \Gamma)$.

For in this case $\sum_{\delta\in\Delta} L_\delta = \sum_{\delta\in\Delta} (o:\Gamma\setminus\{\delta\}) = (o:\bigcap_{\delta\in\Delta}(\Gamma\setminus\{\delta\}) = (o:\Delta\setminus\Gamma)$.

We can generalize 7.28(c) easily to the infinite case:

<u>7.29 THEOREM</u> (M. Johnson (3)). Every $L \trianglelefteq_\ell M_0(\Gamma)$ is the sum of annihilator left ideals.

This is a consequence of 7.25, for $L = \sum_{\ell\in L} (o:Z_\ell)$.

Now we extend our result that $M_G(\Gamma)$ is simple if Γ is finite (7.19).

<u>7.30 THEOREM</u> (Berman-Silverman (2), Nöbauer-Philipp (1)). $M_0(\Gamma)$ is simple for every group Γ.

<u>Proof</u>. We may assume that Γ is infinite. Take $I \trianglelefteq M_0(\Gamma)$, $I \neq \{0\}$. If $m\in M_0(\Gamma)$, call $|\{m(\gamma)|\gamma\in\Gamma\}| =: rk(m)$ the <u>rank of m</u>.

(a) First we remark that for each $\gamma\in\Gamma^*$ there is a maximal set $A\subseteq\Gamma$ with respect to the property that $A \cap (A+\gamma) = \emptyset$.
This follows by Zorn's lemma.

(b) The set A of (a) satisfies $|A| = |\Gamma|$. To see this, consider $A \cup (A+\gamma)$.
If $A \cup (A+\gamma) = \Gamma$, there is nothing to prove.
If $\delta\in\Gamma$ is not in $A \cup (A+\gamma)$, consider $A' := A \cup \{\delta\}$.
Since A is maximal, $A' \cap (A'+\gamma) \neq \emptyset$. But
$A' \cap (A'+\gamma) = (A \cup \{\delta\}) \cap ((A+\gamma) \cup \{\delta+\gamma\}) = A \cap \{\delta+\gamma\}$
by the algebra of sets and the assumptions.
So $\exists\ \alpha\in A$: $\alpha = \delta+\gamma$, whence $\delta\in A-\gamma$.
Thus $\Gamma = A \cup (A+\gamma) \cup (A-\gamma)$. Since $|A| = |A+\gamma| =$
$= |A-\gamma|$ and the fact that Γ is infinite we get
$|\Gamma| = |A|$.

(c) Now we show that I contains some i with $rk(i) = |\Gamma|$. Take $\gamma\in\Gamma^*$ and A as in (a) and (b).

Take $i\varepsilon I^*$ and some $\xi\varepsilon\Gamma$ with $L_\xi \subseteq I$ (7.20).
Let $j\varepsilon L_\xi$ be such that $j(\xi) = -\gamma$.

Define $g,f,h\varepsilon M_o(\Gamma)$ by $g(\delta) := \begin{cases}\gamma & \text{for } \delta\neq o\\ o & \text{for } \delta=o\end{cases}$

$f(\delta) := \begin{cases}\delta & \text{for } \delta\varepsilon A\\ o & \text{for } \delta\notin A\end{cases}$ and $h(\delta) := \begin{cases}\xi & \text{for } \delta\neq o\\ o & \text{for } \delta=o\end{cases}$

Then $i := fo(id+g)-foid\varepsilon I$.

$\forall\ \alpha\varepsilon A\setminus\{o\}: i(\alpha) = f(\alpha+g(\alpha))-f(\alpha) = -\alpha$.
Hence $rk(i) \geq |A\setminus\{o\}| = |A| = |\Gamma|$, thus proving that $rk(i) = |\Gamma|$.

(d) Now we prove the theorem.
Let m be arbitrary $\varepsilon M_o(\Gamma)$. Since $rk(m) \leq |\Gamma| = rk(i)$ (i as in (c)) there is some injective map $f: m(\Gamma) \to i(\Gamma)$. For each $\gamma\varepsilon\Gamma$ choose one $\gamma'\varepsilon\Gamma$ with $i(\gamma') = f(m(\gamma))$. Denote this correspondence $\gamma \to \gamma'$ by g. We may assume that $f(o) = o$ and $i(o) = o$.
Define maps r,s as follows:

$r(\delta) := \begin{cases} m(\gamma) & \text{if } \exists\ \gamma\varepsilon\Gamma: \delta = f(m(\gamma)) \neq o\\ o & \text{otherwise}\end{cases}$

$s(\delta) := \begin{cases} g(\gamma) & \text{if } \exists\ \gamma\varepsilon\Gamma: f(m(\delta)) = i(g(\gamma)) \neq o\\ o & \text{otherwise}\end{cases}$

Then for all δ with $\exists\ \gamma\varepsilon\Gamma: f(m(\delta)) = i(g(\gamma)) \neq o$ we get $(roios)(\delta) =$
$= r(i(s(\delta))) =$
$= r(i(g(\gamma))) =$
$= r(i(\gamma')) =$
$= r(f(m(\gamma))) =$
$= m(\gamma) = m(\delta)$
(since $f(m(\gamma)) = f(m(\delta))$).

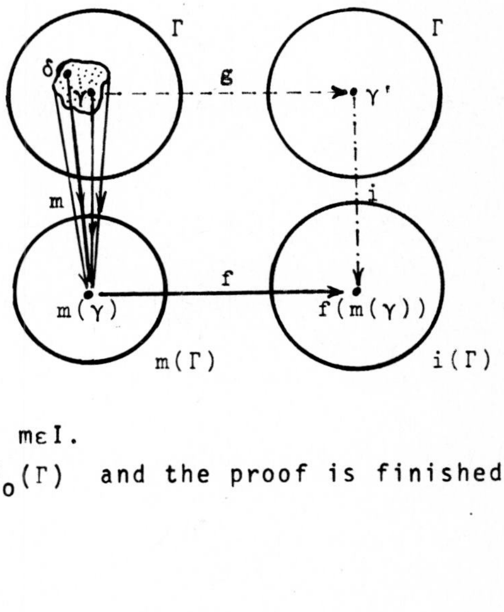

Hence $roios = m$ and $m\varepsilon I$.
This shows that $I = M_o(\Gamma)$ and the proof is finished.

Of course one is lead to the question whether $M(\Gamma)$ is simple or not. We try the simplest example:

7.31 EXAMPLE $M(\mathbb{Z}_2)$ has one non-trivial ideal, namely $M_c(\Gamma)$

But this is the only case that $M(\Gamma)$ is not simple:

7.32 THEOREM (Berman-Silverman (2)).
There are no proper subnear-rings between $M_0(\Gamma)$ and $M(\Gamma)$.

Proof. Let N be a subnear-ring of $M(\Gamma)$ with $N \supset M_0(\Gamma)$. Let n be $\in N \setminus M_0(\Gamma)$. If γ is arbitrary $\in \Gamma$ and if m_γ denotes the map which is constant $= \gamma$ then $\exists\, m_0 \in M_0(\Gamma)$ with $m_0(n(o)) = \gamma$. But $m_0 n =: n' \in N$, and $m_\gamma = -n_0' + n' \in N$ (where $n' = n_0' + n_c'$ as in 1.13), showing that $N = M_c(\Gamma)$.

7.33 THEOREM (Berman-Silverman (2), Nöbauer-Philipp (1)).
$M(\Gamma)$ is simple iff $|\Gamma| \neq 2$.

Proof. Let I be an ideal of $M(\Gamma)$. Then $I_0 := I \cap M_0(\Gamma)$ (2.18) is an ideal in $M_0(\Gamma)$. Hence $I_0 := \{0\}$ or $I_0 = M_0(\Gamma)$. If $I_0 = \{0\}$ then $I \subseteq M_c(\Gamma)$. Examining 1.27 β) shows that $I = \{0\}$ if $|\Gamma| \neq 2$.
If $I_0 = M_0(\Gamma)$ then $I = M_0(\Gamma)$ or $I = M_c(\Gamma)$ by 7.32. Since $M_0(\Gamma)M_c(\Gamma) \subseteq M_c(\Gamma)$, $M_0(\Gamma) \not\trianglelefteq M(\Gamma)$.
It remains $I = M(\Gamma)$.

Other substructures have also been considered. We mention without proof:

7.34 THEOREM (Berman-Silverman (2)). The only two-sided invariant subnear-rings of $M(\Gamma)$ are $M_c(\Gamma) = \{f \in M(\Gamma) \mid rk(f) = 1\}$, $\{f \in M(\Gamma) \mid rk(f) < \aleph_0\}$ and $\{f \in M(\Gamma) \mid rk(f) \leq \aleph\}$ with $\aleph_0 < \aleph \leq |\Gamma|$. The ones of $M_0(\Gamma)$ are just the intersections of the ones above with $M_0(\Gamma)$.

7.35 REMARK See Heatherly (4) for a discussion of the right ideals of $M_0(\Gamma)$.

7.36 COROLLARY By 7.6, the lattices of all subgroups of a finite group Γ and of all right invariant subnear-rings of $M_0(\Gamma)$ are isomorphic (cf. also Heatherly (4)).

7.37 COROLLARY $\mathrm{Aut}_{M_0(\Gamma)}(\Gamma) = \{id\}$.

Proof. 7.13.

The next corollary gives one half of 1.99 once again in a different version.

7.38 COROLLARY $M_0(\Gamma)$ and $M_0(\Delta)$ are isomorphic (say by f) iff there is some isomorphism $S: \Gamma \to \Delta$ with $\forall\ m \in M_0(\Gamma)$: $f \circ m = S \circ m \circ S^{-1}$.

Proof. 7.15.

We now turn to the automorphisms of $M_0(\Gamma)$.

7.39 THEOREM (Ramakotaiah (5)). A homomorphism $f : M_0(\Gamma) \to M_0(\Gamma)$ is an automorphism $\Longleftrightarrow$ there is some automorphism S of $(\Gamma,+)$ with $\forall\ m \in M_0(\Gamma)$: $f \circ m = S \circ m \circ S^{-1}$.
Moreover, $\mathrm{Aut}\,(\Gamma,+) \cong \mathrm{Aut}\,(M_0(\Gamma),+,\cdot)$.

Proof. 7.38 and 7.17 (for $H = \{id\}$).

7.40 REMARKS

(a) See Clark (1) for examples of $M(\Gamma)$ with $\Gamma = \mathbb{Z}_p$, $\Gamma = \mathbb{Z}_{pq}$, $\Gamma = \mathbb{Z}_{p^r}$.

(b) See Blackett (4), (5) and (6) for examples of simple subnear-rings of $M(\mathbb{R})$ and $M(\mathbb{C})$ and for "dense" subnear-rings of $M(\mathbb{R})$.

(c) Beidleman (11) proved for a finite group Γ that Γ is nilpotent (solvable, supersolvable) iff the same applies to $(M_0(\Gamma),+)$.

(d) Malone-McQuarrie (1) showed that if Γ is torsion without elements of order 2 then $(M_0(\Gamma),+)$ is (uniquely) halvable.

(e) M. Johnson (3) characterized the cases when left ideals and normal $M_0(\Gamma)$-subgroups coincide in $M_0(\Gamma)$. We know already that this happens e.g. if Γ is a finite, non-abelian invariantly simple group (6.34 and 6.6(a)). M. Johnson showed that if Γ is finite then left ideals and normal $M_0(\Gamma)$-subgroups coincide in $M_0(\Gamma)$ iff Γ is not abelian.
For infinite groups Γ this happens iff $\exists\ \gamma\in\Gamma$: $[\Gamma : C(\gamma)] = |\Gamma|$, where $C(\gamma)$ is the centralizer of γ.
Or, equivalently: this coinciding happens for some group Γ iff every normal $M_0(\Gamma)$-subgroup of $M_0(\Gamma)$ is the sum of annihilator left ideals of $M_0(\Gamma)$ (7.29).

c) E(Γ), A(Γ) AND I(Γ)

In this section we will examine similar items for the dgnr.'s $E(\Gamma)$, $A(\Gamma)$ and $I(\Gamma)$, generated by all endomorphisms (automorphisms, inner automorphisms, respectively) of Γ, as we did for $M(\Gamma)$ and $M_0(\Gamma)$ (one- and two-sided ideals, simplicity, radicals, automorphisms, etc.). As in b) we will get the best results for the case that Γ is finite.

First we define our objects in question.

7.41 DEFINITION Let Γ be a group. Similar to $E(\Gamma)$ we define $A(\Gamma)$ $(I(\Gamma))$ as the near-rings additively generated by all automorphisms (inner automorphisms) of Γ.

7.42 REMARKS

(a) Let $End(\Gamma)$, $Aut(\Gamma)$ and $Inn(\Gamma)$ denote the semi-groups of all endomorphisms (automorphisms, inner automorphisms) of Γ.

(b) $E(\Gamma)$, $A(\Gamma)$ and $I(\Gamma)$ are dgnr.'s.

(c) Similar to 6.4(d), $A(\Gamma)$ $(I(\Gamma))$ consists of all finite sums $\sum_k \sigma_k \alpha_k$ with $\sigma_k = \pm 1$ and all $\alpha_k \in Aut(\Gamma)$ $(\alpha_k \in Inn(\Gamma))$.

(d) $I(\Gamma) \subseteq A(\Gamma) \subseteq E(\Gamma) \subseteq M_0(\Gamma)$.

6.6(c) gives us

7.43 PROPOSITION If Γ is abelian then $E(\Gamma)$, $A(\Gamma)$ and $I(\Gamma)$ are rings.

Hence we will be interested in non-abelian groups Γ.

7.44 PROPOSITION The $E(\Gamma)$- $(A(\Gamma)$-, $I(\Gamma)$-$)$ subgroups are exactly the fully invariant (characteristic, normal) subgroups of Γ.

Proof. 6.24.

This implies

7.45 COROLLARY

(a) $E(\Gamma)$ is 2-primitive on Γ <=> Γ is invariantly simple.

(b) $A(\Gamma)$ is 2-primitive on Γ <=> Γ is characteristically simple.

(c) $I(\Gamma)$ is 2-primitive on Γ <=> Γ is simple.

Now we characterize in which cases $E(\Gamma)$, $A(\Gamma)$ and $I(\Gamma)$ coincide with $M_0(\Gamma)$ and get as a generalization of results of Johnson (1), Fröhlich (3) and Maxson (14)

<u>7.46 THEOREM</u> Let Γ be a non-abelian, finite group.

(a) $E(\Gamma) = M_o(\Gamma)$ <=> Γ is invariantly simple.

(b) $A(\Gamma) = M_o(\Gamma)$ <=> Γ is characteristically simple.

(c) $I(\Gamma) = M_o(\Gamma)$ <=> Γ is simple.

This result follows from 7.45 and 6.33. Observe that if Γ is invariantly simple then $E(\Gamma)$ is a non-ring: $E(\Gamma)$ is 2-primitive on Γ and can be no ring by 4.8. The same applies to $A(\Gamma)$ and $I(\Gamma)$.

The last theorem can be expressed as a "purely group-theoretic" result:

<u>7.47 THEOREM</u> Let Γ be a finite non-abelian group. Then each map $\Gamma \to \Gamma$ which fixes o can be expressed as a finite sum $\Sigma\sigma_i d_i$ with $d_i \in End(\Gamma)$ $(Aut(\Gamma), Inn(\Gamma))$ <=> Γ is invariantly simple (characteristically simple, simple, respectively).

<u>7.48 COROLLARY</u> Let Γ be as 7.47.

(a) If Γ is characteristically simple then $A(\Gamma) = E(\Gamma) = M_o(\Gamma)$.

(b) If Γ is simple then $I(\Gamma) = A(\Gamma) = E(\Gamma) = M_o(\Gamma)$.

Thus $E(\Gamma)$, $A(\Gamma)$ and $I(\Gamma)$ can be viewed to be characterized by 7b) if Γ is as in the cases of 7.46.
The next step is to consider the case that Γ is a non-abelian group which is the direct sum of fully invariant (characteristic, normal) subgroups. Without taking the hands out of the trouser pockets we deduce from 7.44 the following result:

<u>7.49 COROLLARY</u> Let Γ be the direct sum of minimal fully invariant (characteristic, normal) subgroups. Then Γ is a completely reducible (2.48) $E(\Gamma)$- $(A(\Gamma)$-, $I(\Gamma)$-$)$ group.

Now we will try to decompose E(Γ) (...) itself. We start with E(Γ) and fix our notation. Our discussion follows partly and generalizes Johnson (1).

7.50 NOTATION Let Γ be the direct sum of fully invariant subgroups Φ_i $(i\varepsilon I)$ (don't forget 7.44!). Then each $\gamma\varepsilon\Gamma$ can uniquely be decomposed as $\gamma = \sum_{i\varepsilon I} \phi_i$ with $\phi_i \varepsilon \Phi_i$ (note that this sum is actually finite).

If $m\varepsilon E(\Gamma)$ and $i\varepsilon I$ then
$$m^{(i)}: \Gamma \to \Gamma \quad . \qquad \gamma = \sum_{j\varepsilon I} \phi_j \to m(\phi_i)$$

Finally, let $E^{(i)}(\Gamma) := \{m^{(i)} | m\varepsilon E(\Gamma)\}$.

7.51 PROPOSITION Let Γ be as in 7.50 and let m,n be $\varepsilon E(\Gamma)$ and $i\varepsilon I$.

(a) $m\varepsilon E^{(i)}(\Gamma) \Longleftrightarrow m = m^{(i)}$.

(b) $m^{(i)} = m \circ \pi_i$, where π_i is the projection $\Gamma \to \Phi_i$.

(c) $(m+n)^{(i)} = m^{(i)} + n^{(i)}$ and $(mn)^{(i)} = mn^{(i)}$.

(d) $E^{(i)}(\Gamma) \cong E(\Phi_i)$ by $m^{(i)} \to m^{(i)}/_{\Phi_i} = m/_{\Phi_i}$.

Proof. (a): If $m\varepsilon E^{(i)}(\Gamma)$ then $\exists\, n^{(i)} \varepsilon E^{(i)}(\Gamma): m = n^{(i)}$.

But then for all $\gamma = \sum_{j\varepsilon I} \phi_j \varepsilon \Gamma$ we obtain

$m^{(i)}(\gamma) = m^{(i)}(\Sigma\phi_j) = m(\phi_i) = n^{(i)}(\phi_i) = n^{(i)}(\Sigma\phi_j) =$

$= n^{(i)}(\gamma)$, whence $m^{(i)} = n^{(i)} = m$.

The converse is trivial.

(b) - (d) follow easily.

Now we improve Proposition 5.7 of Johnson (1).

7.52 PROPOSITION $\forall i\varepsilon I: E^{(i)}(\Gamma) \trianglelefteq E(\Gamma)$.

Proof. (a) If $m^{(i)} \varepsilon E^{(i)}(\Gamma)$, consider $m\varepsilon E(\Gamma)$ and decompose m as $m = \sum_k \sigma_k e_k$ with all $e_k \varepsilon \mathrm{End}\,\Gamma$. All $e_k^{(i)} \varepsilon \mathrm{End}(\Gamma)$ by 7.51(b). Using 7.51(c) we obtain

$m^{(i)} = \sum\limits_k \sigma_k e_k^{(i)}$, whence $m^{(i)} \varepsilon E(\Gamma)$.

This shows that $E^{(i)}(\Gamma) \subseteq E(\Gamma)$.

(b) Applying 7.51(c) again shows that $E^{(i)}(\Gamma) \leq E(\Gamma)$.

(c) Now let $n = \sum\limits_{j=1}^{r} \sigma_j f_j \varepsilon E(\Gamma)$ (all $f_j \varepsilon \mathrm{End}\ \Gamma$) and take some $m^{(i)} \varepsilon E^{(i)}(\Gamma)$.

Then $n+m^{(i)}-n = \sigma_1 f_1+\ldots+\sigma_r f_r+m^{(i)}-\sigma_r f_r-\ldots-\sigma_1 f_1$.

But if $\gamma = \sum\limits_{i \varepsilon I} \phi_i = \phi_i+\psi$ where $\psi = \sum\limits_{k \neq i} \phi_k$ then

$(\sigma_r f_r+m^{(i)}-\sigma_r f_r)(\gamma) = \sigma_r f_r(\phi_i+\psi)+m^{(i)}(\phi_i+\psi)-\sigma_r f_r(\phi_i+\psi)=$
$= \sigma_r f_r(\phi_i)+m^{(i)}(\phi_i)-\sigma_r f_r(\phi_i) = (\sigma_r f_r+m^{(i)}-\sigma_r f_r)^{(i)}(\gamma)$,
since $f_r(\psi) \varepsilon \sum\limits_{k \neq i} \Phi_k$ commutes with $m^{(i)}(\phi_i) \varepsilon \Phi_i$.

This shows that $\sigma_r f_r+m^{(i)}-\sigma_r f_r \varepsilon E^{(i)}(\Gamma)$.

Proceeding in this way yields $n+m^{(i)}-n \varepsilon E^{(i)}(\Gamma)$ and $E^{(i)}(\Gamma)$ is normal and by (b) and 6.6(a) we know that $E^{(i)}(\Gamma) \trianglelefteq_\ell E(\Gamma)$.

(d) Finally consider again $n = \sum\limits_j \sigma_j f_j$ and $m^{(i)} = \sum\limits_k \sigma'_k e_k^{(i)}$. Then by 6.4(d),

$m^{(i)} n = \sum\limits_k \sigma'_k \sum\limits_j \sigma_j e_k^{(i)} f_j$. Now if $\gamma = \phi_i+\psi$ as before,

$e_k^{(i)} f_j(\gamma) = e_k^{(i)}(f_j(\phi_i)+f_j(\psi)) = e_k^{(i)}(f_j(\phi_i)) =$
$= (e_k^{(i)} f_j)^{(i)}(\gamma)$, whence $e_k^{(i)} f_j \ \varepsilon\ E^{(i)}(\Gamma)$ and hence $m^{(i)} n \varepsilon E^{(i)}(\Gamma)$.

So $E^{(i)}(\Gamma)$ has no chance any more to escape from being an ideal.

The following result is implicit in Johnson (1).

7.53 THEOREM Let Γ be finite, non-abelian and the direct sum of finitely many minimal fully invariant subgroups Φ_i $(i\varepsilon I)$. Then, as near-rings,

$$E(\Gamma) = \sum_{i\varepsilon I}{}^{\bullet} E^{(i)}(\Gamma) \cong \bigoplus_{i\varepsilon I} E(\Phi_i).$$

Proof. If $m\varepsilon E(\Gamma)$, $m = \sum_{i\varepsilon I} m^{(i)}$, for 2.30 implies that

$\forall\ \gamma = \Sigma\phi_i\varepsilon\Gamma$: $m(\gamma) = m(\Sigma\phi_i) = \Sigma m(\phi_i) = \Sigma m^{(i)}(\phi_i) =$
$= \Sigma m^i(\gamma)$.

If $m^{(i)}\varepsilon E^{(i)}(\Gamma) \cap (\sum_{j\neq i} E^{(j)}(\Gamma))$ then for each $\phi_i\varepsilon\Phi_i$ $m^{(i)}(\phi_i) = o$; consequently $m^{(i)} = \bar{o}$ and the sum is direct. The rest follows from 2.30, 2.28, 7.51(d), 7.43 and 7.46(a).

Using 4.46 and 7.53 one gets

7.54 COROLLARY If Γ is as in 7.53, $E(\Gamma)$ is finitely completely reducible as a near-ring as well as an $E(\Gamma)$-group (2.50). $E(\Gamma)$ is simple iff $|I| = 1$, i.e. iff Γ is invariantly simple.

So the structure of (left) ideals of $E(\Gamma)$ seems to be clear for a group Γ as in 7.53 (see 2.50 and 2.55).

7.55 COROLLARY Let the finite non-abelian group Γ be the direct sum of minimal fully invariant subgroups $\Phi_1,\dots,\Phi_k$. Let $\Phi_1,\dots,\Phi_t$ be abelian and $\Phi_{t+1},\dots,\Phi_k$ not. Then

$$E(\Gamma) \cong \bigoplus_{i=1}^{t} End(\Phi_i) \oplus \bigoplus_{j=t+1}^{k} M_o(\Phi_j)$$

(where $End(\Phi_i)$ denotes the endomorphism ring on Φ_i).

Proof. 7.53, 7.43 and 7.46(a) (observe that the minimal fully invariant subgroups Φ_i are invariantly simple).

It is harder to get similar results for $A(\Gamma)$ and $I(\Gamma)$, since $\alpha\varepsilon Aut(\Gamma)$ does not imply $\alpha^{(i)}\varepsilon E^{(i)}(\Gamma)$ to be an automorphism in general. At least we get the following

7.56 THEOREM Let Γ be the direct sum of non-abelian minimal fully invariant (minimal characteristic, minimal normal) subgroups Φ_j $(j\varepsilon J)$. Then $E(\Gamma) \hookrightarrow \prod_{j\varepsilon J} E(\Phi_j) = \prod_{j\varepsilon J} M_o(\Phi_j)$, $A(\Gamma) \hookrightarrow \prod_{j\varepsilon J} A(\Phi_j) = \prod_{j\varepsilon J} M_o(\Phi_j)$ and $I(\Gamma) \hookrightarrow \prod_{j\varepsilon J} I(\Phi_j) = \prod_{j\varepsilon J} M_o(\Phi_j)$, respectively.

Proof. An embedding map can be

$$h: E(\Gamma) \to \prod_{j\varepsilon J} E(\Phi_j)$$
$$m \longrightarrow (\dots, m/_{\Phi_j}, \dots)$$

and likewise for the other two cases.
Observe that the restriction of an endomorphism is an endomorphism, that of an automorphism to a characteristic subgroup is an automorphism and if Φ_j is simple and non-abelian, this automorphism is inner (cf. e.g. (Rotman), p. 133).

We return repentantly to $E(\Gamma)$. Applying 5.20 or 5.31 to 7.53 yields

7.57 COROLLARY If Γ is as in 7.53 then

$$\mathfrak{J}_2(E(\Gamma)) = \mathfrak{J}_1(E(\Gamma)) = \dots = \mathfrak{P}(E(\Gamma)) = \{0\}.$$

So the situation is quite clear if Γ is the direct sum of finitely many minimal fully invariant subgroups.
If this is not the case, life is much more complicated (e.g. if Γ is the dihedral group with 8 elements).
We mention only two results without proofs (but with hints) in the finite case:

7.58 THEOREM (Johnson (1), (2)). Let Γ be a non-abelian finite group. Then

(a) $E(\Gamma)$ is simple <=> Γ is invariantly simple. (we know <=; conversely, if Δ is a proper fully invariant subgroup of Γ then $(o:\Delta)$ is a proper ideal of $E(\Gamma)$).

(b) $\mathfrak{J}_2(E(\Gamma)) = \mathfrak{J}_1(E(\Gamma)) = \ldots = \mathfrak{P}(E(\Gamma))$ (but not necessarily $= \{0\}$). ($\mathfrak{J}_2(E(\Gamma))$ turns out to be nilpotent. Now apply 5.61(b)).

For more on that see Lyons (5), (6).

We continue with a result due to Fröhlich (3).

7.59 THEOREM (Ramakotaiah (5)). Let Γ be a finite simple non-abelian group. Then $Aut(I(\Gamma)) \cong Aut(\Gamma) \cong Inn(\Gamma) \cong \Gamma$ (as groups) and each $h \in Aut(I(\Gamma))$ is of the form $m \to ama^{-1}$, where $a \in Aut(\Gamma)$.

Proof. 7.39, 7.46(c) and (Rotman) pp. 131 - 133.

If we are going to leave finite groups Γ it becomes pretty dark. In (3), Fröhlich mentioned that one can see (by comparing the cardinal numbers) that if Γ is an infinite simple group then $E(\Gamma) < M_0(\Gamma)$. Introducing a suitable topology in $M_0(\Gamma)$ gives the result that $M_0(\Gamma)$ is the completition of $E(\Gamma)$.

We conclude this section with some remarks.

7.60 REMARKS

(a) If Γ is abelian then $M_0(\Gamma) = E(\Gamma)$ <=> $|\Gamma| \leq 2$. This is trivial.

(b) (Malone-McQuarrie (1)). Let Γ be halvable. If for all $m,n \in E(\Gamma)$ $(A(\Gamma), I(\Gamma))$ $\frac{m+n}{2} - \frac{n}{2} - \frac{m}{2} = 0$ then $E(\Gamma)$ $(A(\Gamma), I(\Gamma))$ is a ring.

(c) (Maxson (14)). If $E(\Gamma)$ is a ring and Γ is fg. then Γ is abelian iff ${}_{E(\Gamma)}\Gamma$ is monogenic. (This can actually happen; so $E(\Gamma)$ can be a ring although Γ is not abelian! - Compare 7.46). If Γ is a finite p-group and $E(\Gamma)$ is a semisimple ring then Γ is abelian.
See also Lyons (5) and Malone (9).

(d) (Chandy (2)). Call a group Γ an <u>L-group</u> iff $I(\Gamma)$ is a ring.
Then Γ is an L-group $\Leftrightarrow$ all conjugated elements commute $\Leftrightarrow$ the centralizer of each $\gamma\varepsilon\Gamma$ is a normal subgroup. If Γ is an L-group and nilpotent of class ≤ 3 then $(I(\Gamma),+)$ is nilpotent of class 1.

(e) (Chandy (3)). $I(\Gamma)$ is a commutative ring iff Γ is nilpotent of class 2.

(f) (Beidleman (11)). If Γ is finite and solvable (nilpotent) then the same applies to $(E(\Gamma),+)$.

(g) (Fain (1), Lyons (3), (4), Lyons-Malone (2), Johnson (1)). If D_n is the dihedral group with $2n$ elements (n odd) then $E(D_n) = A(D_n) = I(D_n)$. See a more detailed description of these dgnr.'s in the references. Lyons-Malone (1) contains also a description of all dgnr.'s definable on D_n. See Meldrum (5) for informations on $E(D)$, where D is the infinite dihedral group.

(h) See Lyons (4), for a proof of the fact that if Δ is a fully invariant abelian direct summand then $E(\Delta)$ embeds in $E(\Gamma)$ as a direct summand.

(i) See Lyons-Malone (1), where $E(S_3)$ is discussed in detail.

(j) Malone (7) does the same for $E(\Gamma)$, where Γ is a (generalized) quaternion group.

(k) Call $e\varepsilon End(\Gamma)$ <u>normal</u> if $\forall\ a\varepsilon Inn(\Gamma)$: $ea = ae$.
Let $N(\Gamma)$ be the nr. generated (additively) by all normal endomorphisms of Γ. By a routine check we get

the following result (Heerema (1), Plotkin (3), 6.1.3.2): $N(\Gamma)$ is a ring.

(l) (B.H. Neumann (2)). Let a be a fixed-point-free automorphism of Γ of order 3 and A_a the subnear-ring of $A(\Gamma)$ generated by a. Then A_a is a ring, too.

(m) See Scott (1) for much information about the invertible elements in certain subnear-rings of $E(\Gamma)$. He also proved that if $I(\Gamma)$ has the DCCL then Γ is finite!

(n) (Scott (7)). Let Γ be finite, $\alpha \varepsilon \mathrm{Aut}(\Gamma)$, α fixed-point-free and N the nr. generated by α. Then $(N,+)$ is nilpotent iff $(\Gamma,+)$ is nilpotent.

(o) See also 7.113(e) and 6.35(d).

d) POLYNOMIAL NEAR-RINGS

1.) POLYNOMIALS AND POLYNOMIAL FUNCTIONS

O.K., you are right: polynomial near-rings on some algebra A are no transformation near-rings on A in general. But they are so close to this subject and necessary to treat near-rings of polynomial functions that we include them into §7.

Polynomials can be defined over any algebraic structure A and any set X (with $A \cap X = \emptyset$) of "indeterminantes" in any variety $\mathfrak{V}$ containing A. Denote the set of these polynomials by $A(X,\mathfrak{V})$. Roughly spoken, $A(X,\mathfrak{V})$ is the "set of all words in $A \cup X$ where equality is defined in accordance with the laws defining $\mathfrak{V}$". In fact, $A(X,\mathfrak{V})$ is defined as a "suitable" factor algebra of the word algebra over $A \cup X$ in $\mathfrak{V}$, hence becoming also an algebra of $\mathfrak{V}$, the "polynomial algebra". Another possible characterization of $A(X,\mathfrak{V})$ is the following: $A(X,\mathfrak{V})$ is the free union of A and the free algebra F over X in $\mathfrak{V}$.
See Lausch-Nöbauer (1) (whole book, particularly pp. 12/13) for a detailed exposition.

In some instances for these polynomials (= equivalence classes of words over $A \cup X$) there exist "normal forms" (see Lausch-Nöbauer (1), pp. 22 ff.), e.g. for the case of commutative rings with identity and for groups.
We will restrict our considerations to the case of a single "variable" $X = \{x\}$ and to two varieties: the one of all groups ($\mathcal{G}$) and the variety $\mathcal{R}$ of all commutative rings with identity 1.
Then we have the following polynomial algebras (in normal forms) (see Lausch-Nöbauer (1), I.8.11 and I.9.11):

<u>7.61 NOTATION</u> If $R \in \mathcal{R}$ and $\Gamma = (\Gamma,+) \in \mathcal{G}$ then

(a) $R[x] := \{ \sum_{i=0}^{n} a_i x^i \mid n \in \mathbb{N}_0, a_i \in R, a_n \neq 0\} \cup \{0\}$.

(b) $\Gamma[x] := \{\gamma_0 + n_1 x + \gamma_1 + n_2 x + \ldots + \gamma_{r-1} + n_r x + \gamma_r \mid r \in \mathbb{N}_0, \gamma_i \in \Gamma,$
$n_i \in \mathbb{Z}^*$ and $\forall\ t \in \{1,\ldots,r-1\}: \gamma_t \neq o\}$.

Our interest in polynomials stems from the easy-to-establish

<u>7.62 PROPOSITION</u> Under addition "+" and substitution "$\circ$" of polynomials (definition as usual, cf. Lausch-Nöbauer (1), p. 77) $(R[x],+,\circ)$ and $(\Gamma[x],+,\circ)$ are near-rings.

We will continue to denote these near-rings by $R[x]$ and $\Gamma[x]$. Throughout this section R will mean an element of $\mathcal{R}$.

<u>7.63 DEFINITION</u>

(a) If $p \in R[x]$ then $\bar{p}: R \to R$, $r \to p \circ r =: p(r)$ is called the <u>polynomial function induced by p</u>.

$\bar{R}[x] := \{\bar{p} \mid p \in R[x]\}$.

(b) Similarly we define $\bar{p}$ for $p \in \Gamma[x]$ and $\bar{\Gamma}[x]$.

Trivial, but good to note is

7.64 PROPOSITION If Γ is abelian then every polynomial function $\overline{p} \in \overline{\Gamma}[x]$ is of the form $\overline{p}: \Gamma \to \Gamma$, $\gamma \to \gamma_0 + n\gamma$ with $\gamma_0 \in \Gamma$ and $n \in \mathbb{N}_0$.

7.65 PROPOSITION The correspondence $h: p \to \overline{p}$ is a near-ring homomorphism. Hence $\overline{R}[x]$ and $\overline{\Gamma}[x]$ are subnear-rings of $M(R)$ ($M(\Gamma)$, respectively).

Proof: straightforward.

7.66 REMARKS

(a) $\overline{R}[x]$ and $\overline{\Gamma}[x]$ are called the near-rings of polynomial functions on R (Γ, respectively).

(b) Throughout this section, h will have the meaning of 7.65. h is not necessarily an isomorphism $R[x] \to \overline{R}[x]$: Take $R =: \mathbb{Z}_p$ and $q := x\cdot(x-1)\cdot\ldots\cdot(x-(p-1))$. Then $p \neq 0$, but $h(p) = \overline{p} = \overline{o}$ (zero function on $\mathbb{Z}_p$). Anyhow, $\operatorname{Ker} h \trianglelefteq R[x]$ will play a decisive rôle. Similar considerations apply to groups (if e.g. $\Gamma = \mathbb{Z}_3$, $q = 3x$ behaves as above). Hence one cannot identify polynomials and polynomial functions (as one is used to do over $\mathbb{R}$) in general. But:

7.67 REMARK If R is torsion-free ((Aczel)) or if R is an infinite field (use the theory of linear equations and the Van-der-Monde-determinant) then h is an isomorphism $R[x] \to \overline{R}[x]$.

Of course, one could also consider $N[x]$ (N a near-ring), $V[x]$ (V a vector space) and so on to get more near-rings. But we remain at $R[x]$ and $\Gamma[x]$ since most phenomena can already be seen there and discuss $R[x]$ first since $R[x]$ seems to be more familiar than $\Gamma[x]$.

2.) R[x]

We start with some elementary, yet fundamental properties. Let, as often, $[p]$ be the degree of p ($[0]:=0$).
Observe that R[x] is an abelian near-ring with identity x.
If we write $p = \sum_{i=0}^{n} a_i x^i$, we allow $a_n = 0$ if $p = 0$ (7.61(a)) to avoid separating cases.

7.68 PROPOSITION (Lausch-Nöbauer (1), p. 134).

(a) $\forall\, p,q \in R[x]$: $[p \circ q] \le [p]\cdot[q]$ (with equality iff R is an integral domain).

(b) If $p \in R[x]$ is invertible (w.r.t. $\circ$) then $[p] = 1$. If R is a field, the converse also holds.

(c) If R is an integral domain then each $p \in R[x]$ with $[p] > 0$ is right cancellable.

Proof. (a) If $p = \sum_{i=0}^{n} a_i x^i$, $q = \sum_{j=0}^{m} b_j x^j$ then

$$p \circ q = \sum_{i=0}^{n} a_i q^i = a_n b_m x^{nm} + \text{summands of lower degree.}$$

(b) follows from (a).

(c) If $f \circ p = g \circ p$ then $(f-g)\circ p = 0$. By (a), $[f-g] = 0$, so $f-g$ is a constant c. Hence $f = g+c$ and so $g\circ p = f \circ p = g \circ p + c$, whence $c = 0$.

7.69 DEFINITION Call $p \in R[x]$ indecomposable if $p = f \circ g$ implies $[f] = [p]$ or $[g] = [p]$.

7.70 PROPOSITION (Lausch-Nöbauer (1)). Let R be an integral domain and $p \in R[x]$.

(a) $[p] \in \mathbb{P} \Rightarrow$ p is indecomposable.

(b) If $[p] > 1$ then there exist indecomposable polynomials $p_1,\dots,p_s \in R[x]$ with $p = p_1 \circ p_2 \circ \dots \circ p_s$.

Proof. (a) is trivial.

(b) follows by induction on $[p]$, using 7.68(a).

7.71 DEFINITION

(a) Let p be called normed if $a_n = 1$.

(b) p is called linear if $[p] = 1$.

Now we can sharpen 7.70(b) and state without proof:

7.72 THEOREM (Lausch-Nöbauer (1)). If R is an integral domain and if $p \in R[x]$ has $[p] > 1$ then

(a) $p = \ell \circ p_1 \circ \ldots \circ p_s$ with ℓ linear, $s \in \mathbb{N}$ and $p_1, \ldots, p_s$ indecomposable and normed.

(b) If $p = m \circ q_1 \circ \ldots \circ q_t$ is another such "prime decomposition" of p then $m = \ell$, $s = t$ and the degrees of the p_i's and q_j's are the same (up to order).

(c) There exist only finitely many decompositions of p of the form (a).

This uniqueness can even be strengthened by the deep theorem 2.46 of Lausch-Nöbauer (1).

7.73 REMARK See Graves-Malone (2) for another result of "prime factor decompositions" in "Euclidean near-domains" (see 9.60 and 9.67(c)).

3.) $\overline{R}[x]$

7.74 DEFINITION Call $R \in \mathcal{R}$ polynomially complete (Nöbauer (6)) or 1-polynomially complete (Lausch-Nöbauer (1)) or primal ((Grätzer)) if $\overline{R}[x] = M(R)$, i.e. if each map $R \to R$ is a polynomial function.

Of course, $\mathbb{Z}$, $\mathbb{Q}$, $\mathbb{R}$ and $\mathbb{C}$ are not polynomially complete. On the other hand, the fact that $\mathbb{Z}_2$ is polynomially complete is of high value in mathematical logic.
Perhaps we now feel which R might be polynomially complete and prove the following interesting result, which will also prove useful in the sequel.

7.75 THEOREM (Nöbauer (6), Lausch-Nöbauer (1)).
R is polynomially complete <=> R is finite and simple (hence a finite field).

Proof. =>: If R is infinite, $|\overline{R}[x]| \leq |R[x]| = |R| < |R|^{|R|} = |M(R)|$, so R cannot be polynomially complete.
Let I be a non-trivial ideal of R. Take some non-zero $i \in I$. If $\overline{p} \in \overline{R}[x]$ then $\overline{p}(i) - \overline{p}(0) \in I$, since $I \trianglelefteq R$.
Let $f \in M(R)$ fulfill $f(i) - f(0) \notin I$. Then $f \notin \overline{R}[x]$ and again R is not polynomially complete.

<=: This follows from Lagrange's interpolation theorem (or from the fact that $\overline{R}[x]$ is 2-primitive on R).

4.) IDEAL THEORY IN R[x]

Of course, R[x] can also be considered as a ring $(R[x], +, \cdot)$, where "$\cdot$" means ordinary multiplication.
There is some connection between the ring- and the near-ring ideals of R[x]:

7.76 PROPOSITION (So (1)). If $p, q, r \in R[x]$ then r divides (w.r.t. "$\cdot$") $p \circ (q+r) - p \circ q$.

Proof. Let $p = \sum_{i=0}^{n} a_i x^i$. Then
$p \circ (q+r) - p \circ q = \sum_{i=0}^{n} a_i (q+r)^i - \sum_{i=0}^{n} a_i q^i$ is a multiple of r.

7.77 COROLLARY (Lausch-Nöbauer (1)). Every ideal of $(R[x],+,\cdot)$ is a left ideal of $(R[x],+,\circ)$.

See 7.94 for a (partial) converse.

Note that if we write $R[x]$ we always mean the near-ring-version $(R[x],+,\circ)$. It's also high time to fix one more notation.

7.78 NOTATION

(a) $(R[x])_o = \{\sum_{i=1}^{n} a_i x^i \mid a_i \in R \wedge n \in \mathbb{N}\} = \{p \mid \bar{p}(0) = 0\} =$ $=: R_o[x]$.

(b) $(R[x])_c = \{\sum_{i=0}^{0} a_i x^i \mid a_o \in R\} = \{p \mid [p] = 0\} =: R_c[x]$.

(Of course, $R_c[x]$ can be identified with R, and we will do so in the sequel.)

(c) The elements of $R_o[x]$ $(R_c[x])$ are called the zero-symmetric (constant) polynomials.

7.79 PROPOSITION (Clay-Doi (2), So (1)). Let $L \trianglelefteq_\ell R[x]$. Then

(a) $R_o[x] \circ L \subseteq L$.

(b) $\forall r \in R \quad \forall \ell \in L$: $r\ell \in L$.

(c) If $x \in L$ then $R_o[x] \subseteq L$.

Proof. (a) follows from 1.34(a).

(b) $r\ell = (rx) \circ \ell \in L$ by (a).

(c) also follows from (a).

7.80 REMARK Observe that R is an $R[x]$- and an $\bar{R}[x]$- group in the natural way ($pr := \bar{p}(r)$ if $p \in R[x]$ and $r \in R$). We will milk this observation in the next number.

Now we take a glance to two interesting ideals of $R[x]$:

7.81 EXAMPLES (Nöbauer (1), Lausch-Nöbauer (1)). Let $J \trianglelefteq R$. Then

(a) $(J) := J[x] \trianglelefteq R[x]$.

(b) $\langle J\rangle := (J:R)_{R[x]} \trianglelefteq R[x]$.

These examples of ideals in $(R[x],+,\circ)$ show up to be also ideals in $(R[x],+,\cdot)$. This leads to the following

7.82 DEFINITION $I \trianglelefteq R[x]$ is called a full ideal (Nöbauer) or T-0-ideal (Menger) or composition ideal (Adler) if also $I \trianglelefteq (R[x],+,\cdot)$.

7.83 REMARK Thus full ideals are exactly the "ideals" of the composition ring (TO-Algebra) (see 1.117) $(R[x],+,\cdot,\circ)$.

In fact, these full ideals of $R[x]$ are much better explored than the "ordinary" ideals of $R[x]$. Not every ideal of $R[x]$ is a full ideal, but this holds in special, important cases (see 7.93).
If e.g. R happens to be a field then all full ideals are principal (a well-known result of ring theory) and so quite easy to overlook. It is not known under which conditions each ideal of $R[x]$ is "principal" (= generated by one single polynomial), or at least f.g. .
The ideals of 7.81 show up to be quite important ones:

7.84 THEOREM (Nöbauer (1), Lausch-Nöbauer (1), Hule (1)). Let I be a full ideal of $R[x]$. Then there exists exactly one ideal J of R with $(J) \subseteq I \subseteq \langle J\rangle$ (namely $J = I \cap R$).

Proof. Let $J := I \cap R$. Then clearly $(J) = J[x] \subseteq I$. Also, if $i \in I$ and $r \in R$ then $\bar{i}(r) = i \circ r \in I \cap R = J$, whence $I \subseteq \langle J\rangle$.
If $J' \trianglelefteq R$ also fulfills $(J') \subseteq I \subseteq \langle J'\rangle$ then $(J') \subseteq \langle J\rangle$, whence $J' \subseteq \langle J\rangle$ and so $J' \subseteq J$. Similarly, $J \subseteq J'$ and hence $J = J'$.

7.85 DEFINITION Let I,J be as in 7.84. Then J is called the "enclosing ideal" of I (J is not ideal in $R[x]$ in general!).

For (much) more information concerning these enclosing ideals see the comprehensive book Lausch-Nöbauer (1). Cf. also Mlitz (1).

We will get more sensational results when R is assumed to be a field:

5.) $F[x]$

Throughout this number, let F denote a commutative field.

7.86 PROPOSITION (Clay-Doi (1), Straus (1)). Let $L \trianglelefteq_{\ell} F[x]$ with $L \cap F \neq \{0\}$ (cf. 7.78(b)).

(a) $F \subseteq L$.

(b) If $|F| > 2$ then $L = F[x]$.

Proof. (a) Let $\ell \neq 0$ be $\in L \cap F$. Then take some $f \in F$. By 7.79(b), $f \cdot \ell^{-1} \cdot \ell = f \in L$.

(b) If char $F \neq 2$, take $f := 2^{-1}$. By (a), $f \in L$ and also $f^2 \in L$. Hence $x^2 \circ (x+f) - x^2 \circ x \in L$, so $x + f^2 \in L$, whence $x \in L$. Use (a) and apply 7.79(c) to get $L = F[x]$.
If char $F = 2$, and $|F| > 2$, then in particular char $F \neq 3$. Hence $x^3 \circ (x+1) - x^3 \circ x = x^2 + x + 1 \in L$, so $x^2 + x \in L$.
Take some $f \in F \setminus \{0,1\}$ and denote $f^{-1} \cdot x^3$ by p.
Then $p \circ (x+f) - p \circ x = x^2 + fx + f^2 \in L$.
Since $x^2 + x \in L$ and $F \subseteq L$, $(f-1)x \in L$, so by 7.79(b) and the fact that F is a field we get $x \in L$ and again $F[x] = L$.

7.87 REMARK Brenner (1) has shown that 7.86(b) does not hold for $F = \mathbb{Z}_2$. See 7.98(b).

7.88 PROPOSITION (Clay-Doi (2)).

(a) $(F,+)$ is an $F[x]$-group and $\overline{F}[x]$-group of type 2.

(b) If h (7.65) is an isomorphism (cf. 7.66(b)) then $F[x]$ is 2-primitive on $(F,+) = F$.

(c) $\overline{F}[x]$ is always 2-primitive on F.

Proof. If $f,f' \in F$, $f \neq 0$, then $\exists\, p_o \in F_o[x]: \overline{p}_o(f) = f'$, namely $p_o = f'f^{-1}x$. The rest is equally obvious.

If the reader is still interested, he is cordially invited to a nearly complete trip to the ideals of $F[x]$ and $\overline{F}[x]$. First we settle the question, for which F $F[x]$ happens to be simple.

7.89 THEOREM (Straus (1)). Let F be infinite. Then $F[x]$ is simple.

Proof. If $I \trianglelefteq F[x]$, $I \neq \{0\}$, take some $i \in I$, $i \neq 0$. By 7.67, $\overline{i} \neq \overline{o}$, so $\exists\, f \in F: \overline{i}(f) = i \circ f \neq 0$. Hence $i \circ f \in I \cap F$ and 7.86(b) implies $I = F[x]$.

So the infinite case is settled and we turn to finite fields. To do so, we first determine all full ideals (7.90) of $F[x]$ and then we will see (7.93) that, if char $F \neq 2$, all ideals of $F[x]$ are full ideals.

7.90 THEOREM (Menger (2), Milgram (1), Lausch-Nöbauer (1), Straus (1)).
Let F be a finite field and I an ideal of $(F[x],+,\cdot)$. Since $(F[x],+,\cdot)$ is a PID, I is some principal ideal (p) of this ring $(F[x],+,\cdot)$. Then the following conditions are equivalent:

(a) $I = (p)$ is a full ideal.

(b) $\forall\, s \in F[x]: p / p \circ s$.

(c) There exist $k \in \mathbb{N}$, $n_1, \ldots, n_k \in \mathbb{N}$ with $1 \leq n_1 < n_2 < \ldots < n_k$ and $m_1, \ldots, m_k \in \mathbb{N}_0$ with

$$p = \text{l.c.m.}\ \{(x^{q^{n_1}} - x)^{m_1}, \ldots, (x^{q^{n_k}} - x)^{m_k}\} \quad \text{where } |F| = q$$

$$(\text{then } I = ((x^{q^{n_1}} - x)^{m_1}) \cap ((x^{q^{n_2}} - x)^{m_2}) \cap \ldots \cap ((x^{q^{n_k}} - x)^{m_k})).$$

Proof. (Straus).

(a) $\Rightarrow$ (b) is trivial.

(b) $\Rightarrow$ (c): Let C be the algebraic closure of the field F. Let c be in C with $\overline{p}(c) = 0$. Let m be the multiplicity of the root c. Then c is a root of multiplicity $\geq m$ of each $p o s$ ($s \in F[x]$). Hence p has a zero of multiplicity $\geq m$ at each element of $F(c)$ (field extension of F by adjunction of c). Applying the theory of finite fields we see that p is divisible by $(x^{q^n} - x)^m$, where $m := [F(c):F]$: $\forall\, d \in F(c)$: $(x-d)^m / p$, hence $\prod_{d \in F(c)} (x-d)^m$ divides p. But $\prod_{d \in F(c)} (x-d)^m = (x^{q^n} - x)^m$. Starting with a root c with maximal $[F(c):F]$ we arrive successively at (c).

(c) $\Rightarrow$ (a): It suffices (7.77) to show that for $n \in \mathbb{N}$ and $s = \sum_{i=o}^{t} a_i x^i \in R[x]$, $(x^{q^n} - x) o s \in (p)$.

In fact, $(x^{q^n} - x) o s = (\sum_{i=o}^{t} a_i x^t)^{q^n} - \sum_{i=o}^{t} a_i x^i =$

$= \sum a_i^{q^n} (x^{i \cdot q^n} - x^i) \in (p)$.

7.91 REMARK The representation in 7.90(c) is moreover unique: see Lausch-Nöbauer (1), Ch. III, 7.21.

7.92 DEFINITION A polynomial p as in 7.90 is called saturated (Milgram (1)).

7.93 THEOREM (Straus (1)). Let F be a finite field. Then: Every ideal of $F[x]$ is a full ideal $\Longleftrightarrow$ char $F \neq 2$.

Proof. $\Rightarrow$: Assume that char $F = 2$. We show the existence of an ideal I of $F[x]$ which is no full ideal:

Let $|F| =: q$ and $F[x^2] := \{\sum_{i=0}^{n} a_{2i}x^{2i} \mid n \in \mathbb{N}_0, a_{2i} \in F\}$.

Consider $I := (x^q+x)^2 \cdot F[x^2] + (x^q+x)^4 \cdot F[x]$.
$I \not\trianglelefteq (F[x],+,\cdot)$ since $(x^q+x)^2 \in I$, but $x \cdot (x^q+x)^2 \notin I$.
But $I \trianglelefteq (F[x],+,\circ)$:

(a) Clearly $(I,+)$ is a subgroup of $(F[x],+)$.

(b) If $(x^q+x)^2 =: p$ then for all $q,r,s \in F[x]$ and all $t \in F[x^2]$ p^2 divides $q\circ(r+pt+p^2s)-q\circ r-pt(q'\circ r)$.
But $q'\circ r$ is a polynomial in q^2, hence in $F[x^2]$.
Thus $q\circ(r+pt+p^2s)-q\circ r \in I$ and I is shown to be a left ideal of $F[x]$.

(c) Finally we have to show that $I \trianglelefteq_r F[x]$. Let $(x^q+x)^2 =: s$. Since s is saturated, it suffices to show that $\forall\, r,t \in F[x]\ \exists\, h \in F[x]$: s^2 divides $(s\circ r)(t\circ r^2)+s(h\circ x^2)$. Since s is easily shown to be distributive, it suffices to show this for $r = x^k$ $(k \in \mathbb{N}_0)$:
If k is even $(k = 2\ell)$ then
$s\circ x^k = (x^q+x)^2\circ x^{2\ell} = (x^{2q}+x^2)\circ x^{2\ell} = x^{4q\ell}+x^{4\ell} =$
$= (x^{4q}+x^4)\circ x^\ell = s^2\circ x^\ell$. Since s^2 is also saturated, $s^2 / s^2\circ x^\ell = s\circ x^k$.
Taking $h = s$, s^2 divides $(s\circ x^k)\cdot(t\circ r^2)+s\cdot(s\circ x^2)$.
If k is odd $(k = 2\ell+1)$ then
$s\circ x^k = x^{2kq}+x^{2k} = (x^{2q}+x^2)^k = x^k\circ((x^{2q}+x^2)+x^2)+x^k\circ x^2 \in I$ (by (b)).
Thus also in this case $I \trianglelefteq_r F[x]$. Hence $I \trianglelefteq F[x]$.

$\Leftarrow$: Let $I \trianglelefteq F[x]$. Then for all $i \in I$ and all $p \in F[x]$ we get $i\cdot p = \frac{1}{2}(x^2\circ(i+p)-x^2\circ p-(x^2\circ(i+0)-x^2\circ 0)) \in I$.

7.94 REMARK The proof of "7.93 <=" also shows:
If F is finite with char $F \neq 2$ then the left ideals of $(F[x],+,\circ)$ are exactly the ideals of $(F[x],+,\cdot)$ (cf. (cf. 7.77).

7.95 REMARK Straus (1) also showed that if F is finite with char $F = 2$ but $|F| > 2$ and if $I \trianglelefteq F[x]$ then $F[x^2]\cdot I \subseteq I$ and I contains an ideal J of $(R[x],+,\cdot)$ which is generated by $\{i^2 = i\cdot i \mid i\in I\}$; J contains all $i_1\cdot i_2$ $(i_1,i_2\in I)$.

7.96 COROLLARY If F is finite then $F[x]$ does not fulfill the DCC.

7.97 COROLLARY If F is a finite field of characteristic $\neq 2$ then $F[x]$ fulfills the ACC, but not the DCC on ideals. Hence $F[x]$ cannot be completely reducible (2.50).

So it remains to consider finite fields F with char $F = 2$ and there in particular $F = \mathbb{Z}_2$. As usual in algebra, characteristic 2 causes a lot of trouble. The ideal structure of $\mathbb{Z}_2[x]$ is much more complicated than that of $F[x]$ in 7.89. or 7.90/7.93.

We get satisfactory results concerning the ideals of $F[x]$ with char $F = 2$ only in the case of maximal ideals:

7.98 THEOREM (Clay-Doi (2), Brenner (1)).

(a) If F is infinite, $\{0\}$ is the only maximal ideal of $F[x]$.

(b) If F is finite, but $\neq \mathbb{Z}_2$ then $\operatorname{Ker} h = \{p\in F[x] \mid \bar{p} = \bar{o}\}$ is the unique maximal ideal of $F[x]$.

(c) $\mathbb{Z}_2[x]$ has exactly two maximal ideals:

$V := \{p\in F[x] \mid \bar{p}$ is constant$\}$ and

$T :=$ the (near-ring) ideal generated by x^3.

Proof. Consider h of 7.65 and the diagram (observe 1.30).

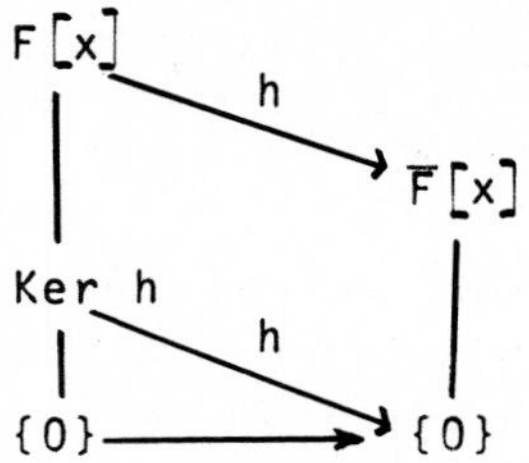

(a) is settled by 7.89. So we may assume that F is finite. By 7.75, $\overline{F}[x] = M(F)$. So if $|F| \neq 2$, $\overline{F}[x]$ is simple and Ker h is maximal. If $|F| = 2$ then $M_c(F)$ is a maximal ideal in $M(F) = \overline{F}[x]$ by 7.31. Hence its pre-image under h (= V) is maximal in $F[x]$. The facts that T is another maximal ideal in $\mathbb{Z}_2[x]$ and the uniqueness statements in (b) and (c) involve some technical reasoning and we overgo the proofs. That of (b) is in Clay-Doi (2), while that of (c) can be found in Brenner (1).

7.99 COROLLARY (Nöbauer (6), Hule (1), Clay-Doi (2), Lausch-Nöbauer (1)).
$F[x]$ is simple $\Leftrightarrow$ F is infinite.

Theorem 7.98 has some applications. We mention

7.100 COROLLARY (Clay-Doi (2)). $\mathbb{Z}[x]$ contains maximal ideals.

Proof. If $p \in \mathbb{P}$, $(p\mathbb{Z})[x] \trianglelefteq \mathbb{Z}[x]$ (7.81(a)), and $\mathbb{Z}[x]/(p\mathbb{Z})[x] \cong \mathbb{Z}_p[x]$, which contains at least one maximal ideal by 7.98. An application of 1.30 gives the statement.

All maximal ideals or all full ideals of $\mathbb{Z}[x]$ are not known (Clay-Doi (1), Lausch-Nöbauer (1), p. 131). One also might raise the question, which $\overline{R}[x]$ happen to be simple.

We are happy to have a full answer:

7.101 THEOREM (Nöbauer (6)).
$\overline{R}[x]$ is simple $\Leftrightarrow$ R is a commutative field with $|R|>2$.

Proof. =>: (a) Let I be a non-trivial ideal of R. Let h be again as in 7.65. By 7.81(a), $I[x] \lhd R[x]$; consequently $h(I[x]) = \overline{I}[x] \trianglelefteq \overline{R}[x]$.
Considering the constant polynomial functions yields $\overline{I}[x] \neq \{\overline{o}\}$ and $\overline{I}[x] \neq \overline{R}[x]$, a contradiction. Since $\overline{R}[x]$ is assumed to be simple, we arrive at a nonsense.

(b) $\mathbb{Z}_2[x] = M(\mathbb{Z}_2)$ is not simple by 7.31.

<=: Let R be a field. If R is finite $\neq \mathbb{Z}_2$ then by 7.75 $\overline{R}[x] = M(R)$ is simple (7.33). If R is infinite then by 7.67 $\overline{R}[x] \cong R[x]$ which is simple by 7.99.

We get only partial results on the radicals of $F[x]$:

7.102 REMARKS

(a) Let F be infinite. Then $\mathcal{J}_2(F[x]) = \ldots = \mathcal{P}(F[x]) = \{0\}$. This holds by 7.88(b).

(b) For any integral domain R, $\mathcal{N}(R[x]) = \mathcal{P}(R[x]) = \{0\}$, for by 7.68(a) $R[x]$ has no nilpotent elements $\neq 0$.

(c) Clay-Doi (1), Mlitz (1) and (3) determined radicals of some $F[x]$'s, which do not always coincide with our $\mathcal{J}_2(N),\ldots,\mathcal{P}(N)$. But one can get instantly results on polynomial near-rings over finite fields F:
If $|F| \neq 2$ then $\mathcal{J}_2(F[x]) \subseteq \operatorname{Ker} h$ (by 7.88(a)).
If char $F \neq 2$ then $\mathcal{J}_{1/2}(F[x]) = \ldots = \mathcal{N}(F[x]) = \{0\}$ (this follows from 7.94).

(d) The situation in the general case does not seem to be clear. Anyhow, we observe two strange phenomena:
(1) The near-ring-radicals might differ substantially from the ring-radicals of $R[x]$ (the latter ones are always $= \{0\}$ if R is a field).
(2) The smaller F, the more complicated is the structure of $F[x]$.

(e) See So (1) for a detailed study of the ideal structure of $R[x]$ and $\overline{R}[x]$.

6.) $\Gamma[x]$ AND $\overline{\Gamma}[x]$

7.103 PROPOSITION (Lausch-Nöbauer (1)). Every normal subgroup D of $\Gamma[x]$ is a left ideal of $\Gamma[x]$.

Proof. Similar to 7.77: if $d \in D \trianglelefteq (\Gamma[x],+)$ and $p,q \in \Gamma[x]$ then with $p = \gamma_0+n_1x+\gamma_1+n_2x+\ldots+\gamma_{r-1}+n_rx+\gamma_r$ we get
$po(d+q)-poq = \gamma_0+n_1(d+q)+\gamma_1+\ldots+\gamma_{r-1}+n_r(d+q)+\gamma_r-$
$-\gamma_r-n_rq-\ldots-\gamma_0 \in D$.

The definition 7.74 of polynomial completeness is carried over to $\Gamma[x]$ in the obvious way. Similar to 7.75 we get

7.104 THEOREM (Lausch-Nöbauer (1)). The polynomially complete groups are exactly $\mathbb{Z}_2$ and all finite, non-abelian simple groups.

Proof. (a) As in 7.75, a polynomially complete group Γ is shown to be finite and simple.
If Γ is abelian and $|\Gamma|\geq 3$ then Γ is some $\mathbb{Z}_p$ ($p \in \mathbb{P}$, $p\geq 3$). Take $f \in M(\mathbb{Z}_p)$ with $f(0) = 0$, $f(1) = 1$ and $f(2) = 0$.
If $\overline{q} \in \overline{\Gamma}[x]$, $\overline{q}$ has the form $\overline{q}: \gamma \to \gamma_0+n\gamma$ (7.64). From $\overline{q}(0) = f(0)$ and $\overline{q}(1) = f(1)$ we get $\gamma_0 = 0$ and $n = 1$. But then $\overline{q}(2) = 2 \neq f(2)$, whence $f \neq \overline{q}$, and Γ is not polynomially complete.

(b) Conversely, $\mathbb{Z}_2$ is easily shown to be polynomially complete, while for finite non-abelian simple groups each $\overline{p}_\delta: \gamma \to \delta+\gamma-\delta$ ($\delta \in \Gamma$) is in $\overline{\Gamma}[x]$. Consequently $I(\Gamma) \subseteq \overline{\Gamma}[x]$. But $I(\Gamma) = M(\Gamma)$ for these groups (7.46).

7.105 REMARK This result is transferred to Ω-groups by Lausch (2).

7.106 PROPOSITION $\Gamma_o[x]: = (\Gamma[x])_o = \{p = \gamma_o+n_1x+\ldots+n_rx+ +\gamma_r \mid \sum_{i=o}^{r} \gamma_i = o\}$. If Γ is abelian then $\Gamma_o[x] = \{nx \mid n\in\mathbb{N}_o\}$.

Proof. trivial.

Again, Γ can be considered as an $\Gamma[x]$- and an $\overline{\Gamma}[x]$-group (cf. 7.80). But in contrast to 7.88 we get

7.107 REMARK Γ is as $\Gamma[x]$- and $\overline{\Gamma}[x]$-group not of type 2 in general. (Consider an abelian, but not simple group and you have a counterexample.)

But the theory of enclosing ideals still works:

7.108 EXAMPLES Let $\Delta \trianglelefteq \Gamma\in\mathcal{G}$. Then

(a) (Δ) (= the ideal of $\Gamma[x]$ generated by Δ) $\trianglelefteq \Gamma[x]$.

(b) $<\Delta>: = (\Delta:\Gamma) \trianglelefteq \Gamma[x]$.

See Lausch-Nöbauer (1) or Hule (1) for a description of (Δ). Completely similar to 7.84 we get

7.109 THEOREM (Lausch-Nöbauer (1)). For each $I \trianglelefteq \Gamma[x]$ there is exactly one $\Delta \trianglelefteq \Gamma$ with $(\Delta)\subseteq I\subseteq<\Delta>$.

7.110 DEFINITION Again, Δ is called the enclosing ideal of I.

Concerning simplicity we get surprisingly

7.111 THEOREM (Hule (1)). $\Gamma[x]$ is never simple (unless $|\Gamma| = 1$).

Proof. Let Γ have at least 2 elements and take some non-zero $\gamma\in\Gamma$. Suppose that $\Gamma[x]$ is simple. Consider the group epimorphism

$$g: \Gamma[x] \longrightarrow \mathbb{Z}_2$$
$$\gamma_o+n_1x+\ldots+n_rx+\gamma_r \to (n_1+\ldots+n_r)\cdot 1$$

Clearly $g \neq \bar{o}$. Hence $\text{Ker } g = \{\bar{o}\}$ and g is an isomorphism. Consequently $|\Gamma[x]| = |\mathbb{Z}_2| = 2$ which is quite hard to fulfill since $\Gamma[x]$ is infinite.

7.112 REMARK See Mlitz (1) for a computation of some radicals in $\Gamma[x]$ (including other ones than ours in §5).

7.) CONCLUDING REMARKS

We close this section with some remarks concerning questions related to polynomial and polynomial-like near-rings.

7.113 REMARKS

(a) Nöbauer (6) remarked that for $R,S \in \mathcal{R}$, $R[x] \cong S[x]$ implies that $R \cong S$.
He also remarked that each subnear-ring of $M(F)$ (F a field) which contains all constant functions is automatically simple.
$\overline{R}[x]$ is directly decomposable iff this applies to R.

(b) If C is a composition ring and D is a map $C \to C$ then D is called a derivation (Menger (3), Müller (1), Lausch-Nöbauer (1), Nöbauer (9)) if for all $a,b \in C$:

(1) $D(a+b) = D(a)+D(b)$ ("sum rule")
(2) $D(a \cdot b) = D(a) \cdot b + a \cdot D(b)$ ("product rule")
(3) $D(a \circ b) = (D(a) \circ b) \cdot D(b)$ ("chain rule")

Clearly the zero endomorphism on C is a (trivial) derivation. $R[x]$ has also a non-trivial derivation, namely the usual one: $D: p \to p'$. All on $R[x]$ are given by $D_r: p \to r \cdot p'$, where $r \in R$ is idempotent (Lausch-Nöbauer (1)).
Nöbauer (6) showed that the composition ring $M(R)$ has no derivations except the trivial one. If R is a finite field, the same applies to $\overline{R}[x]$ (by 7.75). If R is an infinite integral domain then Müller (1) showed e.g. that if $(R,+)$ is torsion-free, the sum-

and the chain rule imply the product rule.
Müller studied also "derivations" in near-rings.

(c) Invertible elements (w.r.t. $\circ$) are studied in Lausch-Nöbauer (1) and Suvak ((1), (2)).
Those $p \in R[x]$ such that $\overline{p}$ is bijective (= invertible) are called permutation polynomials, were considered by many authors and are presented extensively in Lausch-Nöbauer (1).

(d) Clearly $R[x]$ and $\overline{R}[x]$ are in general non-commutative near-rings. Those polynomials which commute with a certain family of others were studied e.g. by Kautschitsch (1) and Lausch-Nöbauer (1).
Call $C \subseteq F[x]$ (F a field) a <u>P-chain</u> ("<u>permutable chain</u>") if $\forall\, c \in C$: $[c] > 0$ and $\forall\, k \in \mathbb{N}\ \exists\, c \in C$: $[c] = k$.

Examples: (1) The <u>P-chain of powers</u> $\{x, x^2, x^3, \ldots\}$.

(2) The P-chain of <u>Čebyshev polynomials</u> $\{t_1, t_2, t_3, \ldots\}$ (where t_n is defined via $\cos n\phi = t_n \circ \cos \phi$ over $F = \mathbb{Q}$ and then transferred to $F[x]$ for an arbitrary field F:

$$
\begin{aligned}
t_1 &= x\\
t_2 &= -1+2x^2\\
t_3 &= -3x+4x^3\\
t_4 &= 1-8x^2+8x^4\\
&\vdots
\end{aligned}
$$

Also, $t_n \circ t_m = t_{nm}$.)

If ℓ is a linear polynomial and C is a P-chain then $C_\ell := \{\ell \circ c \circ \ell^{-1} \mid c \in C\}$ is a P-chain, too, called a <u>conjugate</u> P-chain.
One can see (the proofs are not too easy - see Lausch-Nöbauer (1), p. 156 - 159):

(α) If C is a P-chain then C contains to each $k \in \mathbb{N}$ exactly one c with $[c] = k$.

(β) All P-chains over a field F are conjugates of

either the P-chain of powers or of the P-chain of Čebyshev polynomials.

(e) Lausch-Nöbauer (1), ch. 5, contains more information on $\Gamma[x]$ and $\overline{\Gamma}[x]$. For example, the classes $E_k(\Gamma)$ of all k-place functions generated by all "k-place endomorphisms on Γ" are considered ("k-dimensional composition groups"). These are more examples of dgnr.'s and results similar to our 6.33 and 7.46 are obtained.

(f) Heatherly (8) considered $F_0[x]$ (F a field). This is a near-ring with identity, but without divisors of zero. $F_0[x]$ is also not regular (9.154). The ideals $I_k := \{\sum_{i=k}^{k+n} a_i x^i \mid n \in \mathbb{N}_0,\ a_i \in F\}$ form a strictly descending chain. So $F_0[x]$ does not fulfill the DCC on ideals (cf. 7.97).

(g) Nöbauer (6) also considers the near-rings $R(x)$ and $\overline{R}(x)$ of all "rational polynomials" and "rational polynomial functions". Again $\overline{R}(x) = M(R)$ iff R is a finite field (cf. 7.75). $\overline{R}(x)$ is directly decomposable iff R is it (cf. Remark (a)).

(h) The near-rings $R[[x]]$ of all formal power series over $R \in \mathcal{R}$ were considered by Fröhlich (9), Cartan (1) and others.
Fröhlich (9) studied $M := (R[[x_1,\dots,x_n]])^n$, defined in this set a composition "$\circ$" by $(f \circ g)_i := f_i(g_1,\dots,g_n)$ (where f_i denotes the i-th component of $f \in M$.
If $f \equiv g :\Longleftrightarrow$ all f_i and g_i have the same degree, then one can cefine in $M/\equiv$ an addition "+" in that way that $(M/\equiv,+,\circ)$ is a near-ring of number-theoretic relevance.
Cartan's result was already mentioned.in 1.12.
Graves-Malone (3) looked at the subnear-ring $N = \{\sum_{n\geq 0} a_{2n+1}x^{2n+1} \mid a_{2n+1} \in \mathbb{R}\}$ of $\mathbb{R}[[x]]$.

N satisfies the right Ore condition (1.64) and is integral.

Kautschitsch (1) considered full ideals and Müller (1) studied derivations in $R[[x]]$.

(i) Istinger-Kaiser determined all polynomially complete near-rings in (1) (cf. 7.105).

Anyhow, this section seems to be a wide field for further research.

§ 8 NEAR-FIELDS AND PLANAR NEAR-RINGS

This chapter brings up two important classes of near-rings. We start with perhaps the most important class, the near-fields. A thorough treatment would require nearly a whole book. But there are several excellent presentations of parts of this theory (e.g. Karzel (3), Kerby (9) and Wähling (9)) so that we dare to give the theory partly without proofs. First we characterize those nr.'s which happen to be nf.'s. After showing that the additive group of a nf. is abelian we give a super-sonic trip through the relations between near-fields and geometry (incidence groups, coordinatisation of planes, planar near-fields).

In b) we deal with planar near-rings. Their structure is explored (8.90, 8.96), "blocks" $aN+b$ $(a \neq 0)$ are defined and it is shown that a planar finite near-ring together with its blocks forms a tactical configuration $(N,\mathcal{B})$. The case when $(N,\mathcal{B})$ is a balanced incomplete block design is characterized in 8.118 and several consequences are deduced.

The author thanks Dr. G. Betsch for leaving him unpublished lecture notes concerning this paragraph.

a) NEAR-FIELDS

1.) CONDITIONS TO BE A NEAR-FIELD

We start with (cf. 1.15)

8.1 PROPOSITION If N is a nf. then either $N \cong M_c(\mathbb{Z}_2)$ or N is zero-symmetric.

Proof. If $n_c \in N_c$, $n_c \neq 0$, then $1 = n_c n_c^{-1} = n_c$, whence $1 \in N_c$.
So $\forall\, n \in N^*$: $n = 1n = 1$, hence $N = \{0,1\}$. The rest is obvious.

8.2 CONVENTION In all of our subsequent discussion of near-fields we will exclude this silly nf. $M_c(\mathbb{Z}_2)$ of order 2.

Evidently, every near-field is simple.
We now characterize those near-rings which are near-fields:

8.3 THEOREM (Ligh (2), Maxson (1), Beidleman (1), Fain (1)). Equivalent are for $N \in \mathcal{N}_0$:

(a) N is a near-field.
(b) $N_d \neq \{0\}$ and $\forall\, n \in N^*$: $Nn = N$.
(c) N has a left identity and ${}_NN$ is N-simple.
(d) N has a left identity and N is 2-primitive on ${}_NN$.
(e) N has a left identity and N is 1-primitive on ${}_NN$.

If N is finite one can add:

(f) N has a left identity and N is 0-primitive on ${}_NN$.
(g) N has a left identity and N is simple.

Proof. (a) => (b) is clear.

(b) => (a): $\forall$ $a,b \in N^*$ $\exists$ $a',b' \in N^*$: $b'b = a \wedge a'a = b'$.
Thus $a'(ab) = (a'a)b = b'b = a \neq 0$, so $ab \neq 0$ and N is integral.
Take some $d \in N_d^*$. $\exists$ $e \in N$: $ed = d$.
So $(de-d)d = ded-dd = 0$. From above, we get $de = d$.

Now let n be $\in N^*$.
Then $d(en-n) = den-dn = 0$, whence $en = n$.

Finally, $\forall$ $n \in N^*$ $\exists$ $n' \in N^*$: $n'n = e$.
This shows, that $(N^*, \cdot)$ is a group and $(N,+,\cdot)$ is a near-field.

(a) => (c) <=> (d) <=> (e) are obvious (observe 4.6)

(c) ⇓ (b)

If N is finite, it suffices to apply 4.47(a).

8.4 REMARK (Ligh (2)). Of course, e.g. (c) in 8.3 can be replaced by (c)': "$N_d \neq \{0\}$, $\forall$ $n \in N^*$ $\exists$ $n' \in N^*$: $n'n \neq 0$ and ${}_N N$ is N-simple." (For (c)' => (b) => (c) => (d) => => (c)'!)

Without proof we mention the following results of Ligh (2) and (1):

8.5 THEOREM Let $N \neq \{0\}$ be a dgnr..
N is a skew-field <=> $\forall$ $n \in N^*$ $\exists!$ $n' \in N$: $nn'n = n$ <=> <=> $\forall$ $n \in N^*$: $Nn = N$.

8.6 COROLLARY A finite integral dgnr. is a commutative field.
A dgnr. $N \neq \{0\}$ with left identity is a field iff it is N-simple.

8.7 THEOREM $N \in \mathcal{n}_0 \cap \mathcal{n}_1$ is a nf. <=> every $n \in N$, $n \neq 1$, is qr. (in Beidleman's sense - see 3.37 c)).

8.8 REMARK See Andre (4) for a development of a theory of "linear algebra over near-fields" and "near-vector-spaces" (cf. also Beidleman (1)).

2.) THE ADDITIVE GROUP OF A NEAR-FIELD

Let the characteristic char N of a near-field N be defined as usual - (Wähling (9) defines char N: = char N_d - but this gives the same (see 8.23)).

Then one sees as for fields:

8.9 PROPOSITION Let N be a nf. and o(1) be the order of 1 in (N,+). Then

(a) If o(1) is finite then char N = o(1).
(b) If o(1) is infinite then char N = 0.
(c) char N is either 0 or a prime.

For the following result, cf. and apply 1.5.

8.10 PROPOSITION (Karzel (3), Maxson (1), Ligh-Neal (1)).
Let N be a nf. . Then

(a) $\forall$ $n \in N$: ($n^2 = 1 \iff n \in \{1,-1\}$).

(b) $\forall$ $n,n' \in N$: $n(-n') = (-n)n' = -nn'$.

Proof. (a): "<=" is clear; so let $n^2 = 1$, but $n \neq 1$.
If char N = 2 then (observe that (N,+) is abelian in this case) $(n+1)n = n^2+n = (1+n)\cdot 1$; now $n+1 \neq 0$, whence $n = 1$, a contradiction.
Now let char N be $\neq 2$, and $1+1 =: 2$ ($\neq 0$).
Then $2(-1) = (1+1)(-1) = -1-1 = -(1+1) = (-1)2$.
So $(-2)^{-1} = (2\cdot(-1))^{-1} = (-1)^{-1}\cdot 2^{-1} = -2^{-1}$.
Observe that $(2^{-1}(-1)+1)(-2) = 1-2 = -1$, whence $(2^{-1}(-1)+1) = 2^{-1}$. Let $m: = 2^{-1}(n-1)+1$.
Then $m\cdot n = 2^{-1}\cdot(n-1)\cdot n+n = 2^{-1}(n^2-n)+n = 2^{-1}(1-n)+n-$
$-1+1 = 2^{-1}(-1)(n-1)+(n-1)+1 = (2^{-1}(-1)+1)(n-1)+1 =$
$= (2^{-1})(n-1)+1 = m = m\cdot 1$.
$n \neq 1$ gives $m = 0$, so $n-1 = 2\cdot(-1) = -1-1$, whence $n = -1$.

(b) It suffices to assume $n \neq 0$. But then $(n^{-1}(-1)n)^2 = 1$, so by (a) we get $n^{-1}(-1)n \in \{1,-1\}$. $n^{-1}(-1)n = 1$ implies $-n = n$, so char $N = 2$. But then the result is trivial.
If $n^{-1}(-1)n = -1$, $n(-1) = (-1)n$, so $\forall\, n' \in N$: $n(-n') = n(-1)n' = (-1)nn' = -nn'$.

The following famous result was first shown for finite nf.'s by Zassenhaus in 1936, for infinite nf.'s by B.H. Neumann in 1940. There exist several essentially different proofs. The following (due to Karzel) seems to be the most simple one.

8.11 THEOREM (Dickson (1), Zassenhaus (1), B.H. Neumann (1), Ligh (6), (13), Ligh-McQuarrie-Slotterbeck (1), Karzel (3), Zemmer (3)).
The additive group of a nf. is abelian and characteristically simple.

Proof. By 8.10(b) $\forall\, n \in N$: $n(-1) = -n$. Hence by 1.109(a), $(N,+)$ is abelian.
Consider for $n \in N^*$ the automorphism $\alpha_n: \begin{array}{l} N \to N \\ x \to xn \end{array}$.
If $M \neq \{0\}$ is a characteristic subgroup of $(N,+)$ take $m \in M^*$. Let n' be arbitrary in N^*. Then $\alpha_{m^{-1}n'}(m) = n' \in M^*$, whence $M = N$.

Recall that from 1.88(g) and (a) one sees that every nr. is isomorphic to a nf. of bijective mappings (plus the zero map) on an abelian group Γ.
Standard group theory gives us the structure of the additive groups of nf.'s:

8.12 COROLLARY (Ligh-McQuarrie-Slotterbeck (1)). Let N be a nf.

(a) If char $N = 0$ then $(N,+)$ is torsion-free divisible, so the direct sum of copies of $(\mathbb{Q},+)$.

(b) If char $N = p$ then $(N,+)$ is elementary abelian, so the direct sum of copies of $(\mathbb{Z}_p,+)$.

The orders of finite nf.'s are the same as those of finite fields (cf. also number 4.)), for 8.12(b) implies

8.13 COROLLARY If N is a finite nf. then $\exists\, p\in\mathbb{P}\ \exists\, k\in\mathbb{N} : |N|=p^k$.

8.14 COROLLARY (Heatherly (2), Ligh (2)). Let N be a finite nr. with $N_d \neq \{0\}$ and $(N,+)$ simple.
Then either $N^2 = \{0\}$ or N is a commutative field.

Proof. Let some $n'n''$ be $\neq 0$. Take $d\in N_d^*$.
$(0:n'') \trianglelefteq (N,+)$ implies $(0:n'') = \{0\}$. Hence $dn'' \neq 0$ and $n''\notin D := \{n\in N \mid dn = 0\} \trianglelefteq (N,+)$, so $D = \{0\}$. Consequently, $\forall\, n\in N^*$: $(0:n) = \{0\}$. By 8.4, N is a nf., hence abelian.
But then $\{0\} \neq (N_d,+) \trianglelefteq (N,+)$, so $N_d = N$ and N is a finite field.

8.15 REMARK See Wähling (9), p. 49, for a characterization (due to P. M. Cohn) of those groups which can be the multiplicative group of a near-field.

3.) THE CENTER AND THE KERNEL OF A NEAR-FIELD

8.16 NOTATION Let $C(N) := \{n\in N \mid \forall n'\in N: nn' = n'n\}$ be the center of $(N,\cdot)$ and call N_d the "kernel of N" (Karzel et al.).

8.17 THEOREM (Karzel (3)). The subnear-ring I of the nf. N generated by C(N) consists of all sums of elements of C(N) and is an integral domain. The subnear-field of N generated by C(N) is the field of quotients of I.

Proof: straightforward calculations using 8.10 and 8.11.

8.18 COROLLARY Every nf. N contains a commutative subfield F.

There is a (possibly) different subfield in N:

8.19 THEOREM (Zemmer (1)). If N is a nf. then N_d is a subfield of N.

8.20 REMARK If M is a subfield of a nf. N then N can be considered as a vector space over M. It's dimension will be denoted by $\dim_M N$.

8.21 REMARK Clearly $C(N) \subseteq N_d$. More exactly (but without proof) there is the following relation between center and kernel of a near-field:

8.22 THEOREM (André (2), Wähling (3)). Let N be a nf. which is no proper skew-field. Then

$$C(N) = \bigcap_{n \in N^*} n^{-1}(N_d)n.$$

Moreover, $C(N) = N_d$ iff $\forall\ n \in N^*: n^{-1}N_d n = N_d$.

See more on that e.g. in Wähling (9).

We close this number with the following

8.23 REMARK If N is a nf. then $\operatorname{char} N = \operatorname{char} N_d$ (by 8.9(a)).

4.) DICKSON NEAR-FIELDS

Dickson obtained the first proper near-fields in 1905 by "distorting" the multiplication in a finite field.
We axiomatize this procedure, tracking the presentation of Wähling (9). Proofs (or references where to find them) can be found there.
For this number, confer also the chapter on Dickson near-rings in §9d). Unless otherwise indicated, N will always denote a nf.

8.24 DEFINITION A map $\phi: N^* \to \mathrm{Aut}\,(N,+,\cdot)$, $n \to \phi_n$ is called a coupling map if $\forall\, m,n \in N$: $\phi_n \circ \phi_m = \phi_{\phi_n(m)\cdot n}$.

8.25 EXAMPLE $\phi: n \to id_N$ is a coupling map on N.

8.26 NOTATION If ϕ is a coupling map on N then

$$n \circ_\phi m := \begin{cases} \phi_m(n)\cdot m & \text{if } n \neq 0 \\ 0 & \text{if } n = 0 \end{cases} .$$

8.27 PROPOSITION If ϕ is a coupling map on N then $(N,+,\circ_\phi)$ is again a near-field.
(The "coupling property" in 8.24 is just the restatement of the associativity of $\circ_\phi$).

8.28 DEFINITION $(N,+,\circ_\phi)$ is then called the ϕ-derivation of $(N,+,\cdot)$ and denoted as N^ϕ.
$\{\phi_n \mid n \in N^*\}$ is called the Dickson-group of ϕ.
N is said to be a Dickson near-field if N is the ϕ-derivation of some field F: $F^\phi = N$.

To the author's knowledge, all known near-fields (up to 7 examples - see below) are Dickson near-fields.
We give an example of a class of finite and infinite Dickson near-fields which are not fields:

8.29 EXAMPLE (Zemmer (1)). Let F be a commutative field and F(x) the field of rational functions (7.113(h)).
$\phi: F(x) \to \mathrm{Aut}\, F(x)$, given by $\phi_{\frac{f(x)}{g(x)}}\left(\frac{p(x)}{q(x)}\right) := \frac{p(x+[f]-[g])}{q(x+[f]-[g])}$
is a coupling map on $F(x)$ and $(F(x),+,\circ_\phi)$ is a Dickson near-field.

For "most" finite fields we get important coupling maps. But first we need the following

8.30 DEFINITION $(q,n)\in\mathbb{N}^2$ is called a pair of Dickson numbers if

(a) q is some power p^ℓ of a prime p.

(b) Each prime divisor of n divides $q-1$.

(c) If $q \equiv 3 \pmod n$ then 4 does not divide n.

8.31 THEOREM Let (q,n) be a pair of Dickson numbers. Let F be the (Galois-)field $GF(q^n) = GF(p^{\ell n})$ with q^n elements. $(F^*,\cdot)$ is cyclic. Let g be a generator and let H be the subgroup of $(F^*,\cdot)$ generated by g^n. Let α be the (Frobenius-) automorphism $f \to f^q$ of $(F,+,\cdot)$.

Then F^*/H can be represented as $\{Hg, Hg^{\frac{q^2-1}{q-1}}, \ldots, Hg^{\frac{q^n-1}{q-1}}\}$.

Let $\lambda(Hg^{\frac{q^k-1}{q-1}}) := \alpha^k \in \mathrm{Aut}\,(F,+,\cdot)$. If $\pi: F^* \to F^*/H$ is the canonical epimorphism then $\phi := \lambda\pi$ is a coupling map on F.

$N := F^\phi = (GF(p^{\ell n}),+,\circ_\phi)$ is a nf. with $C(N) = N_d =$
$= \{x\in F \mid \forall f\in F^*: \phi_f(x) = x\} = GF(q)$.

The number of non-isomorphic Dickson near-fields derived in this way (by different choices of g) is $\frac{\Phi(n)}{i}$, where Φ is the Euler-function and i is the order of $p \pmod n$. Their multiplicative groups are isomorphic.

8.32 THEOREM By taking all pairs of Dickson numbers, all finite Dickson near-fields arise in the way described in 8.31.

This makes the question, which (finite) near-fields are Dickson near-fields, even more interesting. Of course, it might be hard to visualize Dickson near-fields with naked eyes. So we use an instrument (see e.g. Wähling (9)):

8.33 THEOREM ("Zassenhaus-criterion"): A finite nf. N is a Dickson nf. iff $G := (N^*,\cdot)$ is metacyclic (i.e. $[G,G]$ and $G/[G,G]$ are cyclic - see Zassenhaus (3), p. 174). In this case, G has two generators a,b with $b^{-1}ab = a^q$, where $q = \mathrm{char}\, N$.

It was Zassenhaus, too, who determined all finite nf.'s:

8.34 THEOREM All finite nf.'s - up to 7 exceptional cases - are Dickson nf.'s.

Now we are going to describe these 7 "outsiders" $N_1,\dots,N_7$ (numbering is the one of Zassenhaus):
All of them are of order p^2 with $p = 5,7,11$ (two cases), 23, 29 or 59.
Since all $N \in \{N_1,\dots,N_7\}$ can be considered as vector spaces over N_d of dimension 2 and since for each $n \in N^*$ the map $x \to xn$ is an element of $\mathrm{Aut}_{N_d}(N)$ it suffices to describe $(N^*,\cdot)$ via 2×2-matrices over $N_d = GF(p)$:

i	p	order of N_i	$(N_i^*,\cdot)$ is generated by $A:=\begin{pmatrix}0&-1\\1&0\end{pmatrix}$ and
1	5	5^2	$\begin{pmatrix}1&-2\\-1&-2\end{pmatrix}$
2	11	11^2	$\begin{pmatrix}1&5\\-5&-2\end{pmatrix}$ and $\begin{pmatrix}4&0\\0&4\end{pmatrix}$
3	7	7^2	$\begin{pmatrix}1&3\\-1&-2\end{pmatrix}$
4	23	23^2	$\begin{pmatrix}1&-6\\12&-2\end{pmatrix}$ and $\begin{pmatrix}2&0\\0&2\end{pmatrix}$
5	11	11^2	$\begin{pmatrix}2&4\\1&-3\end{pmatrix}$
6	29	29^2	$\begin{pmatrix}1&-7\\-12&-2\end{pmatrix}$ and $\begin{pmatrix}-13&0\\0&-13\end{pmatrix}$
7	59	59^2	$\begin{pmatrix}9&15\\-10&-10\end{pmatrix}$ and $\begin{pmatrix}4&0\\0&4\end{pmatrix}$

The smallest Dickson non-field which is no field is given by

$$(GF(3^2),+,\circ) \quad \text{with} \quad x \circ y := \begin{cases} x\cdot y & \text{if } x \text{ is a square in } (GF(3^2),+,\cdot) \\ x\cdot y^3 & \text{otherwise} \end{cases}$$

Its multiplicative group is the quaternion group (of order 8) (see e.g. Zemmer (1), André (1) or (for a slightly different presentation) Karzel-Sörensen-Windelberg (1), p. 22).

8.35 THEOREM (Ligh-Neal (1)). Let N be a finite nf. such that $\forall\, n \in N$: $n^k = n$ where k is of the form $k = p^j+1$ ($p \in \mathbb{P} \setminus \{2\}$, $j \in \mathbb{N}$).
Then N is a field.

5.) NEAR-FIELDS AND DOUBLY TRANSITIVE GROUPS

Near-fields (and some similar structures) can be used to describe "sharply transitive" permutation groups. The following discussion follows Kerby (9).

8.36 NOTATION If M is an arbitrary set, let S_M be the symmetric group on M (i.e. the group of all 1-1-maps $M \to M$). If $k \in \mathbb{N}$, $(m_1,\ldots,m_k) \in M^k$ is called a proper k-tuple if all m_i's are distinct.

8.37 DEFINITION $\Gamma \leq S_M$ is called (sharply) k-transitive (on M) if for all proper k-tuples $(m_1,\ldots,m_k)$, $(m_1',\ldots,m_k') \in M^k$ there is (exactly) one $\gamma \in \Gamma$ with $\gamma(m_i) = m_i'$ for all $i \in \{1,\ldots,k\}$ (cf. 4.26).
1-transitive groups are simply called transitive, the sharply 1-transitive ones are just the regular permutation groups.

8.38 NOTATION Let $(m_1,\ldots,m_k) \in M^k$ be proper and $\Gamma \leq S_M$. Then $\Gamma_{m_1,\ldots,m_k} := \{\gamma \in \Gamma \mid \forall\, i \in \{1,\ldots,k\}: \gamma(m_i) = m_i\}$ denotes the stabilizer (subgroup) of $(m_1,\ldots,m_k)$.

Near-fields are primarily applicable to sharply k-transitive groups. There is no need to consider large k's:

8.39 THEOREM (C. Jordan (1872), M. Hall (1954) and others - see e.g. Kerby (9)):
If $k \geq 4$ then all sharply k-transitive permutation groups are finite and isomorphic either to S_n ($n \geq 4$), A_n ($n \geq 6$) or to the Mathieu groups of degree 11 or 12.

Since regular permutation groups are well-studied (see. e.g. (Wielandt) or (Passman)), we turn our attention to sharply 2- and 3-transitive permutation groups.
We start with the sharply 2-transitive ones. Our interest stems from

8.40 EXAMPLE Let N be a nf.. Then the group $T_2(N)$ of all "affine transformations" (cf. §9c)) $x \to xa+b$ $(a,b \in N,\ a \neq 0)$ is sharply 2-transitive on N.

However, not all sharply 2-transitive groups seem to arise in this way. We have to consider a new algebraic system which is a "little bit" more general than a near-field.

8.41 DEFINITION A near-domain is a set N with two binary operations "+" and "·" subject to the following conditions:

(a) $(N,+)$ is a loop (with zero 0) (see Bruck (1)).

(b) $\forall\ n,n' \in N\colon n+n' = 0 \Rightarrow n'+n = 0$.

(c) $(N^*,\cdot)$ is a group.

(d) $\forall\ n \in N\colon n0 = 0$.

(e) $\forall\ n,n',n'' \in N\colon (n+n')n'' = nn''+n'n''$.

(f) $\forall\ n,n' \in N\ \exists\ d_{n,n'} \in N^*\ \forall\ n'' \in N\colon n+(n'+n'') = (n+n')+d_{n,n'} \cdot n''$.

Near-domains can be viewed as "additively non-associative near-fields" (cf. 8.75):

8.42 REMARK A near-domain with associative addition is a nf..

It is not known if there exist near-domains which are no near-fields. Anyhow, those ones must be infinite:

8.43 THEOREM A finite near-domain is a near-field.

We define for a near-domain N $T_2(N)$ as in 8.40 and get

8.44 THEOREM

(a) For each near-domain N, $T_2(N)$ is sharply 2-transitive.

(b) Conversely, for each sharply 2-transitive permutation group Γ on a set M, M can be made into a near-domain such that $\Gamma = T_2(M)$.

8.45 COROLLARY All finite sharply 2-transitive permutation groups are exactly the $T_2(N)$'s, where N is a finite nf.

So by 8.31, 8.32 and 8.34, all finite sharply 2-transitive permutation groups are determined.

There exist many conditions under which a near-domain is forced to be a near-field. They are excellently presented in Kerby (9). We mention only one:

8.46 NOTATION If Γ is a group then $I_\Gamma := \{\gamma\varepsilon\Gamma | \gamma^2 = 1\}$ denotes the subset of the "involutions" of Γ.
Let $(I_\Gamma)^2 := \{\gamma_1\gamma_2 | \gamma_1,\gamma_2 \varepsilon I_\Gamma\}$.

8.47 THEOREM Let Γ be a sharply 2-transitive permutation group on M and $(M,+,\cdot)$ "it's" near-domain (8.44(b)). Then M is a near-field $\iff (I_\Gamma)^2 \leq \Gamma$.

8.48 REMARK Sharply 3-transitive groups can be characterized in a similar, but more complicated way by groups of things like "fractional affine transformations" on certain near-domains (so-called "Karzel-Tits-fields"). See Kerby (9). See also all S"-labeled items in the bibliography.

6.) NORMAL NEAR-FIELDS AND INCIDENCE GROUPS

In order to be able to formulate the connections between nf.'s and geometry we drive in another country and recall some geometry.

8.49 DEFINITION Let P be a set and $\mathcal{L} \subseteq 2^P$. The pair $(P,\mathcal{L})$ is called an incidence structure. $(P,\mathcal{L})$ is an incidence space provided that

(a) $\forall\ p,q \in P,\ p \neq q\ \exists!\ L \in \mathcal{L}: p \in L \wedge q \in L$.

(b) $\forall\ L \in \mathcal{L}: |L| \geq 2$.

The elements of P are then called "points" and those of $\mathcal{L}$ "lines". L of (a) is called the "line determined by p,q" and denoted by $\overline{pq}$. If $L,M \in \mathcal{L}$, set $L /\!/ M :\Longleftrightarrow (L=M) \vee (L \cap M = \emptyset)$. Call $(P,\mathcal{L})$ degenerated if every set of 3 points is on a common line.

8.50 DEFINITION Two incidence spaces $(P,\mathcal{L})$ and $(P',\mathcal{L}')$ are called isomorphic if $\exists\ h: P \to P'$ with h bijective and $\forall\ M \subseteq P: h(M) \in \mathcal{L}' \Longleftrightarrow M \in \mathcal{L}$. h is then called an isomorphism or (if $P = P'$ and $L = L'$) an automorphism.

8.51 DEFINITION A subset S of an incidence space $(P,\mathcal{L})$ is called subspace if it is "convex", i.e. if $\forall\ s,t \in S,\ s \neq t: \overline{st} \in \mathcal{L}$.

8.52 REMARK The subspaces of an incidence space $(P,\mathcal{L})$ form an inductive Moore-system. Hence it makes sense to speak about the "subspace generated by a subset of P".

8.53 DEFINITION A non-degenerated incidence space $(P,\mathcal{L})$ is called an

(a) affine plane if $\forall\ L \in \mathcal{L}\ \forall\ p \in P\ \exists!\ M \in \mathcal{L}: p \in M \wedge L /\!/ M$.

(b) projective plane if $\forall\ L \in \mathcal{L}: |L| \geq 3$ and $\forall\ L,M \in \mathcal{L}: L \cap M \neq \emptyset$.

Each affine plane can be extended to a projective plane by adding some points. Conversely, one gets an affine plane from a projective one by taking out one line (see e.g. Karzel-Sörensen-Windelberg (1)).

8.54 DEFINITION A subspace of an incidence space $(P,\mathcal{L})$ generable by 3 points is called a plane in $(P,\mathcal{L})$.

8.55 DEFINITION An incidence space $(P,\mathcal{L})$ is called a projective space if each plane in $(P,\mathcal{L})$ is a projective plane.

8.56 DEFINITION Let $(P,\mathcal{L})$ be a projective space. $B \subseteq P$ is called a base of $(P,\mathcal{L})$ if B is a minimal generating set for $(P,\mathcal{L})$.

8.57 THEOREM Each projective space has a (non-empty) base and all bases are equipotent (see. e.g. Karzel-Sörensen-Windelberg (1)).

8.58 DEFINITION If B is a base for the projective space $P := (P,\mathcal{L})$ then $\dim P := |B|-1$ is called the dimension of P.

8.59 PROPOSITION The automorphisms of a projective space P (the "collineations") form a group Coll (P).

8.60 DEFINITION A projective space $(P,\mathcal{L})$ is called Desarguesian if, whenever two "triangles" $\{a_1,a_2,a_3\}$ and $\{b_1,b_2,b_3\}$ $(a_1,a_2,a_3,b_1,b_2,b_3 \in P)$ are "perspective w.r.t. a center $o \in P$" (that means that

$\exists\, L_1,L_2,L_3 \in \mathcal{L}\ \forall\, i \in \{1,2,3\}: o \in L_i \wedge a_i \in L_i \wedge b_i \in L_i$)

then $\overline{a_1a_2} \cap \overline{b_1b_2}$, $\overline{a_1a_3} \cap \overline{b_1b_3}$ and $\overline{a_2a_3} \cap \overline{b_2b_3}$ are in some common line L:

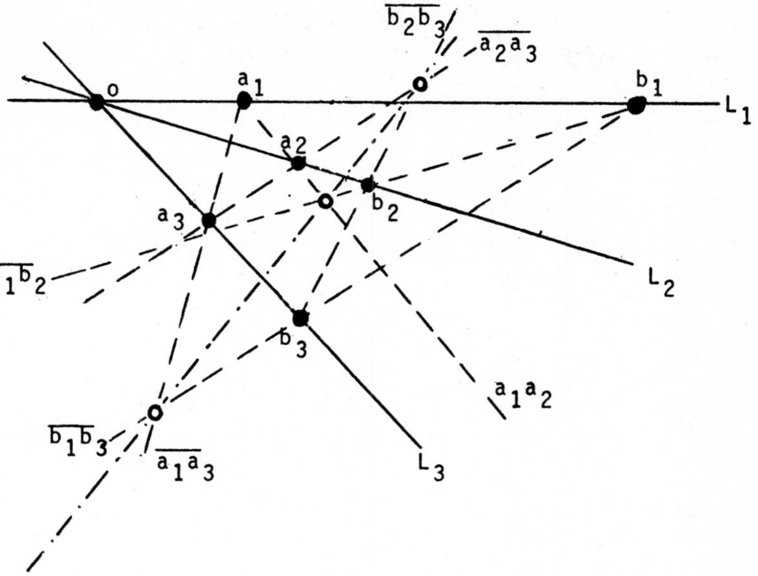

8.61 REMARK Each projective space of dimension ≥ 3 is Desarguesian (Karzel-Sörensen-Windelberg (1)).

8.62 NOTATION If V is a vector space over some field F then $V^*/_{K^*} := \{K^*v \mid v \in V^*\}$ and

$\mathcal{L} := \{K^*L \mid L$ is a subspace of V of (vector-space-) dimension 2$\}$.

8.63 THEOREM (Karzel (3)). In the notation of 8.62, $(V^*/_{K^*}, \mathcal{L})$ is a Desarguesian projective space of dimension dim V-1. Conversely, one gets all Desarguesian projective spaces of dimension ≥ 2 in this way.

8.64 DEFINITION The triple $(P, \mathcal{L}, \cdot)$ is called a (projective) incidence group if

(a) $(P, \mathcal{L})$ is a projective space.

(b) $(P, \cdot)$ is a group.

(c) $\forall\, p \in P$: $c_p: \begin{matrix} P \to P \\ q \to pq \end{matrix}$ $\in \mathrm{Coll}(P)$.

$(P, \mathcal{L}, \cdot)$ is called Desarguesian if $(P, \mathcal{L})$ is it.

8.65 REMARK It seems to be clear how isomorphic projective incidence groups and the dimension dim P of a projective incidence group are defined.

What has all of that to do with near-fields ?

8.66 DEFINITION Let N be a nf. and F a proper subfield of N. N is said to be normal over F provided that

(a) $(F^*,\cdot) \triangleleft (K^*,\cdot)$.

(b) $\forall\, f,f' \in F \ \forall\, n \in N$: $n(f+f') = nf+nf'$.

(N,F) is then called a normal near-field. N can be considered as a vector space over F.

8.67 REMARK If (N,F) is normal and N is a field then by Cartan-Brauer-Hua's theorem (see (Jacobson), p. 186) $F \subseteq C(N)$. This does not hold for general nf.'s (see Wähling (9), p. 76).

The basic fact is in

8.68 THEOREM (Karzel (3), pp. 69 - 73).

(a) Let (N,F) be a normal nf. . Then $(N^*/_{F^*}, \mathcal{L}, \cdot)$ (as in 8.63) is a Desarguesian projective incidence group.

(b) Conversely, all (up to isomorphic copies) Desarguesian projective planes P arise in this way from some normal nf., which (if dim $P \geq 2$) is unique up to isomorphism.

8.69 THEOREM (Karzel (3), pp. 76 and 78). If (N,F) is normal, $N^*/_{F^*}$ commutative and $\dim_F(N) \geq 3$ then N is a commutative field.
If $\dim_F(N) = 2$ and N is finite then either N is a field $GF(p^2)$ or a Dickson nf. of order 9 (see the remarks preceding 8.35) or of order 64.

For generalizations ("normal local near-modules") see e.g. Pieper (1), (3) and Karzel-Meissner (1).
Now we turn to affine planes.

7.) PLANAR NEAR-FIELDS

First again a little bit geometry.

8.70 DEFINITION An automorphism α of an affine plane $(P,\mathcal{L})$ is called dilatation if $\forall\, L \in \mathcal{L}$: $\alpha(L) \parallel L$. A dilatation α is a translatation if $\alpha = id$ or α is fixed-point-free.

8.71 DEFINITION An affine plane $(P,\mathcal{L})$ is called a translation plane if the set (it is a group!) of all translations in $(P,\mathcal{L})$ works transitively on P.

Consider, for a nf. N, in $N^2 =: P$ the "lines" $\{(x,xa+b) \mid x \in N\} =: L_{a,b}$ $(a,b \in N)$. Two of such, $L_{a,b}$ and $L_{a',b'}$ can be considered as "parallel" if $a = a'$. In order to get something like an affine plane, we want two "non-parallel" "lines" to have exactly one common point. This is the case iff every equation $xa = xa'+c$ with $a \neq a'$ has exactly one solution.

8.72 DEFINITION A nf. N is called planar (or projective) if each equation

$$xa = xb+c \qquad (a \neq b)$$

has exactly one solution.

Evidently, every field is planar.

8.73 REMARK (Wähling (9)). It suffices to want $xa = x+1$ to have a unique solution for each $a \neq 1$.

In fact we get

8.74 THEOREM (e.g. Karzel-Sörensen-Windelberg (1)). Let N be a nf., $L_{a,b}$ as in the motivational considerations preceding 8.72 and $L_c := \{(c,x) \mid x \in N\}$ $(c \in N)$ the "vertical lines". Let $\mathcal{L} := \{L_{a,b} \mid a,b \in N\} \cup \{L_c \mid c \in N\}$. Then $(N^2, \mathcal{L})$ is an affine plane $\Longleftrightarrow$ N is planar.
In this case, $(N^2, \mathcal{L})$ is a translation plane.

8.75 REMARK Not all translation planes arise in this way from a planar nf. . One has to use "multiplicatively non-associative planar nf.'s" (cf. 8.41 - 8.42), so called "Veblen-Wedderburn-systems" M to get all translation planes as some $(M^2, \mathcal{L})$ (definition of $\mathcal{L}$ as above) ("Each translation plane can be coordinatised by a Veblen-Wedderburn system".) See e.g. Hall (1), p. 362.
For the more general question, which geometric structures can be coordinatized by which types of algebraic structures, see all G-labeled items in the bibliography.

We look a little bit around to find some planar nf.'s.

8.76 THEOREM (Maxson (10) et al.) A nf. N with $\dim_{N_d}(N)$ finite is planar.

8.77 COROLLARY (e.g. Zemmer (1)). A finite nf. is planar.

So our search is turned around: does there exist non-planar nf.'s at all ?

8.78 EXAMPLES

(a) (Zemmer (1)). The Dickson nf. $(F(x),+,\circ_\phi)$ of 8.29 (char F = 0) is not planar.

(b) There exist planar nf.'s N with $\dim_{N_d}(N)$ infinite (Maxson (10)). Hence the converse of 8.76 does not hold.

The following concept is usually only defined for finite groups. See e.g. (Passman) or Kerby (9).

<u>8.79 DEFINITION</u> $\Gamma \leq S_M$ is called a <u>Frobenius group</u> if

(a) Each $\gamma \varepsilon \Gamma$, $\gamma \neq id$ has at most one fixed-point.

(b) $K_\Gamma := \{\gamma \varepsilon \Gamma \mid \gamma$ is fixed-point-free$\} \cup \{id\}$ is a transitive proper normal subgroup of Γ.

<u>8.80 REMARKS</u>

(a) K_Γ is called the <u>Frobenius-kernel</u> of Γ.

(b) Other characterizations of finite Frobenius groups are e.g.:

(α) $\Gamma \leq S_M$ is a Frobenius-group iff Γ is transitive, but not sharply 1-transitive (= regular) and $\forall (m_1, m_2) \varepsilon M^2$, $m_1 \neq m_2 : \Gamma_{m_1, m_2} = \{id\}$ (8.38).

(β) The Frobenius groups are exactly the semidirect products of a group Δ with a fixed-point-free automorphism group $\neq \{id\}$ of Δ.

Anyhow, if Γ is a finite Frobenius group, K_Γ is characteristic, regular and nilpotent.

(c) The finite sharply 2-transitive permutation groups of degree ≥ 3 are exactly the 2-transitive Frobenius groups.

The connection to planar nf.'s is given by the following two theorems.

<u>8.81 THEOREM</u> (e.g. André (4)). Let N be a planar nf. with $|N| > 2$. Then $T_2(N) =: \Gamma$ (8.40) is a Frobenius group and Γ_K is the set of all mappings $x \to x+b$. Moreover, if

char N = 2 then $I_{\Gamma_K} = \Gamma_K$ (8.46) ...... (*)

char N $\neq$ 2 then $\forall n \varepsilon N \ \exists \gamma \varepsilon I_{\Gamma_K} : \gamma(n) = n$ ....(**).

Conversely, all Frobenius groups with (*) or (**) can be obtained in this way from a planar nf.

Of course one can define Γ_K for every permutation group as in 8.79(b). Then one gets one more characterization of a planar nf.:

8.82 THEOREM (Kerby (9). Let N be a nf. and $\Gamma := T_2(N)$.
N is planar $\Leftrightarrow (I_\Gamma)^2 = \Gamma_K$.

We close our round-up of near-field theory with

8.83 REMARK (Sperner (1), Arnold (1)). Let N be a nf. and $k \in \mathbb{N}$. Consider the N-group N^k and define $\mathcal{L}$ by
$\mathcal{L} := \{(a_1,\ldots,a_k)+N(b_1,\ldots,b_k) \mid (a_1,\ldots,a_k),(b_1,\ldots,b_k) \in N^k\}$.
Then $(N^k, \mathcal{L})$ is a "nearly affine space" as defined by Sperner (1).

For the representation of "affine incidence groups" see Pieper (3).
Do not stop reading now - we are already in the next chapter.

b) PLANAR NEAR-RINGS

1.) THE STRUCTURE OF PLANAR NEAR-RINGS

The "planarity property" 8.72 can of course also be formulated for near-rings. But it is not very wise to do so: a near-ring with this property is a near-field (see 8.91), so we wouldn't get anything new. Therefore we generalize this concept (Anshel-Clay (1)):

8.84 DEFINITION Let N be a nr. and $a,b \in N$.

$a \equiv b :\Leftrightarrow \forall\, n \in N : na = nb$.

In this case, a and b are called (right) equivalent multipliers.

Of course, this is an equivalence relation on N.

8.85 DEFINITION A nr. N is said to be a planar near-ring if $|N/\equiv| \geq 3$ and if every equation

$$xa = xb+c \qquad (a \not\equiv b)$$

has a unique solution (in N).

8.86 NOTATION If $N \in \mathcal{N}$, let $A := \{n \in N \mid n \equiv 0\}$; denote $N \setminus A$ by $N^{\#}$.

By 8.85, $N^{\#}$ has at least two elements.

8.87 PROPOSITION (Anshel-Clay (1)). Every planar nr. is zero-symmetric.

Proof. Take $n \in N$. Let a be $\in N^{\#}$. Then 0 and $n0$ are both solutions of $xa = x0+0$, hence equal.

8.88 PROPOSITION (Anshel-Clay (1)). Let N be planar.

(a) $a \in N$ is a right zero divisor $\Leftrightarrow a \equiv 0 \Leftrightarrow a \in A$.

(b) $\forall\, n \in N^{\#}\ \forall\, m \in N\ \exists\, x \in N$: $xn = m$.

Proof. (a) We only have to show that $na = 0$ $(n \neq 0)$ implies $a \equiv 0$. In fact, $a \not\equiv 0$ implies that 0 and n are solutions of $xa = x0+0$, a contradiction.

(b) If $n \in N^{\#}$, $xn = x0+m$ has a unique solution.

The last result gives rise to the following definition.

8.89 NOTATION For $a \in N^{\#}$ let 1_a be the unique solution of $xa = a$. Let $B_a := \{x \in N^{\#} \mid 1_a x = x\}$.

Evidently, $a \in B_a$, $1_a \in N^{\#}$ and $N^{\#} = \bigcup_{a \in N^{\#}} B_a$.

These B_a's help to clearify the structure of a planar near-ring.

8.90 THEOREM (Anshel-Clay (1)). Let N be a planar nr. . Then

(a) Each $(B_a,\cdot)$ is a group with identity 1_a.

(b) A and the B_a's $(a\in N^{\#})$ form a partition of N.

(c) $\forall\ a\in N^{\#}$: $B_aN^{\#} = B_a$.

(d) If $a,b\in N^{\#}$, then $\phi: B_a \to B_b$, $x \to 1_bx$ is a (group-) isomorphism.

(e) Each 1_a $(a\in N^{\#})$ is a right identity for $(N,+,\cdot)$.

(f) If $S\subseteq N^{\#}$ and $SN^{\#}\subseteq S$ then $S = \bigcup_{a\in S} B_a$.

Proof. $(c)_1$: Let a be $\in N^{\#}$, $n\in N^{\#}$ and $b\in B_a$. Then $1_a(bn) = (1_ab)n = bn$, whence $bn\in B_a$.

(a) By $(c)_1$, $(B_a,\cdot)$ is closed w.r.t. multiplication. Now 1_a1_a and 1_a are both solutions of $xa = x0+a$. So $1_a1_a = 1_a\in B_a$ is a left identity in $(B_a,\cdot)$. If $b\in B_a$, let $\bar{b}$ be the unique element of N with $\bar{b}b = 1_a$ (8.88(b)). Then $\bar{b}$ and $1_a\bar{b}$ solve $xb = 1_a$, whence $1_a\bar{b} = \bar{b}\in B_a$.

$(c)_2$: Conversely, every $b\in B_a$ can be written as $b = 1_ab\in B_aN^{\#}$.

(b): It is enough to show that $\forall\ a,b\in N^{\#}$ either $B_a\cap B_b = \emptyset$ or $B_a = B_b$.

If $n\in B_a\cap B_b$, $1_an = n = 1_bn$, hence 1_a and 1_b are solutions of $xn = x0+n$; so $1_a = 1_b$ and $B_a = B_b$.

(e): For each $n\in N$, $n1_a$ and n solve $x1_a = x0+n1_a$.

(d): By (c), ϕ really goes from B_a into B_b. If $a',a''\in B_a$, $\phi(a'a'') = 1_b(a'a'') = ((1_ba')1_b)a'' =$

$= (1_ba')(1_ba'') = \phi(a')\phi(a'')$, since $1_ba'\in B_b$, where 1_b acts as identity by (a).

If $\phi(a') = \phi(a'')$ then $1_ba' = 1_ba''$. By (e), $a' = 1_aa' = 1_a1_ba' = 1_a1_ba'' = a''$.

If $b'\in B_b$, then $1_a b'\in B_a$ is mapped onto b'.

Hence ϕ is an isomorphism.

(f) is clear.

8.91 COROLLARY (Anshel-Clay (1)). If N is planar and $\equiv$ is discrete (i.e. $\equiv$ is the identity) (N then fulfills 8.72) then N is a nf..

Proof. $\forall\ a,b\in N^{\#}$: $1_a \equiv 1_b$ (by 8.80(e)), so $1_a = 1_b$.

Hence (8.90(b)) $(N^*,\cdot) = (N^{\#},\cdot) = (B_a,\cdot)$ is a group.

8.92 COROLLARY A planar nr. is integral $\Longleftrightarrow N^* = N^{\#} \Longleftrightarrow A = \{0\}$.

8.93 THEOREM (Anshel-Clay (1)). Let N be an integral planar nr. such that each $\overline{B}_a := B_a \cup \{0\}$ $(a\in N^*)$ is a normal subgroup of $(N,+)$. Suppose that no $B_a = N$, but for all $a \neq b$, $B_a + B_b = N$. Then

(a) $(N,+) = (\overline{B}_a,+)\dot{+}(\overline{B}_b,+)$ for all $a,b\in N^*$, $a \neq b$.

(b) Each $(\overline{B}_a,+,\cdot)$ is a nf.

(c) N is abelian.

(d) If $\mathcal{L} := \{n+\overline{B}_a \mid n\in N,\ a\in N^*\}$, $(N,\mathcal{L})$ is an affine plane (8.53).

Proof. (a) is clear from 8.90(b).

(b) follows from 8.90(a).

(c) is a consequence of (a), (b) and 8.11.

(d): If $p,q\in N$, $p \neq q$, take $a\in N^*$ with $p-q\in\overline{B}_a$. Then $\exists\ n\in N$: $p\in n+\overline{B}_a =: L$ and $q\in L$. If $L' = n'+B_{a'}$ also has the property that $p,q\in L'$ then $p-q\in\overline{B}_a \cap \overline{B}_{a'}$. Since $p \neq q$, $\overline{B}_a = \overline{B}_b$ and $L = L'$. Since $0,1_a\in\overline{B}_a$, each $|n+B_a|\geq 2$. Hence $(N,\mathcal{L})$ is an incidence space (8.49). $|N^{\#}| = |N^*/\equiv|\geq 2$ implies that $(N,\mathcal{L})$ is not degenerated.

Now take $L = n+\overline{B}_a\in\mathcal{L}$ and $p\in N$. If $M := p+\overline{B}_a$ then $p\in M$ and $M \parallel L$. If $M'\in\mathcal{L}$ has the same property then

$M' = p+\overline{B}_b$ for some $b \in N^*$.
If $M' = L$ then $p+\overline{B}_b = n+\overline{B}_a$, so $a = b$, whence $M = M'$.
If $M' \neq L$ then $M' \cap L = \emptyset$. If $a \neq b$, $N = B_a + B_b$.
Hence $\exists\, x \in \overline{B}_a\ \exists\, y \in \overline{B}_b$: $n-p = x+y$. So $-x+n =$
$= y+p \in (n+\overline{B}_a) \cap (p+\overline{B}_b) = \emptyset$, a contradiction.
Consequently again $a = b$ and $M = M'$.

<u>8.94 REMARKS</u> It can be shown that the affine plane $(N, \mathcal{L})$ in 8.93(c) can be coordinatized by a skew-field. A similar result can be obtained if the $\overline{B}_a$'s are alternately defined as $\overline{B}_a := B_a \cup \{0\} \cup -B_a$. There is also a close connection to "$\Phi(I,IV)$-groups".
For all of that see Anshel-Clay (2). As Clay points out, there is also some relation to "inverse planes" (cf. Ferrero (12)).

<u>8.95 EXAMPLES</u> (see Anshel-Clay (2) or Clay (11)).

(a) Every planar nf. with more than 2 elements is a planar nr..

(b) Let V be a normed vector space over $\mathbb{R}$. Define $v*w := \|w\|\, v$. Then $(V,+,*)$ is an integral planar non-ring.

(c) Let V be a vector space over $\mathbb{R}$ and $\phi: V \to \mathbb{R}$ have the property that $\exists\, \alpha \in \mathbb{R}^*\ \forall\, t \in \mathbb{R},\ t \geq 0\ \forall\, v \in V$: $\phi(tv) = t^{\alpha}\phi(w)$.
Define $v*w := |\phi(w)|^{1/\alpha} v$. Then $(V,+,*)$ is a planar near-ring.
See Anshel-Clay (2) for the geometric interpretations of the B_a's as lines, rays, hyperbolas etc. .

(d) No $M_0(\Gamma)$ or $M(\Gamma)$ is planar: $\equiv$ is discrete, so planarity would imply that $M_{(0)}(\Gamma)$ is a nf. with more than 3 elements, which is certainly not the case. So in contrast to near-field-theory, a finite nr. is not planar in general (cf. 8.77).

<u>8.96 THEOREM</u> (Ferrero (5), Betsch-Clay (1)).

(a) Let Γ be a group and $G \neq \{id\}$ be a fixed-point-free automorphism group of Γ. If Γ is finite then each $N := (\Gamma,+,\cdot_B)$ of 1.4(b) is a planar near-ring. N is integral iff $\{B_i | i\in I\}$ is the complete set of all non-zero orbits (notation as in 1.4).

(b) Conversely, let N be a planar near-ring. Consider for $a\in N$: $g_a : \begin{matrix} N \to N \\ n \to na \end{matrix}$. Then $G := \{g_a | a\in N^{\#}\}$ is a fixed-point-free automorphism group $\neq \{id\}$ of $(N,+)$. For each $b\in N^{\#}$, $G \cong B_b$.

<u>Proof</u>. (a): Consider again the situation of 1.4(b).

$|N/\equiv| = |G \cup \{\bar{o}\}| \geq 3$, since $\gamma \equiv \delta \iff g_\gamma = g_\delta$.

So it remains to show the "planar property":

Assume that $\xi\cdot\gamma = \xi\cdot\delta+\eta$, $\gamma \not\equiv \delta$. This means that $g_\gamma(\xi) = g_\delta(\xi)+\eta$ or $(-g_\delta+g_\gamma)(\xi) = \eta$ with $g_\gamma \neq g_\delta$ or (with $g := g_\gamma^{-1}g_\delta \neq id$): $(-g+id)(\xi) = -g_\delta^{-1}(\eta)$.

But $-g+id$ is bijective, so this equation has exactly one solution:

Suppose that $(-g+id)(\alpha) = (-g+id)(\beta)$ then $-g(\alpha)+\alpha = -g(\beta)+\beta$ and $g(\alpha-\beta) = g(\alpha)-g(\beta) = \alpha-\beta$. Since g is fixed-point-free and $g \neq id$, $\alpha = \beta$.

Since Γ is finite, $-g+id$ is bijective.

(b): If $a\in N^{\#}$, $\forall\, c\in N\ \exists\, x\in N$: $g_a(x) = xa = c$ by 8.88(b). So $g_a\in Aut(N,+)$ and $G = \{g_a | a\in N^{\#}\}$ is a group.

Consider the map $\psi: \begin{matrix} (B_b,\cdot) \to G \\ a \longrightarrow g_a \end{matrix}$, where $b\in N^{\#}$.

Evidently, ψ is a homomorphism.

If $\psi(a_1) = \psi(a_2)$, then $\forall\, x\in N$: $xa_1 = xa_2$, whence $1_a a_1 = 1_a a_2$, so $a_1 = a_2$, and ψ is shown to be a monomorphism.

Now take some g_c, $c\in N^{\#}$. Since $1_b c\in B_b$ by 8.90(c), $\psi(1_b c) = g_{1_b c} = g_c$ and ψ is an isomorphism.

G is fixed-point-free: let $g_a \in G$ fulfill $g_a(n) = n$ for some $n \in N$, $n \neq 0$, then 0 and n fulfill $xa = x \cdot 1_a + 0$ (8.90(e)). So $a \equiv 1_a$, which means that $\forall\, x \in N$: $g_a(x) = xa = x1_a = x$, from which we deduce that $g_a = id$.

8.97 REMARK (Betsch-Clay (1)). This shows that (similar to the situation in planar near-fields) every finite planar near-ring can be characterized by some pair (Γ, G) of groups, where $G \neq \{id\} \leq Aut\,\Gamma$ is fixed-point-free. So every finite planar near-ring determines a Frobenius group (8.79) and conversely (cf. also Ferrero (5)), and the construction of a planar near-ring on a given additive group Γ is nothing else than the construction of a non-trivial fixed-point-free automorphism group on Γ. Cf. 8.124.

8.98 COROLLARY (Betsch-Clay (1)) Let N be a finite planar near-ring and let G be as in 8.96. Then

(a) $|G|$ divides $|N| - 1$.

(b) $(N,+)$ is nilpotent, but not necessarily abelian.

Proof. (a) is clear from 8.96(b) and 8.90(b), and (b) follows from (Thompson) (cf. 6.33(b) $\Rightarrow$ (d)).

See also 8.124.

The last result is in some other way remarkable: planar near-rings are "not far away from being near-fields" (cf. 8.88(b)). But they are far enough to have non-abelian members in contrast to 8.11. We need

8.99 DEFINITION (Ferrero (5), Szeto (3)). A nr. N is called strongly uniform if $\forall\, n \in N$: $(0:n) = \{0\}$ or $(0:n) = N$, but $\exists\, m \in N$: $(0:m) = \{0\}$.

8.100 THEOREM (Ferrero (5), Clay (10), Szeto (3)).

(a) Let N be a planar nr.. Then N is strongly uniform, the multiplication is not trivial (1.4(b)) and all non-zero orbits of G (see 8.96(b)) are principal (that means that for all x,y in the same non-zero orbit there is exactly one $g \in G$ with $g(x) = y$).

(b) Conversely, if N is a finite nr. which is strongly uniform, has non-trivial multiplication and the property that every non-zero orbit under G (defined as in 8.96(b)) is principal, then N is planar.

Proof. (a) If $a \in A$, $g_a = \bar{o}$ and $(0:a) = N$. If $a \in N^{\#}$, $g_a \in \mathrm{Aut}\, N$ and $(0:a) = \{0\}$, hence N is strongly uniform. Since $|N/\equiv| \geq 3$, the multiplication cannot be trivial.
G is fixed-point-free (8.96), so all orbits are principal (cf. 4.28).

(b) Since N is finite and strongly uniform, all $g_a: x \to xa$ are either $= \bar{o}$ or automorphisms. (Observe that $\mathrm{Ker}\, g_a = (0:a)$). Let G be the group of all those automorphisms. Since all orbits $\neq \{0\}$ are principal, G is fixed-point-free. Since $\cdot$ is not trivial, $G \neq \{id\}$. Now apply 8.96(a).

8.101 REMARK (Szeto (3)). 8.100(b) does not hold in the infinite case: Take $(\mathbb{Z},+) \oplus (\mathbb{Z},+)$ and define $(n,m)*(n',m') := n'(n,m)$. Then $N := (\mathbb{Z}\times\mathbb{Z},+,*)$ is an infinite strongly uniform. $*$ is not trivial and all non-zero orbits are principal.
On the other hand, N is not planar, for $(2,0) \not\equiv (0,0)$, but $x(2,0) = x(0,0)+(1,1)$ has no solution.
Cf. also G. Betsch's revue in the "Zentralblatt für Mathematik".

8.102 REMARK See Szeto (3) for a characterization of these strongly uniform near-rings (similar to 8.90).

Finally, we describe a construction method for finite integral planar near-rings for future use (in the next number).

8.103 THEOREM (Clay (10)). Let F be the field of order p^n ($p\in\mathbb{P}$) and t a non-trivial divisor of p^n-1. $(F^*,\cdot)$ is cyclic of order p^n-1. There is a cyclic subgroup B of F^* of order t. Choose representatives

$u_1 = 1, u_2, \ldots, u_m$ for the cosets $Bu_1, \ldots, Bu_m$ of B in F^*.

Define $f *_t g := \begin{cases} 0 & \text{if } g = 0 \\ f\cdot b_g & \text{if } g \neq 0,\ g\in Bu_i,\ g = b_g u_i \end{cases}$.

Then $(F,+,*_t)$ is an integral planar near-ring.

Proof: straightforward and hence omitted.

2.) PLANAR NEAR-RINGS AND BALANCED INCOMPLETE BLOCK DESIGNS

8.104 DEFINITION Let N be a nr., $a\in N^*$ and $b\in N$. Then the set

$$aN+b$$

is called the block determined by a,b. Blocks of the form aN $(a \neq 0)$ are called basic blocks.

8.105 REMARKS If $G = \{g_x \mid x\in N^{\#}\}$ is as in 8.96(b) and if $G^o := G\cup\{\bar{o}\}$ then $aN+b = G^o a+b$.
The $Ga = aN^{\#} = aN\setminus\{0\}$ form a partitition of N^*, for they are exactly the non-zero orbits of $(N,+)$ under G.

For applications in the (near) future we remark

8.106 PROPOSITION Let N be planar and $a\in N^*$, $b\in N$. Then $|aN+b| = |G^o| = |N/\equiv| \geq 3$.

Proof. $\forall\, n,n'\in N$: $an+b = an'+b \iff an = an' \iff x = a$ and $x = 0$ fulfill $xn = xn' \iff n \equiv n' \iff g_n = g_{n'}$. Observe 8.85.

In order to be able to formulate our principal results we need

<u>8.107 DEFINITION</u> $a\varepsilon N$ is called an "<u>element of the first category</u>" (Ferrero) if $aN = (-a)N+a = -(aN)+a =: -aN+a$.
$C_1(N) := \{a\varepsilon N \mid a \text{ is of the first category}\}$.

Clearly $0\varepsilon C_1(N)$. Much more is in

<u>8.108 THEOREM</u> (Ferrero (8), (19)). Let N be planar and $a\varepsilon N$.
Then $a\varepsilon C_1(N) \iff aN \leq (N,+)$.
In this case and if aN is finite, aN is elementary abelian.

<u>Proof</u>. "$\Rightarrow$": Let a be $\varepsilon C_1(N)$. Let $b\varepsilon aN^{\#}$, $b'\varepsilon aN$.
We want to show that $b-b'\varepsilon aN$. Now $\exists\, n\varepsilon N^{\#}$: $b = an$.
Hence using 8.90(e) and 8.88(b), $bN = aN = aNn =$
$= (-aN+a)n = -aNn+an = -aN+b = -bN+b$.
Now $\exists\, n'\varepsilon N$: $b' = bn'$. Thus we get
$b'-b=bn''-b \;\varepsilon\; bN-b = -bN+b-b = -bN = -aN$. We claim that $aN = -aN$. We may assume that $a \neq 0$.
By 8.106, $\exists\, c,d\varepsilon aN^{\#}$: $c \neq d$. From above we get
$0 \neq c-d \;\varepsilon\; -aN^{\#}$. Hence $\exists\, n''\varepsilon N^{\#}$: $c-d = -an''$.
Therefore $an'' = d-c\varepsilon(aN^{\#}) \cap (-aN^{\#})$ since we have shown above that the difference of any two elements of $aN^{\#}$ is in $-aN^{\#}$. So $(aN^{\#}) \cap (-aN^{\#}) \neq \emptyset$, whence $aN^{\#} = -aN^{\#}$ by 8.105, from which $aN = -aN$.
Now our considerations imply that $aN \leq (N,+)$.

"$\Leftarrow$" is obvious, since $a = a\cdot 1_a \varepsilon aN$ (8.90).

The remark follows from the observation that the automorphisms of G, restricted to aN, form an automorphism group on aN which acts transitively on $aN\setminus\{0\} = aN^{\#}$ and from theorem 11.1 of (Wielandt).

From the first lines of the preceding proof we can deduce

<u>8.109 COROLLARIES</u> (Ferrero (19)). Let N be planar and $a\varepsilon N$. Then

(a) $\forall\ b\varepsilon aN$: $bN = aN$ if $a\varepsilon C_1(N)$.
(b) All $b\varepsilon aN$ are of the same category.
(c) $C_1(N)$ is a union of - say u - orbits of $(N,+)$ under G.
(d) If $\exists\ x\varepsilon C_1(N)$: $xN = N$ then $C_1(N) = N$.

<u>8.110 REMARK</u> See Ferrero (12), (19) for the connection to "difference sets".

Which blocks coincide ?

<u>8.111 THEOREM</u> (Ferrero (19), cf. Ferrero (8) and Clay (10)). Let N be planar, $a,a'\varepsilon N^*$ and $b,b'\varepsilon N$. Then $aN+b = a'N+b' \Longleftrightarrow$ (a) or (b), where

(a) $b = b'$ and $aN = a'N$,
(b) $b \neq b'$, $-aN^{\#} = a'N^{\#}$, $b'\varepsilon aN^{\#}+b$ and $a,a'\varepsilon C_1(N)$.

<u>Proof</u>. $\Longrightarrow$: First let $aN+b = a'N+b'$. If $b = b'$ then $aN = a'N$ and we are in case (a).
So suppose that $b \neq b'$. From $aN = a'N+(b'-b)$ we get some $n\varepsilon N$ with $0 = a'n+(b'-b)$. $b \neq b'$ implies that $n\varepsilon N^{\#}$. So $a'n = b-b'$. Similarly, $\exists\ n'\varepsilon N^{\#}$: $an' = b'-b$.
Hence $0 \neq b-b'\varepsilon(a'N^{\#})\cap(-aN^{\#})$, whence $a'N^{\#} = -aN^{\#}$ by 8.105. So $aN+b = -aN+b'$. Consequently $\exists\ n''\varepsilon N^{\#}$: $b' = an''+b$, so $aN+b = -aN+b' = -aN+an''+b$, whence $aN = -aN+an'' = -aN+g_{n''}(a)$. Applying $g_{n''}^{-1}$ gives $aN = -aN+a$, so $a\varepsilon C_1(N)$. By symmetry, $a'\varepsilon C_1(N)$ and (b) is shown.

$\Longleftarrow$: (a) trivially implies $aN+b = a'N+b'$.
So assume (b). Let $x\varepsilon a'N+b'$. We have to show that $x\varepsilon aN+b$. If $x = b'\varepsilon aN^{\#}+b$, $x\varepsilon aN+b$. If $x \neq b'$, $\exists\ n,n',n''\varepsilon N^{\#}$: $x = a'n+b' = -an'+b' = -an'+an''+b$. Since $aN \leq (N,+)$, by 8.108, $-an'+an''\varepsilon aN$, whence $x\varepsilon aN+b$. The converse inclusion is shown similarly.

These blocks prove useful for constructing block designs. First we define these items.

8.112 DEFINITION An incidence structure $(P,\mathcal{B})$ ($\mathcal{B}\subseteq 2^P$ - 8.49) is said to be a tactical configuration with parameters $(v,b,r,k)\in\mathbb{N}^4$ if

(a) $|P| = v$.
(b) $|\mathcal{B}| = b$.
(c) Each $p\in P$ is in exactly r elements of $\mathcal{B}$.
(d) Each $B\in\mathcal{B}$ contains exactly k elements of P, i.e. $\forall\ B\in\mathcal{B}: |B| = k$.

A tactical configuration is a balanced block design if

(e) Each pair $(p,q)\in P^2$, $p\neq q$, is in exactly λ elements of $\mathcal{B}$

and complete if

(f) $b = \binom{v}{k}$,

otherwise incomplete.
The elements of $\mathcal{B}$ are called blocks. "Balanced incomplete block design" is abbreviated by "BIBD"; (v,b,r,k,λ) are the parameters of the BIBD and $E := \frac{\lambda v}{rk}$ (≤ 1) is called its efficiency.

8.113 EXAMPLES

(a) Let P be a set with v elements, $k\in\mathbb{N}$, $k\leq v$. Let $\mathcal{B} := \{B\subseteq P \mid |B| = k\}$.
Then $(P,\mathcal{B})$ is a tactical configuration with parameters $(v, \binom{v}{k}, \binom{v-1}{k-1}, k)$ and (if $k\geq 2$) a complete balanced block design with $\lambda = \binom{v-2}{k-2}$.

(b) Consider the field $\mathbb{Z}_p$ $(p\in\mathbb{P}\setminus\{2\})$ and the affine plane $(\mathbb{Z}_p^2,\mathcal{L})$ as in 8.74. Then $(\mathbb{Z}_p^2,\mathcal{L})$ is a BIBD with parameters $(p^2,p^2+p,p+1,p,1)$. (For $p = 2$ we get a complete balanced block design.)

For the theory of block designs we refer the reader to (Hall) or (Dembowski), where also 8.113 and 8.114 can be found. The parameters of a BIBD are not independent at all:

8.114 PROPOSITION Let $(P,\mathfrak{B})$ be a BIBD. Then

(a) $kb = vr = |\{(p,B) \mid p \in B, B \in \mathfrak{B}\}|$ (this holds for every tactical configuration!).

(b) $r(k-1) = \lambda(v-1) = |\{p \in P \mid p \neq q \wedge \exists B \in \mathfrak{B}: p \in B \wedge q \in B\}|$, where q is arbitrary in P.

(c) If $b>1$ and $k \leq v-1$ then $b \geq v$ ("Fisher's inequality") and $r \geq k$.

BIBD's are an essential tool in experimental designs. The following example shall illustrate this and provide enough motivation for the reader to endure also the next pages.

8.115 APPLICATION Suppose you have b kinds of fertilizers and want to test some combinations of r fertilizers always on the same number k of experimental fields.
Take some BIBD $(P,\mathfrak{B})$ with parameters (v,b,r,k,λ), and divide the whole experimental area into v parts. Since $|\mathfrak{B}| = b$ = number of fertilizers, $\mathfrak{B}$ can be written as $\mathfrak{B} = \{B_1,B_2,\ldots,B_b\}$. Give the fertilizer number i on every field of the block B_i. Then:
(a) every field contains exactly r different fertilizers,
(b) every fertilizer is applied on exactly k different fields, and
(c) every pair of different fields has exactly λ kinds of fertilizers in common.

8.116 REMARKS Of course, given b,r,k, it is a non-trivial problem how to get a BIBD with suitable parameters.
In general, it is an open question whether for every quintuple (v,b,r,k,λ) of natural numbers which fulfill the conditions of 8.114 there exists a BIBD with these parameters. We will now apply planar near-rings to get new classes of BIBD's.

The efficiency of a BIBD can be interpreted economically in the example above. BIBD's of efficiency $\geq 0{,}75$ are usually considered to be "good". Many of them are listed in (Cochran-Cox).
Balanced complete block designs are usually "rather inefficient". This is the reason for looking at the incomplete ones.

8.117 THEOREM (Ferrero (12)). Let N be planar with $|N| =: v \in \mathbb{N}$. Denote by $\mathcal{B}$ the set of all blocks (8.104). Let α_1 (α_2) be the number of non-zero orbits of $(N,+)$ under G consisting of elements of $C_1(N)$ (not of $C_1(N)$, respectively) (cf. 8.110). Then $(N,\mathcal{B})$ is a tactical configuration with parameters

$$\left(v, \frac{\alpha_1 v}{|G^o|} + \alpha_2 \cdot v, \alpha_1 + \alpha_2 |G^o|, |G^o|\right).$$

Proof. The first parameter is clear.
We compute the number of different blocks and apply 8.111: The number of blocks $aN+b$ with $a \in C_1(N)$ is $\frac{\alpha_1 v}{|G^o|}$, the one of those with $a \notin C_1(N)$ (case (b)) is $\alpha_2 \cdot v$.
Now apply 8.106 to get $k = |G^o|$ in 8.112(d).
Next observe that the number r_n of blocks containing an element $n \in N$ is the same for each $n \in N$, since it equals the number of blocks containing 0. Now we know that $(N,\mathcal{B})$ is tactical and we can apply 8.114(a) to get $r = \frac{kb}{v} = \frac{|G^o| \cdot (\frac{\alpha_1 v}{|G^o|} + \alpha_2 v)}{v} = \alpha_1 + \alpha_2 |G^o|$ (of course, this could be accomplished directly, too).

Observe that $v = (\alpha_1 + \alpha_2)|G| + 1$.
Nothing is more natural now than to ask, under which conditions $(N,\mathcal{B})$ is a BIBD. The next theorem answers this question, thus bringing joy and happiness into our life.

8.118 THEOREM (Ferrero (9) - (12)). Let $(N,\mathfrak{B})$ be as above. $(N,\mathfrak{B})$ is a BIBD $\iff C_1(N) = N$ (then $\lambda = 1$) or $C_1(N) = \{0\}$ (then $\lambda = |G^o|$).

Proof. $\Rightarrow$: It does not seem to be possible to deduce this from the fact that $\frac{r(k-1)}{v-1} \in \mathbb{Z}$ (8.114(b)). So we have to work.

Call (for $a,b \in N$) a,b equivalent if $aN = bN$ (a and b are then in the same orbit under G) and denote this by $a \sim b$. We need a lemma.

Lemma: Let N be planar and n', n'' be $\in N$, $n' \neq n''$. Let $\lambda := \lambda_n$ be the number of blocks B with $n', n'' \in B$ ("blocks through n' and n'' "). Let μ be the number of different representations of $n := n' - n''$ as a difference of two equivalent elements not contained in $C_1(N)$. Then:
If $n \in C_1(N)$ then $\lambda = \mu + 1$.
If $n \notin C_1(N)$ then $\lambda = \mu + 2$.

Proof of the Lemma: First observe that if the block $aN+b$ contains 0 and n (= $n'-n''$) then $aN+(b+n'')$ contains n' and n''. Hence λ is the number of blocks through 0 and n ($\neq 0$).

How many different blocks with $\{0,n\} \subseteq aN+b$ exist? Let $aN+b$ be such a block.

Case (1): If $b = 0$, $n \in aN^{**}$, whence $nN = aN$ by 8.109(a) and there is only one possibility to have $\{0,n\} \subseteq a'N$ for some $a' \in N^*$.

Case (2): $b = n$. Then $0 \in aN+n$, $n \in -aN$, whence $aN = -nN$. So there is again just one block through 0 and n.

Case (3): $0 \neq b \neq n$. $\exists\, n_1 \in N^{**}$: $0 = an_1 + b$. Hence $aN = -bN$, and $aN+b = -bN+b$. So if n is a difference as stated in the lemma, the blocks in consideration

have the form $-cN+c$.
Conversely, for the block $-bN+b$ we get, since $n \in aN+b$, $\exists\, n_2 \in N^{\#}$: $n = -bn_2+b$, which is a representation of n as a difference of two equivalent elements of bN.
If $b \in C_1(N)$, $bN \leq (N,+)$ implies $n \in bN$. Since also $0 \in bN$, we are in case (1), a contradiction to $0 \neq b \neq n$.
So let b be $\notin C_1(N)$. Then $-bN+b$ is neither in case (1) nor (2) nor equal to some other $-b'N+b'$ containing 0 and n, but with $b \neq b'$ by 8.111.
So in case (3) are just as many blocks not in (1) and (2) as there are representations of n of the described kind, namely μ.

So the result follows if one observes that the two blocks in (1) and (2) coincide iff $n \in C_1(N)$.

Proof of the theorem. By the planar property,
$\forall\, n \in N^* \ \forall\, n',n'' \in N^{\#}$, $n' \neq n''$ $\exists!\, x \in N$: $n = xn'-xn''$.
So n has $|G|\cdot(|G|-1)$ such representations (when varying n',n'').
Now take some arbitrary $g \in G$.
Then $n = xn'-xn'' = g_{n'}(x)-g_{n''}(x) = (g_{n'}\circ g)(g^{-1}(x))-(g_{n''}\circ g)(g^{-1}(x))$, providing all other ways to write n as a difference of equivalent elements. So there are just $\frac{|G|\cdot(|G|-1)}{|G|} = |G| - 1$ different ways to write n as such a difference.

(a) If $n \notin C_1(N)$ and $n = a-b$ $(a \sim b)$ then a,b are both $\notin C_1(N)$. For if e.g. $a \in C_1(N)$ then $b \in C_1(N)$, whence $a-b \in C_1(N)$ by 8.108 and 8.110.
By our lemma, $\lambda = \mu+2 = (|G|-1)+2 = |G|+1$.

(b) If $n \in C_1(N)$ then the $|G|-1$ ways to write n as $n = a-b$ with $a \sim b$ are exactly $n = a-(a-n)$ with $a \in nN^{\#}\setminus\{n\}$. For nN is (8.108) an abelian group of order $|G^o|$; so $\forall\, a \in nN^{\#}\setminus\{n\}$; $a-n = -n+a \sim a$. Observe that a and $(a-n)$ are in $C_1(N)$. So none of the $|G|-1$ differences of equivalent elements giving n are as described in the lemma, whence $\mu = 0$, and $\lambda = 1$.

It may happen that (in 8.117 and 8.118) neither $C_1(N) = N$ nor $C_1(N) = \{0\}$ (see Betsch-Clay (1)).
One can even say more (see Ferrero (12)):

<u>8.119 REMARK</u> Let $(N,\mathfrak{B})$ be the BIBD of 8.117/8.118.
If $C_1(N) = N$ then $(N,+)$ is elementary abelian (8.108) and there is some finite field F such that $(N,\mathfrak{B}) = (F^2,\mathfrak{L})$ of 8.74; $(N,\mathfrak{B})$ can be considered as affine space, and the blocks are just the lines of this space.

Looking at the other case (which brings up possibly new designs) yields first

<u>8.120 COROLLARY</u> (Ferrero (12)). Let N be a finite planar nr.
Let $|G^o|$ have not the form p^α, where $p \in \mathbb{P}$ and $p^\alpha/|N|$. Then $(N,\mathfrak{B})$ of 8.117 is a BIBD with $k = \lambda = |G^o|$.

<u>Proof</u>. Assume that $\exists\, n \in N^*$: $n \in C_1(N)$. Then nN is elementary abelian, so $|nN| = p^\alpha$ with $p \in \mathbb{P}$, and $p^\alpha/|N|$ (8.108), a contradiction. Hence $C_1(N) = \{0\}$ and 8.118 gives the result.

See Ferrero (12), Teorema 8 for the connection to finite Möbius planes. Cf. also Anshel-Clay (1).

Another way to reach the case $C_1(N) = \{0\}$ is the following.

8.121 COROLLARY (cf. Ferrero (8)). Let N be a finite integral planar nr. without subnear-fields. Then the same conclusion as in 8.120 holds.

Proof. Suppose that $n\varepsilon N^*$ is in $C_1(N)$. Then nN is an abelian subgroup of $(N,+)$ by 8.108. $(nN)^* = B_n$: $(nN)^* \subseteq B_n$ is clear from 8.89, while every $b\varepsilon B_n$ can be written as $b = 1_n b = nn^{-1}b \varepsilon nN$ by 8.90 (a). Consequently, $((nN)^*,\cdot)$ is a group and nN is a subnear-field of N, a contradiction.
Hence $C_1(N) = \{0\}$ and the result follows from 8.118.

In (8), Ferrero constructs BIBD's from near-rings N with $(|N|,6) = 1$, having parameters $(v, \frac{v(v-1)}{2}, \frac{v-1}{2}, 3, 3)$ (where $|N| = v$).
Both cases in 8.118 can be obtained by the following near-rings:

8.122 COROLLARY (Clay (10)). Consider the planar nr. $(F,+,*_t)$ of 8.103.

(a) If $t = p^m-1$ for some $m\leq n$ then $(F,\mathfrak{B})$ - as in 8.117 - is a BIBD with parameters $(p^n, \frac{p^n(p^n-1)}{p^m(p^m-1)}, \frac{p^n-1}{p^m-1}, p^m, 1)$.

(b) If t is not of the form p^m-1 then $(F,\mathfrak{B})$ is a BIBD with parameters $(p^n, \frac{p^n(p^n-1)}{t}, \frac{(t+1)(p^n-1)}{t}, t+1, t+1)$.

Proof. First observe that $t = |G| = |B|$ (of 8.103), so $t+1 = |G^o|$.

(a) Let t be $= p^m-1$ $(m\leq n)$. Set $\overline{B} := B\cup\{0\}$. Take $a\varepsilon F^*$. Then $a*F = a\overline{B}$ has $t+1 = p^m$ elements and is a subgroup of $(F,+)$: $\overline{B}$ consists of all $x\varepsilon F$ with $x^{p^m} = x$, hence being a subgroup of $(F,+)$. This is easily transferred to $a\overline{B}$. Now apply 8.118.

(b) follows from 8.120.

8.123 REMARK Observe that one can get BIBD's out of 8.122 with efficiency $E = \frac{p^n - p^{n-m}}{p^n - 1}$ (in (a)) and $E = \frac{p^n \cdot t}{p^{n-1}(t+1)}$ (in (b)) arbitrary close to 1.

8.124 REMARK BIBD's can also be constructed from non-abelian finite planar near-rings (see Clay (1)):
Define on $\mathbb{Z}_7 \times \mathbb{Z}_7 \times \mathbb{Z}_7$ an addition "$\oplus$" by

$$(a,b,c) \oplus (a',b',c') := (a+a',\ b+b',\ c+c'+a'\cdot b).$$

Let $g: N \to N$ be defined via $g(a,b,c) := (2a,2b,4c)$. Then (B.H. Neumann (2)) $(N,\oplus)$ is a non-abelian group and $G := \{id, g, g^2\}$ is a fixed-point-free automorphism group of $(N,\oplus)$.
8.96(a) gives some planar near-ring $(N,\oplus,\cdot_B)$. Clay goes on to prove that $(N,\mathcal{B})$ is a BIBD, of course with $k = \lambda$ and $C_1(N) = \{0\}$ (this follows from 8.120).
Clay also generalizes this example.

See Betsch-Clay (1) for an excellent summary of the theory of planar near-rings together with new results (e.g. connections to partially balanced incomplete block designs) and hints for further research.

§ 9 MORE CLASSES OF NEAR-RINGS

a) contains commutativity theorems similar to the "n(x)-theorem" of Jacobson and the "n(x,y)-theorem" of Herstein in ring theory. Our discussion is done in the world of IFP-near-rings (that are nr.'s N where $ab = 0$ implies $anb = 0$ for all $n \in N$). A dgnr. with the "n(x,y)-property" is a commutative ring. p-near-rings and Boolean nr.'s are also considered (as special cases).

Next, we study nr.'s without nilpotent elements. They are (if in $\mathcal{N}_0$) subdirect products of integral nr.'s which are studied in part 2) of b). The finite integral near-rings are planar iff they are not "trivial". Special integral nr.'s are called "near-integral domains". Their characteristic is zero or a prime.

c) contains a discussion of affine nr.'s (i.e. a generalization of nr.'s of type $M_{aff}(V)$). We examine the ideal structure, the radicals and nr.'s constructed out of affine nr.'s. Fundamental for these nr.'s is the fact that N_0 is a ring and N_c an ideal of N.

d) brings (for certain classes of groups) answers to the questions, which nr.'s (nr.'s with identity, ...) are definable on a given additive group. For instance, every nr. with identity on a cyclic group is a commutative ring. Several groups are explored which cannot be the additive group of a nr. with identity.

Finally, we discuss ordered nr.'s in e) (and discover that very few nr.'s can be fully ordered) and several other topics in f).

a) IFP - NEAR-RINGS

In ring theory, the following two theorems are certainly among the most famous commutativity theorems (see e.g. (Procesi)):

THEOREM 1: Let R be a ring with $\forall x \in R\ \exists n(x) \in \mathbb{N}\setminus\{1\}: x^{n(x)} = x$.

Then R is commutative.

THEOREM 2: Let R be a ring with

$$\forall x,y \in R\ \exists n(x,y) \in \mathbb{N}\setminus\{1\}: (xy-yx)^{n(x,y)} = xy-yx.$$

Then R is commutative.

(The first one was obtained by N. Jacobson; the second one is due to I.N. Herstein.)
We will generalize these results to certain classes of near-rings (including the dgnr.'s) using subdirect decompositions. In order to get a satisfactory treatment we start with a more general class of near-rings:

1.) IFP-NEAR-RINGS

9.1 DEFINITION A nr. N is said to fulfill the insertion-of-factors-property (IFP) provided that

$$\forall a,b,n \in N: (ab = 0 \Rightarrow anb = 0).$$

N has the strong IFP if every homomorphic image of N has the IFP.

IFP-nr.'s are called "pseudo-integral domains" in Ramakotaiah-Rao (2).
The next is an intrinsic characterization of the strong IFP:

<u>9.2 PROPOSITION</u> (Plasser (1)). N has the strong IFP: <=>
<=> $\forall$ $I \trianglelefteq N$ $\forall$ $a,b,n \in N$: $(ab \in I \Rightarrow anb \in I)$.

The proof is straightforward and hence omitted.

We will soon get examples of IFP-near-rings. But before we characterize these near-rings.

<u>9.3 PROPOSITION</u> (Bell (1), Plasser (1)). The following assertions are equivalent:

(a) N has the IFP-property.

(b) $\forall$ $n \in N$: $(0:n) \trianglelefteq N$.

(c) $\forall$ $S \subseteq N$: $(0:S) \trianglelefteq N$.

Again, the proof is obvious.

Observe that every IFP-near-ring N with left identity e is in $\mathcal{N}_0$, for $e0 = 0$ implies that $en0 = 0$, whence $n0 = 0$ for all $n \in N$.

<u>9.4 DEFINITION</u> Consider the following properties:

(P_0): $\forall$ $x \in N$ $\exists$ $n(x)>1$: $x^{n(x)} = x$.

(P_1): (P_0) and N is $\in \mathcal{N}_0$.

(P_2): $\forall$ $x,y \in N$ $\exists$ $n(x,y)>1$: $(xy-yx)^{n(x,y)} = xy-yx$ and $N \in \mathcal{N}_0$.

(P_3): $\forall$ $x,y,z \in N$: $xyz = xzy$ ("<u>weak commutativity</u>").

(P_4): $\forall$ $x,y \in N$ $\forall$ $I \leq N$: $xy \in I \Rightarrow yx \in I$.

<u>9.5 REMARKS</u>

(a) The "$x^{n(x)} = x$"-property does not imply that $N \in \mathcal{N}_0$, for every $N \in \mathcal{N}_c$ fulfills it. Nr.'s with (P_1) are called "<u>L-near-rings</u>" in Ligh (11). See Szeto (6), (8) for a characterization via sheaf representations.

(b) Abelian nr.'s N with $\forall$ $x \in N$: $x^2 = x$ and (P_3) were studied by Ratliff (1) and Subrahmanyam (1)

("Boolean semirings"). Abelian nr.'s with (P_3) are called "semirings" there. The nr.'s N with $\forall\ x\varepsilon N: x^2 = x$ and (P_3) are the "β-near-rings" of Ligh (14).

(c) (P_4) was considered by Bell (1), (2) and Plasser (1). Every nr. with (P_4) is in $\mathcal{N}_0$. But, on the other hand, every constant nr. has (P_3).

9.6 PROPOSITION (Bell (2), Ligh (16)).

(a) $(P_1) \Rightarrow (P_2) \Rightarrow (P_4)$.

(b) Each one of (P_1) to (P_4) implies the strong IFP-property.

Proof. (a): $(P_1) \Rightarrow (P_2)$ is immediate. Assume (P_2) and $xy\varepsilon I$. Then $yx-xy \equiv yx \pmod I$ and $\exists\ n\varepsilon \mathbb{N}\setminus\{1\}$: $yx-xy = (yx-xy)^n \equiv (yx)^n = yxyx\ldots yx \equiv 0 \pmod I$. Hence $yx\varepsilon I$.

(b): Since (P_1) - (P_3) are inherited to homomorphic images it suffices to show the IFP-property in this case. By (a), we only have to look at (P_3) and (P_4).

(P_3): If $ab = 0$ and $n\varepsilon N$ then $anb = abn = 0n = 0$.

(P_4): If $ab\varepsilon I$ and $n\varepsilon N$ then $ba\varepsilon I$, hence $b(an)\varepsilon I$, whence $anb\varepsilon I$ by (P_4).

See e.g. Ligh (16) and (Thirrin) for the connection to "duo rings" (i.e. rings, in which every one-sided ideal is two-sided). Clearly each duo ring is a strong IFP-nr. (but not conversely). For a detailed study of "duo-near-rings" see Choudhari (1), ch. VIII.
For easy reference, it rewards to define for this chapter

9.7 DEFINITION Let a nr. N be of

type I if $N\varepsilon\mathcal{N}_0$, N simple and strongly uniform (8.99).
type II if $N\varepsilon\mathcal{N}_0$ is not simple, but the intersection P of all non-zero ideals contains no non-zero idempotent.

type III if $N \notin \mathcal{N}_0$, $N \notin \mathcal{N}_c$ and if P (as above) has a non-zero idempotent then $P = N_0$.

type IV if $N \in \mathcal{N}_c$.

type V if $\forall\ x,y \in N$: $xy = 0$.

The structure of strong IFP-near-rings is given by

9.8 THEOREM (Ligh (16)). Every strong IFP-near-ring N is a subdirect product of subdirectly irreducible IFP-near-rings of type I,II,III,IV or V.

Proof. Let N be the subdirect product of some subdirectly irreducible near-rings N_i ($i \in$ some index set I) (1.62(a)). The N_i's have the IFP-property by 9.1.

(a) If $N_i \in \mathcal{N}_0$ and N_i is simple, use 9.3 to get N_i into type I or type V.

(b) Now let $N_i \in \mathcal{N}_0$ be not simple and P be as in 9.7. By 1.60(c), $P \neq \{0\}$. Assume that P contains the idempotent $e \neq 0$.
If $\exists\ x \in N_i$: $xe \neq x$ then $0 \neq xe - x \in (0:e) \trianglelefteq N_i$, so $P \subseteq (0:e)$, $e \in (o:e)$ and $e = e^2 = 0$, a contradiction. Hence e is a right identity, contained in P, whence $P = N_i$, a contradiction.

(c) If N_i is neither $\in \mathcal{N}_0$ nor $\in \mathcal{N}_c$ then $\{0\} \neq (N_i)_0 = (0:0) \triangleleft N_i$, so $P \subseteq (0:0)$. As in (b), every idempotent $e \neq 0$ is a right identity in N_i. If $x \in (0:0)$, $x = xe \in P$, hence $P = (0:0) = (N_i)_0$.

In special cases one gets more out of 9.8:

9.9 COROLLARY (Ratliff (1), Ligh (16)). A nr. with (P_0) and the strong IFP is a subdirect product of subdirectly irreducible near-rings $N_i \neq \{0\}$ with right identity of type I (in which case N_i is simple and integral), type III (in which case the annihilator ideals are exactly $\{0\}$ and $(N_i)_0$) or type IV.

Proof. Clearly, type V cannot occur. Suppose that N_i has type II. By the subdirect irreducibility, $P \neq \{0\}$. If $x \in P^*$, $x^{n(x)-1}$ is a non-zero idempotent in P, a contradiction. By the same argument, we get $P = N_0$ in the case that N_i is of type III. If then $(0:n) \neq \{0\}$, we deduce from 9.3 that $N_0 \subseteq (0:n)$. If $m = m_o + m_c \in (0:n)$, m_c is zero. Hence $(0:n)$ is either $\{0\}$ or $= (N_i)_o$ and the same follows for all $(0:S)$ $(S \subseteq N_i)$.
Now let N_i be of type I. (P_o) forces N to be integral. Now pick up some $e \neq 0$. $e^{n(e)-1} =: r$ is a right identity, for $\forall\ z \in N_i$: $(zr-z)e = zre-ze = = ze^{n(e)}-ze = ze-ze = 0$, whence $zr = z$.

9.10 COROLLARY If N has (P_1) then N is isomorphic to a subdirect product of simple integral near-rings $\in \mathcal{N}_0$ with a right identity.

Corollary 9.9 cries for

9.11 DEFINITION (Ratliff (1)). A nr. N is called almost small if N has at most 2 different annihilator ideals.

9.12 COROLLARY (Ratliff (1), Ligh (11), (16)). Every strong IFP-nr. with (P_o) is representable as subdirect product of almost small near-rings.

See also Ligh (11) and Szeto (1) for more detailed versions of 9.8 for near-rings with (P_o) or (P_3).

Now we turn to (P_2). First we need

9.13 PROPOSITION Let $N = N_0$ be subdirectly irreducible.

(a) N has the IFP, but no nilpotent elements beside 0 $\Rightarrow$ $\Rightarrow$ N is integral.

(b) If N has (P_2), is integral and has $N_d \neq \{0\}$ then N fulfills both cancellation laws, is abelian and either commutative or $\in \mathcal{N}_1$.

Proof. We may assume that $N \neq \{0\}$.

(a) Consider any $x \in N^*$. The semigroup $(\{x^k \mid k \in \mathbb{N}\},\cdot)$ does not contain 0 and is contained (Zorn !) in a multiplicative semigroup M_x maximal for not containing 0. Consider $I_x := \bigcup_{m \in M_x} (0:m)$. The IFP implies that $I_x \trianglelefteq N$. Since $x \notin I_x$ (M_x is closed w.r.t. multiplication!), $\bigcap_{z \in N^*} I_z = \{0\}$.

Since N is subdirectly irreducible, $\exists\, y \in N^*: I_y = \{0\}$. If $n \in N$ is not in M_y, the subsemigroup generated by M_y and n contains 0. So some product containing at least one times n (and possibly elements of M_y) must be zero. Such a product has one of the following forms:

$m_1 n m_2 = 0$, $nm' = 0$, $m''n = 0$, $n = 0$ $(m_1, m_2, m', m'' \in M_y)$.

An application of (P_4) yields $n = 0$ or $\exists\, m \in M_y: nm = \{0\}$, in which case $n \in I_y = \{0\}$, so again $n = 0$.

Thus $M_y = N^*$ and N is integral.

(b) 1) If $\exists\, x,y \in N: xy-yx \neq 0$, take $k \in \mathbb{N}\setminus\{1\}$ with $(xy-yx)^k = xy-yx$. $(xy-yx)^{k-1} =: e$ is a non-zero idempotent. Let $d \in N_d$ be $\neq 0$ and let n be arbitrary $\in N$. Then $(ne-n)e = 0$, so $ne = n$ and $d(en-n) = den-dn = dn-dn = 0$, whence $en = n$. So N has an identity 1 and each non-zero idempotent = 1.

2) Now let $a,b,c \in N$ with $ab = ac$, $a \neq 0$. If a is central, we get $b = c$. If a is not central, $\exists\, f \in N^*: af-fa \neq 0$, hence $(af-fa)a \neq 0$. Let $\ell \in \mathbb{N}\setminus\{1\}$ be such that $(a(fa)-(fa)a)^{\ell} = a(fa)-(fa)a$. Then $(a(fa)-(fa)a)^{\ell-1} = 1$ by 1) and a has a left inverse which again results in $b = c$.

3) We now show that $(N,+)$ is abelian. Let $1+1 =: 2$. If $2 = 0$, each element of N is of order 2 and N is abelian.

If $2 \neq 0$, but 2 is central then expanding $(n+m)(1+1)$ in both ways gives $n+m = m+n$ for all $m,n \in N$, so again N is abelian.

If $2 \neq 0$ and 2 is not central, we have to examine the conditions of 1.109(c). By the considerations above, N is $\epsilon \mathcal{N}_1$ and $n(-1) = n = n\cdot(1)$ yields $n = 0$.

2 has a left inverse u (say). Then u is a right inverse, too, for $(1-2u)2 = 2-2u2 = 2-2\cdot 1 = 2-2 = 0$ implies $2u = 1$. Let $r \epsilon N$ be arbitrary; call $h := u\cdot r$. Then $h+h = ur+ur = (u+u)r = (2u)r = 1r = r$.

Finally, $r = h+h = h'+h'$ gives $2h = 2h'$, whence $h = h'$.

An application of 1.109(c) shows that $(N,+)$ is abelian.

<u>9.14 THEOREM</u> (Bell (1), (2), Ligh (12), Ligh-Luh((1))).
Let N be a dgnr. . N has (P_2) <=> N is commutative.

<u>Proof</u>. "=>": Decompose, as in 9.8, N into subdirectly irreducible nr.'s $N_i \neq \{0\}$. Consider some N_i. N_i has also (P_2) and $(N_i)_d \neq \{0\}$.

(a) We first show that each nilpotent element is central (i.e. in C(N) - 8.16). We will accomplish this by induction on the degree k of nilpotence. $k = 1$ is trivial, but we also need $k = 2$:

Suppose that $n^2 = 0$. Then $\forall x \epsilon N$: $(xn-nx)xn = xnxn-nxxn = 0-0 = 0$, since $nn = 0$ implies $nxn = 0$ and $nxxn = 0$ by the IFP. Similarly, $(xn-nx)nx = 0$, so $(xn-nx)^2 = 0$, whence $xn-nx = 0$ by (P_2).

Now assume that $\forall n \epsilon N$: $n^{k-1} = 0 \Rightarrow n \epsilon C(N)$, and take $m \epsilon N$ with $m^k = 0$. Then $(m^{k-1})^2 = 0$, so $m^{k-1} \epsilon C(N)$; hence $\forall x \epsilon N$: $0 = m^k x - x m^k = mxm^{k-1} - xmm^{k-1} = (mx-xm)m^{k-1} = (IFP!) = (mx-xm)m(xm-xm)m = \ldots = (mx-xm)m = ((mx-xm)m)^{k-1} = (m(xm)-(xm)m)^{k-1}$.

Applying (P_2) again yields $m(xm)-(xm)m = 0$.

As above, it turns out that $(mx-xm)^2 = 0$, whence $mx-xm = 0$.

(b) From (a) and the IFP-property one gets (as for rings) that the set $Npt(N_i)$ of all nilpotent elements of N_i forms an ideal.

(c) If $Npt(N_i) = N_i$, (P_2) instantly results the commutativity of $(N,\cdot)$.

(d) If $Npt(N_i) = \{0\}$, N_i is integral by 9.13(a), hence abelian by 9.13(b), consequently a ring (6.9(c) and 6.6(c)) and therefore a commutative one (Theorem 2).

(e) If $\{0\} \neq Npt(N_i) \neq N_i$, consider $\overline{N}_i := N_{i/Npt(N_i)}$. $\overline{N}_i$ has no non-zero nilpotent elements, but is again dg. with (P_2). By (d), $\overline{N}_i$ is a commutative ring. So for all $n', n'' \varepsilon N$, $n'n''-n''n' \varepsilon Npt(N_i)$, from which we get $n'n'' = n''n'$ by (P_2).

(f) Since all N_i are commutative near-rings, the same applies to N.

(g) "<=" is trivial.

9.15 REMARK (Ligh-Luh (1)). The assumption in 9.14 that "N is dg." can be relaxed by "N is a D-nr." which means that each homomorphic image $\overline{N}$ of $N \varepsilon \mathcal{N}_0$ has $\overline{N}_d \neq \{0\}$ and is either non-abelian or a ring. Clearly each dgnr. is a D-nr., but there exist others, too (see Appendix, number 6 of the nr.'s on S_3).

9.16 COROLLARY (Bell (2), Ligh (12), (16)). Let N be a dgnr. with $N^2 = N$. Then N has (P_2) iff N is a commutative ring.

Proof: apply 9.14 and 1.107(c).

9.17 COROLLARY (Bell (1), Ligh (7), (11)). Let N have IFP, (P_0) and non-zero distributive elements in every non-zero homomorphic image. Then N is $\varepsilon \mathcal{N}_0$ and a subdirect product of nf.'s (hence abelian).

Proof. $N_0 = (0:0) \trianglelefteq N$ by the IFP. Now N/N_0 is constant, so $(N/N_0)_d = \{0\}$, whence $N/N_0 = \{0\}$ and $N = N_0 \in \mathcal{N}_0$. Thus we have (P_1) and hence (P_2) and the strong IFP available.
By 9.10, N is the subdirect product of simple integral near-rings N_i with right identities and no nilpotent elements. By 9.13(b), every N_i is abelian and either commutative (then a simple commutative ring with identity, hence a commutative field) or $\in \mathcal{N}_1$ (then (P_0) implies that N_i is a nf.).

Observe from 2.52(b) that the DCCI in N will turn N into a finite direct sum of nf.'s.

9.18 COROLLARY (Bell (1), Ligh (7), (11), (16)). Every dgnr. with (P_0) is a subdirect product of commutative fields (by 9.17 and 9.14) and hence a commutative ring.

9.19 COROLLARY (Ratliff (1)). A nr. with (P_0) and (P_3) is a subdirect product of nr.'s $N_i \neq \{0\}$ of type I, III or IV. If N_i is of type I then N_i is a commutative field or has more than one right identity.

Proof. According to 9.9, N cannot be of type II. If N_i is of type I and has just one right identity e then $\forall\, x \in N_i^* : x^{n(x)-1} = e$ (as in the proof of 9.13(b)). Hence $\forall\, x,y \in N_i^* : xy = x^{n(x)-1}xy = exy = eyx = $ $= y^{n(y)-1}yx = yx$. So N_i is commutative, e is an identity and N_i is (by 9.13(b)) a simple integral domain, hence a field.

9.20 COROLLARY (Ratliff (1)). Let N have (P_0), (P_3) and non-zero distributive elements in every non-zero homomorphic image (this happens e.g. if $N \in \mathcal{N}_1$ or if N is dg.). Then N is a subdirect product of commutative fields and hence a commutative ring.

Proof. By 9.17, $N \in \mathcal{N}_0$, so in the subdirect decomposition of 9.9 all N_i are of type I.
By 9.13(b), $N \in \mathcal{N}_1$. Due to 9.19, N_i is a commutative field.

9.21 REMARKS Now we mention (without proof, but with reference) other commutativity theorems.

(a) (Bell (5), (6)). Let $N \in \mathcal{N}_1$ be a dgnr. with $\forall\ x \in N\ \exists\ n(x) \in \mathbb{N} : x - x^{n(x)} \in C(N)$ (8.16). Then N is a commutative ring.

(b) (Ligh-Luh (1)). A finite D-nr. with identity in which all nilpotent elements are central is a commutative ring.

(c) (Ligh (11)). A finite dgnr. without nilpotent elements is a commutative ring.

(d) (Ligh (12)). A finite dgnr. is commutative iff all zero divisors are central.

(e) (Plasser (1)). If N has (P_0) and a left identity e. Then $N \in \mathcal{N}_0 \iff \forall\ x \in N : e - x0$ is idempotent $\iff$ all idempotents are distributive $\iff$ all idempotents are central $\iff$ N has $(P_1) \implies$ N is subdirect product of nf.'s. Anyhow, $(N,+)$ is torsion and each element has a square-free order in $(N,+)$.

(f) (Ligh (8)). Call $N \in \mathcal{N}$ an <u>α-nr.</u> if $d \in N_d$ implies $-d \in N_d$. Every α-nr. without nilpotent elements is a ring. Each α-nr. with (P_0) is a commutative ring.

(g) (Ligh (15)). Each nr. N with (P_3) fulfills $\forall\ n \in \mathbb{N}\ \forall\ x,y \in N : (xy)^n = x^n y^n$. Every α-nr. N with this property (or with $\forall\ n \in \mathbb{N}\ \forall\ x,y \in N: (x+y)^n = x^n + y^n$) has only nilpotent commutators of $(N,\cdot)$.
A nr. $N \in \mathcal{N}_1$ without nilpotent elements and with $\forall\ x,y \in N: (xy)^2 = x^2 y^2$ is abelian.

(h) (Ligh-Utumi (1)). N is a <u>C_1-nr.</u> (<u>C_2-nr.</u>) if $\forall\ n \in N: nN = nNn$ ($Nn = nNn$, respectively). Neither one implies the other:

If F is a field then $M_{aff}(F)$ (1.4) is a C_1-, but not C_2-nr. .
A finite integral nr. has C_2, but not C_1.
Every C_2-nr. (but not every C_1-nr.) has the IFP.
N is C_1- and C_2-nr. iff N is C_1-nr. and every idempotent is central.
See this paper for decomposition theorems for C_1- and C_2-nr.'s with finiteness conditions.

(i) A ring R is called a <u>P_1-ring</u> if for all $r \in R$ there is a central idempotent r^o with $rr^o = r$ and $\forall\ e^2 = e \in R : (er = re \Rightarrow r^o e = r^o)$ (this P_1 has nothing to do with our (P_1)).
See Plasser (1) for a similar concept for near-rings.

(j) For more results see Ramakotaiah-Rao (2).

2.) p-NEAR-RINGS

<u>9.22 DEFINITION</u> Let p be a prime. A nr. N is called a <u>p-near-ring</u> provided that $\forall\ x \in N$: $x^p = x \wedge px = 0$.

Evidently, every p-nr. has property (P_o).

<u>9.23 PROPOSITION</u> (Plasser (1)). A p-nr. with left identity is zero-symmetric.

<u>Proof.</u> Let e be the left identity. Then it is easily shown by induction that $\forall\ x \in N\ \forall\ k \in \mathbb{N} : (e+x0)^k = e+k(x0)$. Hence $e+x0 = (e+x0)^p = e+p(x0) = e$, whence $x0 = 0$.

<u>9.24 REMARK</u> (Plasser (1)). 9.23 does not hold for general nr.'s with (P_o).

<u>9.25 COROLLARY</u> (Ratliff (1)). A p-nr. with (P_3) and non-zero distributive elements in every homomorphic image is isomorphic to a subdirect product of copies of the field $\mathbb{Z}_p$, hence a p-ring.

<u>Proof</u>. By 9.20, N is a subdirect product of simple commutative p-rings N_i with identity. Ring theory tells us that $N_i \cong \mathbb{Z}_p$.

<u>9.26 THEOREM</u> (Plasser (1)). A finite p-nf. N is isomorphic to the field $\mathbb{Z}_p$.

<u>Proof</u>. N is a Dickson nf., for N cannot be one of the 7 exceptional cases (8.34 and the subsequent discussion), since in each one of these cases $A^5 = A$, but $B^5 \neq B$.
Now $\exists\, q \in \mathbb{P}\ \exists\, n \in \mathbb{N}: |N| = q^n$ by 8.13.
Since $(N,+)$ is a finite p-group, $|N|$ is some power of p, consequently $q = p$.
Now $(N^*,\cdot)$ has generators a,b with $b^{-1}ab = a^q = a^p$ (8.33). Thus $ab = ba^p = ba$, N is commutative and hence has (P_3).
Now the result follows from 9.25.

<u>9.27 REMARKS</u>

(a) Cf. also 8.35.

(b) The finiteness condition in 9.26 is indispensable, for there exist infinite p-fields.

An application of 2.52(b) gives with 9.17 and 9.26 the following

<u>9.28 COROLLARY</u> (Plasser (1)). Let N be a finite p-nr. with IFP and with non-zero distributive elements in every non-zero homomorphic image. Then N is isomorphic to a (finite) direct sum of copies of $\mathbb{Z}_p$, hence a finite p-ring.

<u>9.29 REMARK</u> Ratliff (1) studied p-nr.'s N (especially for $p=3$ and $p=5$), which can be derived from a p-ring R in a way that $(N,+) = (R,+)$ and the product in N is defined via a fixed polynomial function over R. The nr.'s considered in this dissertation fulfill (P_0) and (P_3).

3.) BOOLEAN NEAR-RINGS

It does not seem to be quite clear how to define a Boolean near-ring. So we take what seems to be the most general possible definition.

9.30 DEFINITION A nr. N is Boolean: $\Longleftrightarrow \forall\, x \in N : x^2 = x$.

Hence a Boolean nr. is a (P_0)-near-ring with $n(x) = 2$ for all x.

9.31 REMARKS

(a) Every constant nr. is a Boolean nr. with (P_0) and (P_3), but no 2-nr. in general.

(b) A Boolean nr. with (P_3) (a β-nr.) and non-zero distributive elements in every non-zero homomorphic image is a subdirect product of copies of $\mathbb{Z}_2$. This result of Ligh (14) follows from 9.25.

(c) (Ligh (5), (14), (8), (10), Heatherly (8)). The same assertion holds for dg. Boolean nr.'s. Of course, this follows from 9.18, but there is also a direct elementary proof in Ligh (10).

(d) See p.346/48 for a list of all Boolean nr.'s definable on the two non-abelian groups of order 8.

(e) Ferrero-Cotti (2), (3) considered nr.'s with the identities abc = acbc = abac. These are those ones which contain an ideal I with $I^3 = \{0\}$ and N/I is a Boolean ring.

9.32 COROLLARY (Heatherly (8)). A Boolean nr. $\in \mathfrak{N}_0$ with DCCI is a finite direct sum of ideals which are integral simple nr.'s with the trivial multiplication $xy = \begin{cases} x & y \neq 0 \\ 0 & y = 0 \end{cases}$.

Proof. Apply 9.10, 2.52(b) and the fact that every non-zero element is (as an idempotent - see the proof of 9.13(b)) a right identity.

9.33 EXAMPLES

(a) (Clay-Lawver (1)). Let $(B,+,\wedge)$ be a Boolean ring with identity 1. Let $a' := a+1$ and $a \vee b := (a' \wedge b')'$. If $x \in N$, define for $a,b \in B$ $a *_x b := a \wedge (b \vee x)$. Then $(B,+,*_x)$ is a Boolean nr. with (P_3) which is a ring iff $x = 0$.
Nr.'s derived from Boolean rings are called "special Boolean near-rings" in this paper. Their ideal structure is considered.

(b) Subrahmanyam (1) called an abelian Boolean nr. with (P_3) "Boolean semiring". Every P_1-ring (9.21(i)) $(B,+,\cdot)$ gives rise to a Boolean semiring $(B,+,*)$, where $a*b := ab^o$.
Every constant abelian nr. is a Boolean semiring.
A Boolean semiring can be represented as a disjoint union of "nearly distributive" lattices.
See this paper for more details.

b) NEAR-RINGS WITHOUT

1.) NEAR-RINGS WITHOUT NILPOTENT ELEMENTS

Nr.'s without non-zero nilpotent elements came up at several different places in our discussion of near-rings. We collect some of the results concerning these near-rings.

9.34 REMARKS Let N be a nr. without non-zero nilpotent elements. Then

(a) N has no nil(potent) subsets (2.96).

(b) If $N \in \mathfrak{N}_0$ has DCCN then every non-zero N-subgroup contains a non-zero idempotent (3.51). Moreover, $\mathfrak{J}_{1/2}(N) = \mathfrak{J}_0(N) = \mathfrak{N}(N) = \mathfrak{P}(N) = \{0\}$ in this case (5.40).

(c) In any case, $\mathfrak{N}(N) = \mathfrak{P}(N) = \{0\}$.

9.35 EXAMPLES

(a) Every constant nr. has no non-zero nilpotent elements.

(b) Every integral nr. (hence every nf.) has this property, too.

The connection to the previous chapter is given by

9.36 THEOREM (Bell (1), Marin (1), Ramakotaiah-Rao (2)). Let N be zero-symmetric. Equivalent are:

(a) N has no non-zero nilpotent elements.

(b) N is a subdirect product of integral nr.'s.

Proof. (a) => (b) is nothing else than in the proof of 9.13(a): N has a family of ideals I_x $(x \in N^*)$ with zero intersection and each N/I_x is integral.

(b) => (a): If $x^n = 0$, in each component $\pi_i(N)$ of the subdirect representation of N we get $\pi_i(x)=0$, whence $x = 0$.

Hence we will devote the next number to integral near-rings. But before, some more results might be appropriate.

9.37 PROPOSITION (Bell (1), Heatherly (8), Marin (1)). A nr. A nr. $N \in \mathfrak{N}_0$ without non-zero nilpotent elements is an IFP-nr.

Proof. If $xy = 0$ $(x,y \in N)$ then $yxyx = y0x = 0$, whence $(yx)^2 = 0$, so $yx = 0$. Now $\forall$ $n \in N$: $xny = (ny)x = n(yx) = n0 = 0$, so N has the IFP.

9.38 COROLLARY (Heatherly (8)). Every subdirectly irreducible nr. $N \in \mathcal{N}_0$ without non-zero nilpotents is integral. Every non-zero idempotent is a right identity.

Proof. The first assertion holds by 9.13(a).
If $e \neq 0$ is idempotent, $\forall$ $x \in N$: $(xe-x)e = 0$, whence $xe = x$.

To get more, we have to impose some finiteness conditions on N.

9.39 PROPOSITION (Heatherly (8)). Let $N \in \mathcal{N}_0$ be a subdirectly irreducible nr. $\neq \{0\}$ with DCCN and without non-zero nilpotent elements. Then

(a) N is integral and 2-primitive on N.
(b) N has a right identity.
(c) $N_d \neq 0 \Rightarrow N$ is a nf. .
(d) If N is dg. then N is a field.

Proof. (a) Consider, for $x \in N^*$, the chain $Nx \supseteq Nx^2 \supseteq Nx^3 \supseteq \ldots$ There is some $n \in \mathbb{N}$ with $Nx^n = Nx^{n+1} = \ldots$. N is integral by 9.37, so $Nx^n = (Nx)x^n$ implies $N = Nx$. Therefore N is 2-primitive on N.

(b) holds by 4.46 or by 9.38.

(c) By the same argument as in the proof of 9.13(b), N contains an identity. Now apply 4.47(a) and 9.17.

(d) is obvious.

9.40 REMARK (Heatherly (8)). There exist even finite simple abelian nr.'s N with (P_1) and without non-zero nilpotent elements, which are no nf.'s.

We can reduce the theory of near-rings with DCCN and no non-zero nilpotent elements to that of 9.39:

<u>9.41 THEOREM</u> (Heatherly (8)). Let $N \in \mathcal{N}_0$ have DCCN and no non-zero nilpotent elements. Then N has a right identity, is 2-semisimple and the finite direct sum of nr.'s which fulfill all conditions of 9.39. If every non-zero homomorphic image of N has non-zero distributive elements then N is a finite direct sum of nf.'s; if N is dg. then N is a finite direct sum of fields.

<u>Proof</u>. Decompose N into subdirectly irreducible integral nr.'s N_i (9.36). In fact, N is a finite direct sum of these ones (2.52(b)). Now apply 9.39(b), 9.39(a) and 5.49.

<u>9.42 COROLLARY</u> (Heatherly (8)). If $N \in \mathcal{N}_0$ is a finite nr. without non-zero nilpotent elements. Then N has (P_1).

This is clear by 9.38 and 9.41 (Heatherly goes on to show that $n(x)$ (9.4) can be chosen to be constant for all $x \in N$. Cf. also Ligh (11)).

Moreover, we have some information concerning the near-rings in discussion, which guys belong to the center $C(N)$ of N:

<u>9.43 PROPOSITION</u> (Bell (1)). Let $N \in \mathcal{N}_0$ have no non-zero nilpotent elements. Then

(a) Every distributive idempotent is central.
(b) If $N \in \mathcal{N}_1$, all idempotents are in $C(N)$.

<u>Proof</u>. First we show that for each idempotent e, $\forall\, x \in N$: $ex = exe$. Now $(ex-exe)e = 0$, so (9.36) $e(ex-exe) = 0$ and $ex(ex-exe) = 0$ (IFP). Hence $(-exe)\cdot(ex-exe) = (-ex)0 = 0$. Therefore $(ex-exe)^2 =$ $= ex(ex-exe)+(-exe)(ex-exe) = 0+0 = 0$, whence $ex-exe = 0$.

(a) If $e \in N_d$, $\forall\, x \in N$: $e(xe-exe) = exe+e(-exe) =$ $= exe-exe = 0$, hence $(xe-exe)e = 0$, whence $xe = exe = ex$.

(b) If N has an identity 1, consider again some idempotent e. $(1-e)e = 0$, so $\forall\, x\in N$: $(1-e)xe = 0$. Also, $(xe-exe)e = xe-exe$ and $(1-e)xe = xe-exe$, therefore $(xe-exe)^2 = (xe-exe)e(xe-exe) =$ $= (xe-exe)(1-e)xe = 0$, so $xe = exe = ex$ for all $x\in N$.

9.44 REMARKS

(a) See Marin (1) for characterizations of those near-rings without non-zero nilpotent elements which are (finitely or not) completely reducible into certain other near-rings.

(b) Recall 9.21(f).

(c) Again, let $N\in\mathfrak{N}_0$ have no non-zero nilpotent elements. Then (Bell-Ligh (1)):

α) If N is dg. with finitely many subnear-rings, N is a finite commutative ring.

β) If N has at most 2 idempotents and no proper (finitely many) subnear-rings, N is a finite field (a near-field, respectively).

(d) Don't forget to observe 9.54.

2.) NEAR-RINGS WITHOUT ZERO DIVISORS (INTEGRAL NEAR-RINGS)

9.45 EXAMPLES

(a) Every constant nr. is integral.

(b) If $(\Gamma,+)$ is any group, also $\gamma*\delta := \begin{cases} \gamma & \ldots\ \delta \neq 0 \\ 0 & \ldots\ \delta = 0 \end{cases}$

defines an integral nr. $(\Gamma,+,*)$ (cf. 1.4(b)).

So one can say nothing about the additive group of an integral near-ring. To overcome this we will give the following

9.46 DEFINITION An integral nr. N is non-trivial if its multiplication is not one of 9.45(a) or (b).

These non-trivial integral nr.'s are sometimes called "near-integral domains" (see Ligh (13), Heatherly-Olivier (1), (2), Adams (1), (2)). But they are not always embeddable into a near-field, so we reserve this distinguishing name to a more special class of non-trivial integral near-rings (see 9.52 and 9.65).

9.47 PROPOSITION (Clay (8), Heatherly-Olivier (1), Plasser (1)). If N is integral then $N \in \mathcal{N}_0$ or $N \in \mathcal{N}_c$.

Proof. Suppose that $\exists\ x \in N$: $x0 \neq 0$. Then for all $n \in N$ we have $(nx0-n)x0 = nx0-nx0 = 0$, whence $nx0 = n$. Hence $n0 = nx00 = nx0 = n$, and N is constant.

Thus every non-trivial integral near-ring is zero-symmetric. Integral near-rings also appear in previous chapters. In order to present a good aerial view on this topic we compile these facts:

9.48 REMARKS Recall that an integral near-ring N has the following properties:

(a) N has the right cancellation law (1.111(a)).

(b) If N is finite and non-abelian then each element of N has a unique square-root (1.112).

(c) N is a prime ring (2.66).

(d) If N is non-trivial in the sense of 9.46 and has the DCCI then N is subdirectly irreducible (2.107). Applying 9.39 we get: If N has moreover the DCCN then there exists a right identity, $N_d \neq \{0\}$ implies that N is a nf. (Ligh-Malone (1)), N is dg. implies that N is a field (cf. also 6.14(b)), N is 2-primitive on N and so N is simple (Heatherly (8)). See also Graves-Malone (1).

(e) On the whole, 9b)1) is applicable, for N has no non-zero nilpotent elements. So if N is non-trivial, it has the IFP.

<u>9.49 REMARK</u> to 9.48(d). Without chain conditions one cannot conclude that an integral nr. N with $N_d \neq \{0\}$ is a near-field: take a field F and form $N := F_0[x]$ (7.78). N is integral (7.68(c), 1.111(a)), each ax $(a \in N)$ is in N_d, but N is no nf. (7.68(b)). In fact, for every $k \in \mathbb{N}$, $I_k := \{a_k x^k + a_{k+1} x^{k+1} + \ldots \mid a_k, a_{k+1}, \ldots \in F\}$ is an ideal and $I_1 \supset I_2 \supset I_3 \supset \ldots$ is a strictly descending chain (Heatherly (8)). Cf. also Graves-Malone (1).

<u>9.50 THEOREM</u> (Ferrero (8)). Let N be a finite integral near-ring. N is non-trivial $\Longleftrightarrow$ N is planar.

<u>Proof</u>. $\Rightarrow$: Consider G of 8.96(b). Since $\forall\, n \in N^*$: $Nn = N$ (9.48(d)). Each g_a $(a \in N^*)$ is a monomorphism since N is integral. N is finite, so $G \leq \mathrm{Aut}\,(N,+)$. N is non-trivial, so $G \neq \{id\}$. G is also fixed-point-free (Heatherly-Olivier (1)): Let g_a $(a \in N^*)$ have a fixed-point $n_0 \neq 0$. Let x be arbitrary in N. Since $Nn_0 = N$, $\exists\, y_x \in N$: $x = y_x n_0$. Hence $g_a(x) = xa = y_x n_0 a = y_x g_a(n_0) = y_x n_0 = x$, so $g_a = id$. Since N is trivially strongly uniform we may apply 8.100 and are through.

$\Leftarrow$: is immediate, since $|N/\equiv| \geq 3$ (8.85).

We apply 8.98 to get

<u>9.51 COROLLARY</u> (Ligh (13)). The additive group of a finite non-trivial integral nr. is nilpotent, but not necessarily abelian.

Mathematics is a crazy job: the additive groups of these nr.'s without nilpotent elements is nilpotent.

Anyhow, we can use 8.11 to get:

9.52 COROLLARY Not every non-trivial integral nr. can be embedded into a near-field.

9.53 REMARKS (Betsch). If one recalls $(\Gamma,+,\cdot_B)$ of 1.4(b), 1.15 and 8.97, the following results are simple corollaries from group theory, since a group G is the additive group of a non-trivial integral near-ring iff G has a non-tivial group of fixed-point-free automorphisms.

(a) (Adams (1), (2)). For each $k\in\mathbb{N}$ there exist both finite and infinite non-trivial integral nr.'s N such that $(N,+)$ is nilpotent of degree k (see (Huppert), p. 499).

(b) (Ligh (13), Heatherly-Olivier (2)). Neither the commutativity nor the nilpotency of $(N,+)$ force N to be integral.

(c) (Heatherly-Olivier (2), Adams (1), (2)). If N is infinite and integral, $(N,+)$ is not necessarily nilpotent (not even for dgnr.'s with both cancellation laws).

The following result is somewhat nostalgic in nature, for it concerns nr.'s of 9b)1).

9.54 COROLLARY (Heatherly-Olivier (2)). Let $N\in\mathfrak{N}_0$ be a finite nr. without non-zero nilpotent elements, such that no homomorphic image is trivial (9.46). Then $(N,+)$ is nilpotent.

Proof. 9.36 and 9.51.

9.50, 9.51 and 9.53 exclude many groups from being additive groups of non-trivial integral nr.'s:

9.55 REMARKS

(a) (Ligh-Malone (1)). Complete groups, the dihedral group of order 8 and the quaternion group cannot be the additive group of a non-trivial integral nr..

(b) (Betsch, Heatherly-Olivier (1)). The same applies to all finite groups Γ of order 2n, where n is an odd integer > 1. For Γ is nilpotent in this case, having an element of order 2 which is fixed by all automorphisms of Γ. Hence there is no non-trivial fixed-point-free automorphism group on Γ.

(c) If $|(\Gamma,+)| = p+1$ $(p \in \mathbb{P}\setminus\{2\})$ then either there is no non-trivial integral nr. definable on $(\Gamma,+)$ or the only one is $GF(2^n)$, where $2^n = p+1$.

9.56 EXAMPLES For non-trivial integral nr.'s on $\mathbb{Z}_5$, $\mathbb{Z}_7$ see Clay (6), for ones on $\mathbb{Z}_{31}$, $\mathbb{Z}_{121}$ consult Ferrero (8), for some on $\mathbb{Z}_9$, $\mathbb{Z}_3 \oplus \mathbb{Z}_3$ and $\mathbb{Z}_{15}$ see Heatherly (6), for all of them Heatherly-Olivier (2). Whittington (1) gives a computer-aided description of all non-trivial integral nr.'s on groups of "low" order. Additional information can be found in Adams (1), (2). See also p. 346 and p. 348.

Clay (6) raised the question, if every non-trivial integral nr. has as characteristic zero or a prime (cf. 8.9). Some of the examples above answer this negatively. So we give a condition where this is the case:

9.57 THEOREM (Heatherly-Olivier (1)). Let N be an integral near-ring with $N_d \neq 0$. Then the characteristic of N is either 0 or a prime.

Proof. Take $d \in N_d^*$. Let $x \in N$ have a finite order, say $p \cdot j$, where $p \in \mathbb{P}$, $j \in \mathbb{N}$. Then $0 = d0 = d(pjx) = (pd)(jx)$, whence $pd = 0$. So $d(px) = (pd)x = 0x = 0$ and hence $px = 0$.

9.58 REMARK (Heatherly-Olivier (1)). The same conclusion holds if the non-trivial integral nr. N has no non-trivial right ideals.
Confer also 9.17.

9.59 REMARKS

(a) (Heatherly-Olivier (2)). If a non-trivial integral nr. has the property that $\exists\, n\in N^* \ \forall\, m\in N$: $n(-m) = -nm$ or if N is finite with $|N| = p_1 \dots p_n\, q_1^2 \dots q_m^2\, p^3\, 2^k$ ($p_1,\dots,p_n$, $q_1,\dots,q_m$, p distinct odd primes, where p is of the form 2^r+1 and 2^k-1 is a prime or $k = 0$) then N is abelian.
The same follows for integral nr.'s N with a non-zero idempotent e such that $\exists\, h\in N$: $h+h = e$ (B.H. Neumann (1)).

(b) (Ferrero (8)). Given a prime power p^b, there exists a non-trivial integral nr. on $\mathbb{Z}_{p^b}$ if there is some $k\in\mathbb{N}$ such that the smallest of all numbers k^x-1 which are divisible by p is also divisible by p^b.
Ferrero called N $\mathbb{Z}$-distributive if $\forall\, a,b\in N\ \forall\, z\in\mathbb{Z}$: $a(zb) = z(ab)$. A finite non-trivial integral $\mathbb{Z}$-distributive nr. has an elementary abelian additive group.

(c) See Heatherly-Olivier (2), Szeto (2) and Ramakotaiah-Reddy (1) for a description of the multiplicative semigroup $(N^*,\cdot)$ of a non-trivial integral near-ring N.

Having 9.52 in mind we look for integral nr.'s which can be embedded into a near-field.

9.60 DEFINITION A nr. N is called near-integral domain (nid.) if N fulfills the left cancellation law and the left Ore condition (1.64).

9.61 WARNING Nid.'s are called "near-domains" by Graves-Malone (1) - (3). Since this collides heavily with 8.41, we will not use this name.

Near-integral domains are really integral:

9.62 THEOREM (Graves-Malone (1)). Let $N \neq \{0\}$ be a nid. . Then

(a) N is integral.

(b) $N \in \mathcal{N}_0$ and if $|N| > 2$, N is not trivial.

(c) If $N \neq \{0\}$ has the DCCN then N is a near-field.

Proof. Suppose that $n0 = n_c \neq 0$ for some $n \in N$. Since $n_c n_c = n_c = n_c 0$ we get $n_c = 0$ by the left cancellation law, a contradiction. Hence $N \in \mathcal{N}_0$. If $nm = 0$, but $n \neq 0$, we again use left cancellation to get $m = 0$ out of $nm = 0 = n0$.

So if N is trivial, $\forall\, n,m \in N$: $nm = \begin{cases} n & m \neq 0 \\ 0 & m = 0 \end{cases}$ (9.45).

If $|N| \geq 3$, take $n, m_1, m_2 \in N^*$ with $m_1 \neq m_2$. But then $nm_1 = n = nm_2$ is again resulting in a contradiction $m_1 = m_2$.

To show (c), observe that for $|N| = 2$ we get $N = \mathbb{Z}_2$. Otherwise N is non-trivial. 9.48(d) tells us that N has a right identity e. If e' is another one and $x \in N^*$ then $xe = x = xe'$ implies $e = e'$. By 1.112, N is a nf. .

9.63 REMARKS

(a) Observe that we did not use the Ore condition in 9.62.

(b) (Graves-Malone (1)). 9.62(c) does not hold without a chain condition: Let $(\Gamma,+)$ be the free group on two generators x,y. For $n \in \mathbb{N}_0$ let T_n be the map $\Gamma \to \Gamma$ sending a word $w(x,y)$ in x and y into $w(nx,ny)$. Let N be the dgnr. generated additively by the set $\{T_n \mid n \in \mathbb{N}_0\}$. Then $N \in \mathcal{N}_1$, N has the left cancellation law but neither the left nor the right Ore property. N is no near-field.

<u>9.64 THEOREM</u> (Graves-Malone (1)). Let N be a nid. and $S := (N^*,\cdot)$. Then the nr. N_S of left quotients of N w.r.t. S is a nf. .

<u>Proof</u>. S fulfills the conditions of 1.65, so N_S exists. In N_S, each non-zero element is invertible, hence N_S is a nf.

Now we get two corollaries due to Graves-Malone (1) (see 8.9, 8.10 and 8.11).

<u>9.65 COROLLARY</u> Every nid. can be embedded into a near-field.

<u>9.66 COROLLARY</u> Let N be a nid.. Then

(a) N is abelian.

(b) $\forall\ n,m\in N$: $n(-m) = -nm$.

(c) N has as characteristic either zero or a prime.

This is a satisfactory result, which may conclude our considerations of integral near-rings. We only make some remarks:

<u>9.67 REMARKS</u>

(a) (Graves-Malone (1)). If I is an ideal of the nid. N then the near-fields of quotients of I and N coincide.

(b) Berman-Silverman (1) defined a nr. N to be a <u>D-ring</u> if N is integral and $\forall\ n\in N\ \exists\ n_e, n_r \in N^*$: $n_e n \in C(N)$ and $n n_r \in C(N)$. ($C(N)$ is again the center of N). Graves-Malone showed in (1) that each D-ring is a nid. and in (3) that the converse does not hold by looking at the nr. N of formal power series over $\mathbb{R}$ with $a_{2n} = 0$ for all $n\in\mathbb{N}_0$. N is a nid., but no D-ring.

(c) See Graves-Malone (2) for a discussion of nid.'s which have an Euclidean algorithm ("Euclidean near-domains"). Each such nid. N has the ACCN, an identity, only monogenic N-subgroups and a unique factorisation into units and primes for every non-zero element of N. Confer 7.72.

c) AFFINE NEAR-RINGS

Now we study a class of near-rings which are in a certain sense the "most elementary non-zero-symmetric near-rings". The dominant property is that the constants form an ideal.
Let F be a field, V a vector space over F and $M_{aff}(V)$ the near-ring of affine transformations on V as in 1.4(c).

9.68 PROPOSITION (Blackett (2)). $M_{aff}(V) =: N$ has the following properties:

(a) N is abelian.
(b) $N_o = N_d$.
(c) $N_c \trianglelefteq N$.
(d) $N_o \cong N/N_c$ is a ring.

The proof is straightforward and omitted. Observe that all sub-near-rings of $M_{aff}(V)$ also fulfill these properties.
Now we consider "affine transformations" over groups:

9.69 NOTATION Let Γ be an abelian group.

$M_a(\Gamma) := \mathrm{Hom}(\Gamma,\Gamma)+M_c(\Gamma)$.

9.70 PROPOSITION $M_a(\Gamma) \leq M(\Gamma)$.

Again, the proof is obvious.

$M_a(\Gamma)$ also enjoys the properties (a) - (d) of 9.68. This gives motivation for an axiomatic treatment:

9.71 DEFINITION (Gonshor (1)). A nr. N is called an abstract affine near-ring (=: a.a.n.r.) provided that

(a) N is abelian.
(b) $N_o = N_d$.

Evidently, $M_a(\Gamma)$, $M_{aff}(V)$, every ring and every nr. of linear polynomials (linear polynomial functions) are examples of a.a.n.r.'s. See also 9.81.
There is no need to postulate (c) and (d) of 9.68 since

9.72 PROPOSITION Let N be an a.a.n.r. . Then

(a) $(N,+) = (N_0,+)\dot{+}(N_c,+)$.

(b) $N_c \trianglelefteq N$.

(c) $N_0 \cong N/N_c$ is a ring.

The proof of (a) is done by remembering 1.13, the rest is established by straightforward computations (for (c) one can use 2.8).

The main types of substructures of an a.a.n.r. are easily characterized:

9.73 PROPOSITION (Gonshor (1)). Let N be an a.a.n.r.. Then

(a) All N-subgroups S of N are of the form $S = S_o\dot{+}N_c$ with $S \leq_{N_0} N_0$.

(b) All right ideals R of N are $R = R_o\dot{+}R_c$ where $R_o \trianglelefteq_r N_0$ and $(R_c,+) \leq (N_c,+)$.

(c) All ideals I of N are given by $I = I_o\dot{+}I_c$ with $I_o \trianglelefteq N_0$, $N_0I_c \subseteq I_c$, $I_oN_c \subseteq I_c$.

(d) All two-sided invariant subgroups T of N are $T = T_o\dot{+}N_c$, where $T_o \trianglelefteq N_0$.

Again, the proof consists only of standard arguments (observe 2.18). The formulation of (a) - (e) is meant in this way that every $S_o\dot{+}N_c$ with $S \leq_{N_0} N_0$ is an N-subgroup of N, and so on.

9.74 COROLLARY Every two-sided invariant subgroup of an a.a.n.r. is an ideal, and every N_0-subgroup is a left ideal (and conversely, of course).

In the case of $M_{aff}(V)$ one can improve 9.73 (see 9.76).

<u>9.75 REMARK</u> Not every left ideal of an a.a.n.r. can be directly decomposed similar to 9.73: take for instance $N = M_{aff}(\mathbb{R}^2)$ and $L := ((0,0):(0,1))$. $\ell \in N$, defined by $\ell(x,y) := (x+y-1, x+y-1)$ is in L, but ℓ_o and ℓ_c are not in L, for $\ell_c(x,y) = (-1,-1)$ and $\ell_o(x,y) = (x+y, x+y)$.

<u>9.76 PROPOSITION</u> (Wolfson (1)). For each ordinal $\lambda \geq 0$ let $T_\lambda := \{h \in Hom_F(V,V) \mid \dim \operatorname{Im} h < \aleph_\lambda\}$. Let $T_{-1} := \{\bar{o}\}$. Then

(a) All ideals of $M_{aff}(V)$ are given by $T_\lambda + M_c(V)$ $(\lambda \geq -1)$; hence the ideals are $M_c(V) \subset T_0 + M_c(V) \subset T_1 + M_c(V) \subset \ldots \subset M_{aff}(V)$.

(b) For $\nu \geq -1$, $M_{aff}(V)/_{T_\lambda + M_c(V)} \cong Hom_F(V,V)/_{T_\lambda}$, so every homomorphic image $\neq M_{aff}(V)$ is a ring.

(c) In particular, $\dim_F V < \infty$ implies that $M_c(V)$ is the unique ideal of $M_{aff}(V)$.

Essentially, this follows from 9.73(c) and the fact that every ideal in $Hom_F(V,V)$ is some T_λ (see e.g. (Baer)).

<u>9.77 THEOREM</u> Let N be an a.a.n.r., $\mathcal{J}(N_o)$ the Jacobson-radical of the ring N_o and $\mathcal{J}({}_{N_o}N_c)$ the radical of the N_o-module N_c (= the intersection of all maximal N_o-subgroups of N_c). Then $\mathcal{J}_0(N) = \ldots = \mathcal{J}_2(N) = \mathcal{J}(N_o) + \mathcal{J}({}_{N_o}N_c)$.

<u>Proof.</u> By 9.74, $\mathcal{J}_0(N) = \ldots = \mathcal{J}_2(N)$. If $n \in \mathcal{J}_2(N) \cap N_o$, take some N_o-group Γ of type 2. Since $N_o \cong N/N_c$, Γ is an N-group of type 2 by 3.14 (b). Hence $n \in (o:\Gamma)$, so $n \in \mathcal{J}_2(N_o) = \mathcal{J}(N_o)$. Therefore $\mathcal{J}_2(N) \cap N_o = \mathcal{J}(N_o)$ by 5.25. The rest follows from 9.74 and the observations in 5.67 (u).

<u>9.78 REMARK</u> It is not known if one can obtain similar results for the prime and nil radicals of an a.a.n.r.. Concerning prime ideals one can say that if $P_o \trianglelefteq N_o$ then P_o is prime in N_o iff $P_o + N_c$ is prime in N.

To get a theorem on constructions of a.a.n.r.'s it will be convenient to have

9.79 DEFINITION For any nr. N and $n,n',n''\in N$ let

$$D(n;n',n''):=n(n'+n'')-nn'-nn''$$

denote the "distributor of n w.r.t. n' and n''".

9.80 THEOREM Let N be a nr. Then the following statements are equivalent:

(a) N is an a.a.n.r. .
(b) N is abelian and $(N,+)=(N_d,+)\dot{+}(N_c,+)$.
(c) $(N_c,+)$ is an N_o (ring-) module (where n_on_c is defined as in N).
(d) There exists an abelian group Γ with $N\hookrightarrow M_a(\Gamma)$.
(e) N is abelian and $\forall\, n,n',n''\in N$: $D(n;n',n'')=-n_c=-n0$.

Proof. (a) => (b) => (c) is obvious.

(c) => (d) (Gonshor (1)): Extend ${}_{N_o}N_c$ to a faithful N_o-module Γ (this can always be done). Consider $h: N\to M_a(\Gamma)$, where $h(n)=h(n_o+n_c)=:f_n:\ \Gamma\to\Gamma,\ \gamma\to n_o\gamma+n_c$.

It is easy to see that $\forall\, n\in N$: $f_n\in M_a(\Gamma)$ and that h is a nr.-homomorphism. h is injective since $f_n=f_m$ implies $\forall\,\gamma\in\Gamma$: $f_n(\gamma)=f_m(\gamma)$; taking $\gamma=o$ yields $n_c=m_c$ and this in turn that $n_o=m_o$, since Γ is faithful.

(d) => (e) Suppose that $N\leq M_a(\Gamma)$. Then N is abelian and $D(n;n',n'')=n(n'+n'')-nn'-nn''=n_o(n'+n'')+n_c-$ $-n_on'-n_c-n_on''-n_c=n_on'+n_on''-n_on'-n_on''-n_c=-n_c$.

(e) => (a): It suffices to show that $N_o\subseteq N_d$. Take $n_o\in N_o$. Then for all $n',n''\in N$ we get $D(n_o;n',n'')=-n_o0=0$, so $n\in N_d$.

(c) in 9.80 has some sort of a converse:

<u>9.81 PROPOSITION</u> Let R be a ring and ${}_R M$ an R-module. Then there is exactly one way to extend the multiplication $\cdot: R\times M \to M$ to a multiplication "$\circ$" in $(N,+) := (R,+) \oplus (M,+)$ such that $(N,+,\circ)$ is a nr. with $N_d = N_o = R\oplus\{0\}$ and $N_c = \{0\}\oplus M$, namely

$$(r,m)\circ(r',m') = (rr',rm'+m) \quad \text{(cf. Clay (1) and 9.78).}$$

Moreover, $(N,+,\circ)$ is then an a.a.n.r. and all a.a.n.r.'s arise in this way.

<u>Proof</u>. It is a routine check to see that $(N,+,\circ)$ is an a.a.n.r. with the indicated properties.
Suppose that $(N,+,*)$ is a nr. with $(N,+) = (R,+)\oplus(M,+)$, $N_d = N_o = R\oplus\{0\}$ and $N_c = \{0\}\oplus M$.
Then $\forall\ (r,m),(r',m')\in N$: $(r,m)*(r',m') = ((r,0)+(0,m))*((r',0)+(0,m')) = (r,0)*(r',0)+(r,0)*(0,m')+(0,m)*(r',m') = (rr',0)+(0,rm')+(0,m) = (r,m)\circ(r',m')$.
Since $N_d = N_o$, N is an a.a.n.r.
The rest follows from 9.80(c).

<u>9.82 COROLLARY</u>

(a) The class of all a.a.n.r.'s is a variety.
(b) Subnear-rings, homomorphic images and direct products of a.a.n.r.'s are again a.a.n.r.'s.
(c) Every a.a.n.r. is a subdirect product of subdirectly irreducible a.a.n.r.'s.

<u>Proof</u>. 9.80(e) shows that the class of all a.a.n.r.'s is equationally definable, hence a variety. (b) and (c) are well-known consequences.

The last assertion of the preceding result cries for a description of subdirectly irreducible a.a.n.r.'s:

<u>9.83 THEOREM</u> Let N be an a.a.n.r. with $N \neq N_o$. N is subdirectly irreducible $\Leftrightarrow$ N_c has a smallest under all N_o-subgroups $\neq \{0\}$.

Proof. By 1.60, N is subdirectly irreducible iff N has a smallest non-zero ideal I. By 9.73(c) I has the form $I = I_o \dot{+} I_c$ with $I_o \trianglelefteq N_0$, $N_0 I_c \subseteq N_c$ and $I_o N_c \subseteq I_c$. Since $N_c \trianglelefteq N$, $I \subseteq N_c$, whence $I_o = \{0\}$. So I is just an N_0-subgroup of N_c and of course the smallest one.

Conversely, let N_c have a smallest non-zero N_0-subgroup M, and let J be an ideal $\neq \{o\}$ of N. Again, $J = J_o + J_c$, whence $M \subseteq J_c \subseteq J$. By 9.73, M is an ideal of N and hence the smallest under all non-zero ones.

9.84 REMARKS

(a) Neither "$N_c \trianglelefteq N$" nor "N_0 is a ring" alone imply that a nr. N is an abstract affine one (take e.g. any non-abelian constant near-ring as counterexample).

(b) Also, not every a.a.n.r. N can be embedded into some $M_{aff}(V)$ (each element of any $M_{aff}(V)$ has as characteristic 0 or a prime; this does not necessarily hold in the $M_a(\Gamma)$'s.).

We will discuss these problems now.

9.85 THEOREM Let N be a nr. with $N_0 N_c = N_c$ and N_c a base. Then:

$$N \text{ is an a.a.n.r.} \iff N_c \trianglelefteq N.$$

Proof. $\Rightarrow$ holds because of 9.72(b).

To prove "$\Leftarrow$", let N_c be an ideal.

(a) First we show that N is abelian. By $N_c \trianglelefteq N$ and 1.13, the elements of N_0 and N_c commute.

By $N_0 N_c = N_c$, $\forall\, n_c \in N_c\ \exists\, n_c' \in N_c\ \exists\, n_0 \in N_0$: $n_c = n_0 n_c'$.

So $\forall\, n_c, n_c'' \in N_c$: $n_c + n_c'' = n_0 n_c' + n_c'' = (n_0 + n_c'') n_c' =$

$= (n_c'' + n_0) n_c' = n_c'' + n_0 n_c' = n_c'' + n_c$, proving that $(N_c, +)$ is abelian.

Since N_c is a base, we get $N \hookrightarrow M(N_c)$. Hence N is abelian.

(b) Now we show that $N_o \subseteq N_d$ ($N_d \subseteq N_o$ is always true). Take $n_o \in N_o$.
$\forall\, n_o' \in N_o \ \forall\, n_c \in N_c$: $x := n_o(n_o'+n_c)-n_on_o' \in N_c$, since $N_c \trianglelefteq N$.
Hence $x = x0 = n_o(n_o'0+n_c0)-n_on_o'0 = n_on_c$. Therefore $n_o(n_o'+n_c) = n_on_o'+n_on_c$.
Furthermore, let n_c, n_c' be arbitrary $\in N_c$. Then $\exists\, n_o' \in N_o \ \exists\, n_c'' \in N_c$: $n_c' = n_o'n_c''$. Thus $n_o(n_c+n_c') =$

$= n_o(n_c+n_o'n_c'') = n_o(n_c+n_o')n_c'' = (n_on_c+n_on_o')n_c'' =$

$= n_on_c+n_on_o'n_c'' = n_on_c+n_on_c'$.

Consequently, for all $n_o', n_o'' \in N_o$ and $n_c \in N_c$ we get $(n_o(n_o'+n_o''))n_c = n_o(n_o'n_c+n_o''n_c) = n_on_o'n_c+n_on_o''n_c =$

$= (n_on_o'+n_on_o'')n_c$.

Since N_c forms a base, $n_o(n_o'+n_o'') = n_on_o'+n_on_o''$.
Plugging these results together yields

$\forall\, n', n'' \in N \ \forall\, n_o \in N_o$: $n_o(n'+n'') = n_o((n_o'+n_c')+(n_o''+n_c'')) =$

$= n_o((n_o'+n_o'')+(n_c'+n_c'')) = n_o(n_o'+n_o'')+n_o(n_c'+n_c'') = n_on_o'+$

$+n_on_o''+n_on_c'+n_on_c'' = n_o(n_o'+n_c')+n_o(n_o''+n_c'') = n_on'+n_on''$.

Hence n_o is distributive and $N_o = N_d$.

If instead of $N_oN_c = N_c$ more is postulated, one gets much more information out of 9.85 concerning 9.84(b):

<u>9.86 THEOREM</u> (cf. Heatherly (3)). Let N_c be a base and ideal of a nr. N, such that $\forall\, n_c \in N_c^*$: $N_on_c = N_c$. We may assume that $N \subseteq M(N_c)$ (1.96). Then there exists a field D making N_c into a vector space over D, N is 2-primitive and dense in $M_{aff}(N_c)$ and N_o is a primitive ring, dense in $\mathrm{Hom}_D(N_c, N_c)$.

<u>Proof</u>. 9.85 tells us that N is an a.a.n.r., so N_o is a ring. The assumptions imply that N_o is a primitive ring, hence a dense subring of $\mathrm{Hom}_D(N_c,N_c)$, where D is the centralizer $\mathrm{Hom}_{N_o}(N_c,N_c)$. An application of 4.27(a) finishes the proof, since $N_c = M_c(N_c) \subseteq N$.

We close this section with theorem 9.88 for which we need

9.87 DEFINITION A set N together with two binary operations +,• is called a generalized ring (Beaumont (1)) if

(a) (N,+) is an abelian group.

(b) (N,•) is a semigroup.

(c) $\exists\ r,s\in\mathbb{N}\setminus\{1\}\ \forall\ n_1,\ldots,n_r,n_1',\ldots,n_s'\in N$: $(\sum_{i=1}^{r} n_i)(\sum_{j=1}^{s} n_j') = \sum_{i=1}^{r}\sum_{j=1}^{s} n_i n_j'$.

9.88 THEOREM (Beaumont (1) - Ferrero (3)). Let N be a nr. with bounded order of the elements of $(N_c,+)$. Then:

N is an a.a.n.r. <=> N is a generalized ring.

Proof. =>: Let N be an a.a.n.r. . Then, by 9.80(e),

$\forall\ n,n',n''\in N$: $n(n'+n'') = nn'+nn''-n0$. Hence

$\forall\ k\in\mathbb{N}\ \forall\ n_1,\ldots,n_k\in N$: $n(\sum_{i=1}^{k} n_i) = \sum_{i=1}^{k} nn_i-(k-1)\,n0$.

Let s be the l.c.m. of the orders of the elements of $(N_c,+)$, and set $s'+1 =: s$. Then $s\geq 2$ and

$\forall\ n,n_1,\ldots,n_s\in N$: $n(\sum_{i=1}^{s} n_i) = \sum_{i=1}^{s} nn_i$.

From this one gets (c) in 9.87 for arbitrary $r\in\mathbb{N}\setminus\{1\}$.

<=: From 9.87(c) we can conclude that

$\forall\ n,n',n''\in N$: $n(n'+n'') = \underbrace{(n+0+\ldots+0)}_{r\text{-summands}}\ \underbrace{(n'+n''+0+\ldots+0)}_{s\text{-summands}} =$

$= nn'+nn''+(s-2)\,(n0)+(r-1)\,(0\,(n'+n'+0+\ldots+0)) =$

$= nn'+nn''+(s-2)\,(n0)$.

But $n0 = (n+0+\ldots+0)\,(0+0+\ldots+0) = n0+(s-1)\,(n0)$.

Thus $(s-1)(n0) = 0$ and $(s-2)(n0) = -n0$.

Consequently we get for all $n,n',n''\in N$:

$D(n;n',n'') = nn'+nn''-n0-nn'-nn'' = -n0$.

By 9.80, N is an abstract affine near-ring.

9.89 REMARKS

(a) (Maxson (1)). $M_{aff}(V)$ has a unique maximal ideal.

(b) (Heatherly (3)). $Hom_F(V,V)$ is a maximal subnear-ring of $M_{aff}(V)$.

(c) Malone (5) describes how automorphisms of N_o and N_c (where N is an a.a.n.r.) can be "mated" to give an automorphism of N (cf. 1.114).

(d) See Blackett (3), (4) for matrix representations of affine transformations over a finite-dimensional vector space V.

(e) Observe the connections to near-fields and doubly transitive groups (see e.g. 8.40).

d) NEAR-RINGS ON GIVEN GROUPS

1.) MULTIPLICATIONS ON A GROUP

9.90 DEFINITION Similar to 8.24 - 8.28, we can define coupling maps on a nr. N as maps $\phi: N \to \text{End}\,(N,+,\cdot)$ with $\phi_n \circ \phi_m = \phi_{\phi_n(m)\cdot n}$ for all $n,m \in N$.

As in 8.26/8.27 one gets a new nr. $(N,+,\circ_\phi) =: N^\phi$, again called the ϕ-derivation of N, where $n \circ_\phi m := \phi_m(n)\cdot m$. If for a nr. N there exists a ring R and a coupling map ϕ on R such that $R^\phi = N$, we call N a Dickson near-ring.

More on Dickson near-rings can be found in 9.153. We only need

9.91 PROPOSITION Let $(\Gamma,+)$ be a group and define the "constant multiplication" $*$ as in 1.4(b). Then $(\Gamma,+,*) \in \mathcal{N}_c$. Every nr. $(\Gamma,+,\cdot)$ on $(\Gamma,+)$ can be coupled to this $(\Gamma,+,*)$ by $\phi: \Gamma \to \text{End}\,(\Gamma,+,*)$, given through

$$\gamma \to \phi_\gamma: \begin{array}{l}\Gamma \to \Gamma \\ \delta \to \delta\cdot\gamma\end{array} .$$

Since End $(\Gamma,+,*)$ = End $(\Gamma,+)$, we get

9.92 THEOREM (Clay (2), (4), (7), (8)). There is a 1-1-correspondence between all near-ring multiplications on Γ (that are binary operations "$\cdot$" making $(N,+,\ )$ into a near-ring) and all maps $\phi: \Gamma \to \text{End } \Gamma$, $\gamma \to \phi_\gamma$ with $\phi_\gamma \circ \phi_\delta = \phi_{\phi_\gamma(\delta)}$.

9.93 NOTATION

(a) If ϕ is defined by $\cdot$ we write $\phi^\cdot$.
(b) If $\cdot$ is defined via ϕ we write $\cdot_\phi$ (as in 8.26).

With straightforward proofs one gets

9.94 PROPOSITION (Clay (4)). Let Γ,ϕ be as before.

(a) $\cdot_\phi$ gives a distributive nr. $\iff \forall\ \gamma,\delta\in\Gamma : \phi_{\gamma+\delta} = \phi_\gamma+\phi_\delta$.
(b) $\cdot_\phi$ is commutative $\iff \forall\ \gamma,\delta\in\Gamma : \phi_\gamma(\delta) = \phi_\delta(\gamma)$.
(c) $\cdot_\phi$ produces a zero-symmetric nr. $\iff \phi_0 = \bar{o}$.

We have to look, which multiplications yield isomorphic nr.'s:

9.95 DEFINITION Two multiplications $\cdot_1$ and $\cdot_2$ are similar if there is some $\alpha\in\text{Aut }\Gamma$ with $\gamma,\delta\in\Gamma$: $\alpha(\gamma\cdot_1\delta)=\alpha(\gamma)\cdot_2\alpha(\delta)$.

9.96 PROPOSITION (Clay (4)). The following conditions are equivalent:

(a) $\cdot_1$ and $\cdot_2$ are similar.
(b) $(\Gamma,+,\cdot_1) \cong (\Gamma,+,\cdot_2)$.

Moreover we evidently get

9.97 PROPOSITION (e.g. Clay (8)).

(a) γ is a zero divisor in $(\Gamma,+,\cdot_\phi) \iff \phi_\gamma$ is not injective.
(b) $(\Gamma,+,\cdot_\phi)$ is integral $\iff$ all ϕ_γ are injective $(\gamma\in\Gamma^*)$.
(c) If Γ is finite, $(\Gamma,+,\cdot_\phi)$ is integral iff $\forall\ \gamma\in\Gamma^*: \phi_\gamma\in\text{Aut }(\Gamma,+)$.

At this place it might be appropriate to remark the construction method of 1.4(b) of a nr. on $(\Gamma,+)$ via a fixed-point-free automorphism group. (cf. also Theorem G of Clay (8)).

We also mention without the evident proof

9.98 PROPOSITION (Clay (4)). The multiplication $\cdot$ on $(\Gamma,+)$ is trivial (in the sense of 1.4(b)) $\Leftrightarrow \forall\ \gamma\in\Gamma$: $\phi^{\bullet}_{\gamma}\in\{\bar{o},id\}$.

9.99 THEOREM (Clay (4)). $(\Gamma,+,\cdot)$ is constant iff $\phi^{\bullet}_{o}\in Aut(\Gamma)$.

This holds by 9.91. In this case, $\phi^{\bullet}_{o} = id$.

9.100 REMARKS

(a) See the appendix for all nr.'s on groups of order ≤ 7 and of many ones of order 8.
Examples of nr.'s on non-abelian groups of order 12 and 18 are in Malone (1). All composition rings of small order are listed in Wiesenbauer (1).

(b) There are groups (Z_p^{∞}, for instance), on which no rings except the zeroring is definable. Abelian groups with this property are called "nil groups". Lawver (1),(2) has shown that there might exist non-trivial near-rings on nil groups.
On the other hand, there are also groups on which only the trivial nr. multiplications can be defined. Ligh-Malone (1) have shown that complete groups belong to this class of groups (a group is said to be complete if its center $= \{o\}$ and if all automorphisms are inner; all S_n ($n \neq 2,6$) are complete).

(c) Lawver (3) studied nr.'s on free groups and on direct sums of groups. Cf. 1.22(a).

(d) Clay (1),(7) studied the multiplication on an abelian group by giving them a group structure.

2.) NEAR-RINGS ON SIMPLE AND ON CYCLIC GROUPS

9.101 PROPOSITION (e.g. Heatherly (13)). Let $(\Gamma,+)$ be a simple group and $(\Gamma,+,\cdot) =: \Gamma$ a near-ring on Γ. Then

(a) $\Gamma \in \mathcal{N}_0 \vee \Gamma \in \mathcal{N}_c$.

(b) $\forall\ \gamma \in \Gamma : (\dot{\phi}_\gamma = \bar{o}) \vee (\dot{\phi}_\gamma$ is a monomorphism).

(c) Γ finite $\Rightarrow \forall\ \gamma \in \Gamma : (\dot{\phi}_\gamma = \bar{o}) \vee (\dot{\phi}_\gamma \in \mathrm{Aut}(\Gamma,+))$.

Proof. (a) follows from the fact that $(\Gamma_0,+) \trianglelefteq (\Gamma,+)$.

(b) and (c) result from considering $\mathrm{Ker}\ \dot{\phi}_\gamma$.

9.102 THEOREM (Heatherly (1),(2)). Let $\Gamma = (\Gamma,+,\cdot)$ be a nr. on the finite simple group $(\Gamma,+)$. Then Γ falls into one of the following disjoint classes:

(a) $\forall\ \gamma \in \Gamma : \dot{\phi}_\gamma = \bar{o}$ (in this case, $\cdot$ is the "zero multiplication").

(b) $\Gamma_d = \{o\}$ and Γ has a right identity.

(c) $\Gamma_d \neq \{o\}$ and Γ has an identity.

Proof. Suppose that $\cdot$ is not the zero multiplication. By 9.101(c), $\exists\ \delta \in \Gamma : \dot{\phi}_\alpha : \begin{array}{l}\Gamma \to \Gamma \\ \gamma \to \gamma\alpha\end{array} \in \mathrm{Aut}(\Gamma,+)$. Now $\exists\ k \in \mathbb{N} : (\dot{\phi}_\alpha)^k = id_\Gamma$. $\forall\ \gamma \in \Gamma : \gamma = id(\gamma) = (\dot{\phi}_\alpha)^k(\gamma) = \gamma\alpha^k$, and α^k is a right identity.

If $\Gamma_d \neq \{o\}$, take $\delta \in \Gamma_d^*$. Consider ${}_\delta\psi : \begin{array}{l}\Gamma \to \Gamma \\ \gamma \to \delta\gamma\end{array} \in \mathrm{End}(\Gamma,+)$.

Since $\cdot$ is not the zero multiplication, $(o:\Gamma) = \{o\}$ for $(o:\Gamma) \trianglelefteq (\Gamma,+)$. Hence ${}_\delta\psi \neq \bar{o}$ and (as above) ${}_\delta\psi \in \mathrm{Aut}(\Gamma,+)$ and some power of δ is a left identity, hence the identity.

Observe that 9.109 implies that in case (c) of 9.102, Γ has to be a finite prime field.

We now turn to cyclic groups.

9.103 THEOREM (Heatherly (1) and others). Let N be a nr. on $\mathbb{Z}_n$ or $\mathbb{Z}$ with a generator $g \in N_d$ then N is a commutative ring.

Proof. In this case, N is an abelian dgnr., hence a ring. Every ring on a cyclic group is commutative (see (Beaumont)).

9.104 COROLLARY (Heatherly (1)). If N is a nr. on $\mathbb{Z}_p$ $(p \in \mathbb{P})$ or $\mathbb{Z}$ with $N_d \neq \{0\}$ then N is a commutative ring and there is some $x \in N$ with $\forall\, n,n' \in N : nn' = n \bullet n' \bullet x$ (usual product in $\mathbb{Z}_p$ or $\mathbb{Z}$).

Proof. If $d \in N_d$ is $\neq 0$, a short calculation (cancel d!) shows that 1 is also $\in N_d$; now we may apply 9.103 to get the first assertion.
Let $1 \cdot 1 =: x$ and $n,n' \in N$. Then

$$nn' = \underbrace{(1+\ldots+1)}_{n\text{-summands}}\underbrace{(1+\ldots+1)}_{n'\text{-summands}} = n \bullet n' \bullet (1 \cdot 1) = n \bullet n' \bullet x.$$

9.105 REMARK The same result as in 9.104 holds in every $\mathbb{Z}_n$ if $1 \in N_d$. On the other hand, Heatherly (1) gives an example of a nr. on $\mathbb{Z}_4$ which is not a ring (in fact, 1 and 3 are not distributive).

For the next result, let $C(k,j)$ be the number of combinations of k elements to the class j). Without proof we state

9.106 THEOREM (R. Jacobson (1)). The number of different nr.'s definable on $(\mathbb{Z}_p,+)$ $(p \in \mathbb{P})$ is given by

$$2 + \sum_{k/p-1} \left(\sum_{j=1}^{k} C(k,j)\left(\frac{p-1}{k}\right)^j \right).$$

3.) NEAR-RINGS WITH IDENTITIES ON GIVEN GROUPS

We start with

9.107 PROPOSITION (Clay (4)). Let $N = (\Gamma,+,\cdot_\phi)$ be a nr. on Γ. Then $1\in N$ is an identity of N iff $\phi_1 = id$ and $\forall\ \gamma\in\Gamma: \phi_\gamma(1) = \gamma$.

The proof is obvious.

Out of 9.103 and 9.104 we get (observe that under the given assumptions, x of 9.103 is invertible in N):

9.108 COROLLARY (Clay-Malone (1)). If N is a nr. $\in\mathfrak{N}_1$ on the cyclic group $(N,+)$ then N is a commutative ring. All nr.'s on $(N,+)$ are isomorphic. There are $\Phi(n)$ (Φ the Euler function) ones on $(\mathbb{Z}_n,+)$ and 2 on $\mathbb{Z}$.

9.109 COROLLARY There are exactly $p-1$ nr.'s with identity definable on $(\mathbb{Z}_p,+)$; all of them are isomorphic to the field $\mathbb{Z}_p$ and hence all are finite prime fields.

This result was obtained by Malone, Clay, Maxson and Heatherly under different circumstances.

Observe that if in $(\Gamma,+,\cdot)\in\mathfrak{N}_1$ $(\Gamma,+)$ is abelian with exactly one proper subgroup then $(\Gamma,+) \cong \mathbb{Z}_{p^2}$ and $(\Gamma,+,\cdot)$ is a commutative ring by 9.108 (Ligh (9)). But there do exist non-rings with identity on groups of order p^2 (cf. also 9.115(c)):

9.110 PROPOSITION (Maxson (1)). For each $p\in\mathbb{P}$ there exists a group Γ of order p^2 and a non-ring with identity on Γ.

The proof is established by defining a multiplication on $(\Gamma,+) := (\mathbb{Z}_p,+)\oplus(\mathbb{Z}_p,+)$ in an appropriate manner (see Maxson (1) for details.)

Now we study nr.'s of square-free order. First we need

9.111 PROPOSITION (Clay-Malone (1), Maxson (1)). Let $(\Gamma,+,\cdot)$ be a nr. with identity 1 on the finite group Γ. Let $ord(\gamma)$ be the order of $\gamma\varepsilon\Gamma$. Then

$$ord(1) = l.c.m.\{ord(\gamma)\,|\,\gamma\varepsilon\Gamma\} =: \ell .$$

Proof. If $\gamma\varepsilon\Gamma$, $o = o\gamma = (ord(1).1)\gamma = ord(1)\cdot\gamma$, so $ord(\gamma)/ord(1)$. Hence $\ell/ord(1)$. But $1\varepsilon\Gamma$, hence $ord(1)/\ell$ whence $ord(1) = \ell$.

9.112 THEOREM (Maxson (1),(2)). Let $(\Gamma,+,\cdot)\varepsilon\mathcal{N}_1$ have finite square-free order. Then $(\Gamma,+)$ is cyclic, and $(\Gamma,+,\cdot)$ is a commutative ring.

Proof. Let $|\Gamma| = p_1p_2\ldots p_r$, where $p_1,\ldots,p_r$ are distinct primes. Using the Sylow theory we get for each $i\varepsilon\{1,\ldots,r\}$ some $\gamma_i\varepsilon\Gamma$ of order p_i. Hence
$|G| \geq ord(1) = l.c.m.\{ord(\gamma)\,|\,\gamma\varepsilon\Gamma\} \geq$
$\geq l.c.m.\{ord(\gamma_1),\ldots,ord(\gamma_r)\} = |G|$.
So $ord(1) = |G|$ and G is cyclic. Now use 9.108.

Several groups cannot bear a nr. with identity (call a subset P of a partially ordered set an antichain if no distinct elements are comparable):

9.113 THEOREM (Krimmel (1),(2)). Let $(\Gamma,+)$ be a group having elements $\gamma_1,\ldots,\gamma_r$ of distinct prime orders $p_1,\ldots,p_r$ $(r \geq 2)$. If every antichain in the lattice of normal subgroups of Γ has cardinality $< r$ then $(\Gamma,+)$ cannot be the additive group of a nr. with identity.

Proof. Suppose that $(\Gamma,+,\cdot)$ is a nr. with identity 1. If there are $i,j\varepsilon\{1,\ldots,r\}$ with $(o:\gamma_i) \subseteq (o:\gamma_j)$ then
$$h\colon (\Gamma\gamma_i,+) \to (\Gamma\gamma_j,+) \qquad \gamma\gamma_i \to \gamma\gamma_j$$
is a well-defined group-homomorphism. But $h(\gamma_i) = h(1\gamma_i) = 1\gamma_j = \gamma_j$, whence

$p_j = \mathrm{ord}(\gamma_j)/\mathrm{ord}(\gamma_i) = p_i$, so $i = j$. Hence $\{(o:\gamma_1),\ldots,(o:\gamma_r)\}$ is an antichain with r elements, a contradiction.

Observe that we didn't use associativity of $\cdot$; 1 could have been only a left identity.

9.114 COROLLARY (Clay-Malone (1)). A nr. with identity on a finite simple group Γ is a finite prime field.

Proof. Γ cannot have a composite order by 9.113. Hence Γ is a simple p-group, thus coinciding with it's center. So Γ is isomorphic to $\mathbb{Z}_p$ and we can apply 9.109.

9.115 COROLLARIES (Krimmel (2), Clay-Malone (1), Clay-Doi (1), Ligh (9)).
The following groups Γ cannot be the additive groups of near-rings with identity:

(a) groups of composite order in which the lattice of normal subgroups is linearly ordered (e.g. S_n $(n\geq 3)$),

(b) simple groups of composite order (e.g. A_n $(n\geq 4)$),

(c) finite non-abelian groups with exactly one proper normal non-zero subgroup,

(d) non-cyclic groups of square-free order.

Proof. Evidently 9.114 => (a) => (b) and 9.112 => (d). In (c), Γ must be of composite order since otherwise Γ is a non-abelian p-group, hence of order p^k with $k\geq 3$. In this case, Γ has at least two non-trivial normal subgroups (see e.g. (Rotman), Cor. 5.5 and Ex. 5.2).

We now mention without proof some more results on this subject.

9.116 THEOREM

(a) (Ligh (9)). There is no nr. with identity definable on a torsion divisible group.

(b) (Clay-Doi (1)). The same holds for $S_\infty := \bigcup_{n\in\mathbb{N}} S_n$ and $A_\infty := \bigcup_{n\in\mathbb{N}} A_n$.

(c) (Clay-Maxson (1)). There are also no nr.'s $\in\mathfrak{N}_1$ definable on generalized quaternion groups.

(d) (Ligh (13)). There do exist nr.'s with identity on perfect groups (that are groups coinciding with its commutator subgroup) (cf. Ligh (9)).

(e) See Johnson (4) for the nr.'s on the dihedral groups D_{2n} of order 2n. There are no nr.'s $\in\mathfrak{N}_1$ on D_{2n} for odd n (this follows from 9.111), for the only ones exist on D_{4p} ($p\in\mathbb{P}$). They are zero-symmetric and normal N-subgroups and left ideals coincide (and all left ideals are annihilator left ideals).
There are (up to isomorphism) 7 nr.'s with identity on D_8 (p. 345) and (again up to isomorphism) just one on D_{4p} ($p\in\mathbb{P}\setminus\{2\}$).

(f) (Clay-Maxson (1)). All nr.'s with identity definable on p-groups with exactly one subgroup of order p are commutative rings.
(This follows from 9.108 and (c) since a group as described above is either cyclic or a generalized quaternion group (Hall (1), Theorem 12.5.2.).

4.) NEAR-RINGS WITH OTHER PROPERTIES ON GIVEN GROUPS

Now we briefly study nr.'s with special properties (other than having an identity) on some group $(\Gamma,+)$.
We will only cite the results or even only the papers which are concerned with these topics. See also the chapters concerning the types of near-rings in discussion. For example,

there are no near-fields definable on non-abelian groups (8.11), and so on.

We start with nr.'s with chain conditions. Generalizing 9.102 one gets

9.117 THEOREM (Ligh (3)). Let N be a nr. with DCC on monogenic N-subgroups on the simple group (N,+) such that $N_d \neq \{0\}$. Then N is either the zero-nr. or a field.

9.118 REMARK For a detailed study of nr.'s N on a group which fulfill the DCC on monogenic N-subgroups and the "ACC on principal annihilator left ideals" (i.e. each $(0:x) \subseteq \subseteq (0:x^2) \subseteq (0:x^3) \subseteq \ldots$ terminates) see Ligh-Ramakotaiah-Reddy (1).

9.119 THEOREM (Timm (3)). $(\Gamma,+)$ is the additive group of a (not necessarily associative (!)) near-ring in which every non-zero element has a right inverse iff Γ is invariantly simple and every $\gamma \in \Gamma$ has (the same) prime order.

The question concerning the additive group of near-fields is settled by the following theorem.

9.120 THEOREM (Timm (3)). The follwoing conditions on a group $(\Gamma,+)$ are equivalent:

(a) Γ is the additive group of a near-field.
(b) Γ is abelian and the additive group of a nr. with right cancellation law.
(c) Γ is the additive group of a vector space over some field.
(d) Γ is the additive group of a commutative field.
(e) Γ is the additive group of an alternative field.
(f) There is some $p \in \mathbb{P}$ such that Γ is the direct sum of the groups $(\mathbb{Z}_p,+)$ or Γ is a direct sum of copies of $(\mathbb{Q},+)$.
(g) Γ is abelian and either each element has the same prime order or Γ is torsionfree divisible.

Finally, we consider the additive group of dgnr.'s.

9.121 THEOREM

(a) (Ligh (10)). There are just 3 non-isomorphic dgnr.'s on S_6 and at least 3 on S_n $(n \geq 5,\ n \neq 6)$.
There are precisely 3 non-isomorphic dgnr.'s definable on D_{2p} $(p \in \mathbb{P})$.

(b) (Ligh (13)). The additive group of a simple dgnr. is perfect.

(c) (Malone (7)). There are exactly 16 dgnr.'s on a generalized quaternion group. All of them are distributive.

e) ORDERED NEAR-RINGS

9.122 DEFINITION A nr. N is called partially (fully) ordered by $\leq$ if

(a) $\leq$ makes $(N,+)$ into a partially (fully) ordered group.

(b) $\forall\, n,n' \in N$: $(n \geq 0 \wedge n' \geq 0 \Rightarrow nn' \geq 0)$.

"Ordered" means "partially ordered".

9.123 REMARKS

(a) Thus an ordered near-ring is a nr. where $(N,\leq)$ is an ordered set such that $n \geq 0$, $n' \geq 0$ implies $n+n' \geq 0$ and $nn' \geq 0$.

(b) The standard work on ordered algebraic systems (semigroups, groups, rings and fields) is (Fuchs).

(c) For an ordered near-ring we will write $(N,+,\cdot,\leq)$ or simply $(N,\leq)$.

(d) Our discussion up to 9.132 is implicit in (Gabovich).

9.124 NOTATION We adopt the usual conventions to write $n<n'$, $n\geq n'$, $n>n'$, $n \| n'$ (n and n' are incomparable, i.e. neither $n\leq n'$ nor $n'\leq n$ holds).
"Partially ordered" will be abbreviated by "p.o.", "fully ordered" by "f.o.".

Just as in the theory of ordered groups or rings, it is more convenient to work with the set of "positive" elements instead of the order relation itself:

9.125 THEOREM

(a) Let $(N,+,\cdot,\leq)$ be a p.o. nr. ; then the "positive cone" $P_{\leq} := P := \{n\in N \mid n\geq 0\}$ fulfills

(α) $P+P = P$.

(β) $P\cap(-P) = \{0\}$, where, as usual, $-P := \{n \mid n\leq 0\}$.

(γ) $\forall\, n\in N$: $n+P = P+n$.

(δ) $P\cdot P \subseteq P$.

(b) Conversely, for every subset P of a nr. N fulfilling (α) - (δ) we get an ordered nr. $(N,\leq_P)$ via $n \leq_P n' :\Longleftrightarrow n'-n\in P$.

(c) This correspondence between order relations and subsets with (α) - (δ) is 1-1, that means that $\leq_{P_{\sqsubseteq}} = \sqsubseteq$ and $P_{\leq_{P'}} = P'$.

The proof of 9.125 is easy and left to the reader.

9.125 enables us to say that "the nr. N is ordered by P". The following result is obvious.

9.126 PROPOSITION Let N be ordered by P.

(a) $\leq_P$ is a full order $\Longleftrightarrow P\cup(-P) = N$.

(b) $\leq_P$ is trivial (i.e. $n \leq_P n' \Longleftrightarrow n = n'$) $\Longleftrightarrow P = \{0\}$.

There is no place for finite near-rings in this section:

9.127 PROPOSITION Every non-trivially ordered near-ring is infinite.

Proof. $\exists\, n \in N$: $n>0$. But then $n<n+n = 2n<3n<\dots$.

9.128 DEFINITION Let N,N' be nr.'s ordered by P,P', respectively. $f: N \to N'$ is order-preserving: $\iff$ $\iff f(P) \subseteq P'$.
If there is an order-preserving mono-(iso-)morphism $f: N \to N'$ we write $N \hookrightarrow_0 N'$ ($N \cong_0 N'$, respectively) (for isomorphisms f we also want f^{-1} to be order-preserving since the category of ordered near-rings and order-preserving homomorphisms is not balanced).

9.129 DEFINITION A subset T of an ordered nr. N is called convex if

$$\forall\, t_1,t_2 \in T \ \ \forall\, n \in N : (t_1 \leq n \leq t_2 \Rightarrow n \in T).$$

Similar to ring theory (see (Fuchs)) one can easily prove the following two results. Again, they are corollaries of theorems of (Gabovich).

9.130 THEOREM A subset I of an ordered nr. N is the kernel of an order-preserving nr.-homomorphism from N to some ordered nr. N' iff I is a convex ideal.

9.131 THEOREM Let N,N' be nr.'s, ordered by P,P'. Let $h: N \twoheadrightarrow N'$ be an order-preserving epimorphism (i.e. $h(N) = N'$ and $h(P) = P'$).
If $I' \trianglelefteq N'$ is convex then $h^{-1}(I') =: I$ is a convex ideal of N and $N/_I \cong_0 N'/_{I'}$.

Convexity is quite trivial, so we won't prove it.

9.132 PROPOSITION Every order P in N can be extended to a maximal order $\overline{P}$.

The proof is accomplished by an application of Zorn's Lemma.

9.133 PROPOSITION To every abelian ordered nr. N there exists an (abelian) ordered nr. $\hat{N}$ with identity such that $N \hookrightarrow_0 \hat{N}$.

Proof. By 1.86, we can find a group Γ with $N \hookrightarrow M(\Gamma)$ (say by h). If N is ordered by P, take $\hat{P} := h(P)$. Then α), β), δ) and (since N is abelian) also γ) of 9.125 are clearly fulfilled, hence $\hat{N} := M(\Gamma)$ is ordered by $\hat{P}$ and h is an order-preserving monomorphism.

9.134 REMARKS

(a) Not every ordered nr. can be embedded into a f.o. nr. with identity (this follows from 9.137).

(b) If N contains an identity 1 then $1 \geq 0$ or $1 \| 0$, for $1 < 0$ implies $(-1) > 0$, hence $(-1)(-1) \geq 0$, whence $1 \geq 0$, a contradiction to the assumption $1 < 0$.

In some instances we can describe $\overline{P}$ of 9.132 explicitely:

9.135 PROPOSITION Let N be a nr. such that N_c is ordered by P_c and such that P_c forms a base (1.91). Then $P := \{n \in N \mid \forall p \in P_c : np \geq 0\}$ is the unique maximal extension of P_c to an order of N.

Uniqueness is clear for $P = (P_c : P_c)$, and α) - δ) of 9.125 are easily verified. So the proof is easy. Nevertheless, this proposition has heavy consequences, e.g. that in general $\overline{R}[x]$ cannot be fully ordered: each $\overline{p} \in \overline{R}[x]$ would have to have only positive or only negative values at $\{r \in R \mid r \geq 0\}$!
If N is complete (i.e. if $\forall\, n \in N\ \forall\, k \in N_c\ \exists\, \overline{n} \in N\ \forall\, c \in N_c : \overline{n}c = n(c+k)$) then 9.135 implies that one cannot get full orders except $N = N_0$ or $N = N_c$ (see Pilz (1),(3),(6)).

Fundamental for the following is

9.136 THEOREM Let N be f.o. (by P) and $N_c \neq \{0\}$. Then

$$\forall\, n \in N\ \forall\, c \in P_c := P \cap N_c : \quad nc = n0.$$

Proof. Suppose there are some $n\in N$ and $c\in P_c$ with $nc \neq n0$. W.l.o.g. we may assume that $n0\geq 0$ (otherwise change to $-n$). Now $n_0c = (n-n0)c = nc-n0 \neq 0$. Hence $n_0 \neq 0$.
If n_0 would be >0 then $0\leq n_0c \neq 0$, whence $n_0c>0$. Consider $\ell := n-nc+n_0$; ℓ fulfills $\ell c = n_0c>0$, whence $\ell>0$. On the other hand, $\ell 0 = n0-nc =$ $= -n_0c<0$ implies $\ell<0$ and we arrive at a contradiction.
The assumption $n_0<0$ leads to the same disaster.

9.137 COROLLARY If N has a right identity and is fully ordered then $N\in\mathcal{N}_0$.

9.138 REMARK If under the circumstances of 9.136 N_c forms a base and N is considered as subnear-ring of $M(N_c)$ (1.96) then 9.137 tells us that each $n\in N$ is constant on all positive elements of M_c. We will see that in fact f.o. nr.'s are closely related to constant near-rings (9.141(a)).

9.139 DEFINITION An ordered nr. N is called archimedian ordered if $(N,+)$ happens to be this (see e.g. (Fuchs)).

If a f.o. n.r. N is archimedian, $(N,+)$ is abelian and $(N,+) \hookrightarrow_0 (\mathbb{R},+)$. If N is not archimedian then there exist pairs (a,b) of N^2 with $k\cdot|a| = |a|+|a|+\ldots+|a|<|b|$ for all $k\in\mathbb{N}$ (where $|a|$ has the usual meaning - see (Fuchs)). In this case we call a "small w.r.t. b" and write $a\ll b$. If $A\subseteq N$, "$A\ll b$" and similar notations are then clear.

9.140 DEFINITION Let N be ordered and $n\in N$.

n is nearly constant: $\Longleftrightarrow n_0 \ll N_c^*$.

9.141 THEOREM Let N be fully ordered and $N_c \neq \{0\}$. Then

(a) Every $n\in N$ is nearly constant.
(b) $N_0 \neq \{0\} \Rightarrow$ the order is non-archimedian.

Proof. (a) If $N_c \neq \{0\}$, take any $n \in N$ with $n_c > 0$.
For $k \in \mathbb{N}$, let $a_k := k\cdot(-n_c+n)-n_c$, $b_k := (k-1)\cdot n_0+n$.
Then $a_k 0 = -n_c < 0$, hence $a_k < 0$ and $b_k 0 = n_c > 0$, whence $b_k > 0$, so $\forall k \in \mathbb{N}: a_k < 0 < b_k$.
Thus $-n_c+n-n_c+\ldots+n-n_c < 0 < n-n_c+n-n_c+\ldots-n_c+n$. So $-n_c+k\cdot n_0 < 0 < (k-1)n_0+n$. From the first inequality we get $k\cdot n_0 - n_c < 0$. Hence $k\cdot n_0 - n_c < 0 < (k-1)n_0+n$, so $-n_c < -k\cdot n_0 < -n_0+n = n_c$ and $k\cdot|n_0| < n_c$ for all $k \in \mathbb{N}$.
Now let $x \in N$ and $c \in N_c^*$ be arbitrary, but $c > 0$.
Let $m := x_o + c$. Since $m_c = c > 0$, we can apply our considerations above to m and get $\forall k \in \mathbb{N}: k\cdot|m_o| < m_c$.
Since $m_o = x_o$, we see that $x_o \ll N_c^*$.

(b) follows at once from (a).

It is high time for examples.

9.142 EXAMPLES

(a) Non-zero near-rings of the kind $M(\Gamma)$, $M_{cont}(\Gamma)$ (1.4(a)), $R[x]$ and $M_{aff}(V)$ cannot be fully ordered since they contain an identity and have non-zero constants.

(b) Let R be a fully ordered commutative nr. with identity. Then $R_o[x]$ with $P := \{\sum_{i=0}^{k} a_i x^i \mid a_k > 0\} \cup \{\bar{o}\}$ ("lexicographic order") and $R_o[[x]] := (R[[x]])_o$ with $P := \{\sum_{i \geq k} a_i x^i \mid a_k > 0\} \cup \{\bar{o}\}$ ("antilexicographic order") supply non-trivial examples of ordered near-rings $\in \mathcal{N}_o$.
In $R_o[x]$, for example, we get the following "archimedian classes" $\{p \mid \deg p = 1\} \ll \{p \mid \deg p = 2\} \ll \ll \ldots$.

(you may multiply a linear polynomial by any natural number you want and you will never arrive at a quadratic one).

(c) Let R be a ring, f.o. by P. Take an arbitrary, but fixed subset $Q \subseteq -P^*$, and form the near-ring

$N_Q := \{r_1 + r_2 \cdot \chi_Q \mid r_1, r_2 \in R\}$ (χ_Q the indicator function of Q).

Ordering N_Q lexicographically gives a f.o. n.r. with $\{0\} \neq (N_Q)_c \neq N_Q$, in which all elements are nearly constant.

9.143 REMARK 9.141 shows that one cannot expect for "real" near-rings to get any "neat" full order.
This is not very surprising: one has some subnear-ring of some $M(\Gamma)$ in hand, whose cardinality is, in general, "much bigger" than those of Γ. Anybody, who ever had to bring order in a large storehouse knows: the larger the set, the harder is it to get a full order.
Also, one might expect that well-ordered near-rings are quite special:

9.144 THEOREM (Maxson (11)). Let $N \in \mathfrak{N}_1$ be well-ordered. Then $N \cong_o (\mathbb{Z},+,\cdot,\leq)$ ($\leq$ the usual order in $\mathbb{Z}$).

Proof. Again, let P be the positive cone of N. Let a be the smallest element of P^* and A the cyclic subgroup of $(N,+)$ generated by a.
Suppose that $A \neq N$. Let b be the smallest positive element of $U := N \setminus A$.
Consider a-b. If $a-b>0$ then $a-b = a$, a nonsense.
Hence $b-a>0$. $b-a \in U \Longrightarrow b-a = b$, which is a contradiction, too.
Hence $U = \emptyset$ and N is cyclic and infinite (9.127).
The map $h: N = A \to \mathbb{Z}$, $za \to z$ is an order-isomorphism between the additive groups.

By 9.137, N is isomorphic to the ring $\mathbb{Z} = (\mathbb{Z},+,\cdot)$ (usual multiplication) or $\mathbb{Z}' := (\mathbb{Z},+,*)$ with $z*z' := -z\cdot z'$. $\mathbb{Z}$ and $\mathbb{Z}'$ are order-isomorphic via $z \to -z$. Hence in any case $N \cong_o \mathbb{Z}$.

In an ordered nr. N one can ask, how $|nn'|$ and $|n||n'|$ might be related. In general there is no direct relationship, but for $R[x]$ of 9.142(b) we get the following result which we state without proof.

<u>9.145 THEOREM</u> (Pilz (4)). In $(R[x],+,\circ)$ we have for all $p,q\in R[x]$:

(a) $|p\circ q| = |p|\circ|q| \iff (q\geq 0) \vee$ (p contains only even or only odd degrees).

(b) $|p\circ q| \leq |p|\circ|q| \iff (q<0) \wedge$ (the coefficients of the greatest even and greatest odd degree of p have the same sign).

(c) $|p\circ q| \geq |p|\circ|q| \iff (q<0) \wedge$ (the coefficients of the greatest even and greatest odd degree of p have opposite sign).

J. Zemmer has shown that a direct sum of f.o. rings can be f. ordered iff all but at most one of the summands are zerorings. We now obtain a similar result for nr.'s implying some statements on the structure of f.o. nr's.

<u>9.146 THEOREM</u> If $N = \bigoplus_{i=1}^{s} N_i$ is f.o. then in all but at most one of the N_i's all positive elements (in the order induced by N via the projection maps) annihilate N_i from the right.

<u>Proof</u>. Assume that $\exists\ i,j\in\{1,\ldots,s\},\ i \neq j$
$\exists\ 0\leq n_i\in N_i\ \ \exists\ 0\leq n_j\in N_j:\ N_in_i \neq \{0\} \wedge N_jn_j \neq \{0\}$.
Then one can choose $n_i'\in N_i$ and $n_j'\in N_j$, both positive, such that $n_i'n_i>0$ and $n_j'n_j>0$.
If $n_i'<n_j'$ then $0<n_i'n_i\leq n_j'n_i = 0$ (2.27 and 2.6(b)), a contradiction. $n_i'>n_j'$ yields the same, so $n_i' \parallel n_j'$, and we have no full order.

9.147 COROLLARY Let $N = \bigoplus_{i=1}^{s} N_i$ be fully ordered and all $N_i \neq \{0\}$.
If N is either strictly ordered (i.e. $n>0,\ n'>0 \Rightarrow nn'>0$ — $R[x]$ of 9.142(b) is strictly ordered if R is integral) or if N contains a left identity then $s = 1$.

Proof. For strict orders this is immediate from 9.146.

If N contains a left identity e, let $e = \sum_{i=1}^{s} e_i$ with $e_i \in N_i$. As in 3.43, e_i is a left identity in N_i and again we can employ 9.146 to get $s = 1$.

9.148 COROLLARY Every 1-semisimple f.o. nr. $N \in \mathcal{N}_1$ with DCCL is simple.

Proof: by 9.137, 5.31 and 9.147.

9.149 REMARK One cannot improve 9.146 to get the exact analogue of Zemmer's result: take for N_1, N_2 any f.o. constant non-zero near-rings and use the lexicographic order.

Examining abstract affine near-rings gives a strange result which shows that "nearly no" a.a.n.r. can be fully ordered:

9.150 THEOREM Let N be an a.a.n.r. such that N_0, N_c are fully ordered. Then N can be f.o. $\Leftrightarrow N_0 N_c = \{0\} \wedge (N_c = \{0\} \vee N_0$ is a zeroring).

Proof. $\Rightarrow$: (a) First we show that $(N,+) = (N_c,+) \dot{+} (N_0,+)$ (9.74(a)) must have the lexicographic order.
If $n \geq 0$ then $n_c = n0 \geq 0$; likewise $n \leq 0$ implies $n_c \leq 0$.
If $n_c = 0$ then $n>0 \Leftrightarrow n = n_0 > 0$.
So we get for $n \in N$: $n \geq 0 \Leftrightarrow (n_c > 0) \vee (n_c = 0 \wedge n_0 \geq 0)$, i.e. the lexicographic order.

(b) Assume now that $N_0N_c \neq \{0\}$. Since $N_0 = N_d$, we can find $n_0 \in N_0$ and $N_c \in N_c$ with $n_c > 0$ and $n_0 n_c < 0$. Then $n := 2n_0 - n_0 n_c > 0$ by (a). But $nn_c = n_0 n_c < 0$, a contradiction. Hence $N_0 N_c = \{0\}$.

(c) If $N_0^2 \neq \{0\}$ and $N_c \neq \{0\}$ $\exists\, n_o, n_o' \in N_o$: $n_o > 0 \wedge$ $\wedge\, n_o' < 0 \wedge n_0 n_o' < 0$. Also, $\exists\, n_c \in N_c$: $n_c > 0$. Let $n := n_o > 0$ and $n' := n_0' + n_c > 0$. Then $nn' = n_o n_o' +$ $+ n_o n_c = n_0 n_0' < 0$ (use (b)!), a contradiction. Hence either $N_0^2 = \{0\}$ or $N_c = \{0\}$.

<=: The multiplication rule in N is
$nn' = (n_o + n_c)(n_0' + n_c') = n_0 n_0' + n_c$.
If $N_c = \{0\}$ we get $N = N_0$ as a f.o. ring.
If $N_c \neq \{0\}$ then $N_0^2 = \{0\}$ and all $nn' = n_c$.
In this case it is easily verified that the lexicographic order in $N_c + N_0 = N$ makes N into a f.o.nr.

<u>9.151 COROLLARY</u> No a.a.n.r. $N \neq N_c$ in which N_c forms a base (this happens e.g. in $M_{aff}(V)$ and $M_a(\Gamma)$) can be fully ordered.

<u>Proof</u>. If N is f.o., the same can be said about N_0 and N_c. $N_0 N_c = \{0\}$ implies $N_0 = \{0\}$, hence $N = N_c$.

<u>9.152 REMARKS</u>

(a) See Kerby (1),(3),(5) for a theory of ordered near-fields with some geometric interpretations.

(b) (Pilz (1),(4)). $n \in N$ is called <u>even</u> <u>(odd)</u> if $\forall\, n' \in N$: $n(-n') = nn'$ ($n(-n') = -nn'$, respectively). For instance, $f \in M(\mathbb{R})$ is even (odd) iff f is an even (odd) function. A nf. contains only odd elements (8.10(b)); the same applies to rings.
N is said to be <u>cleavable</u> if each $n \in N$ is the sum of an even and an odd element. $R[x]$, $R[[x]]$, $R_o[x]$ and the subnr. N generated by id, sin and cos in $M(\mathbb{R})$ are examples of cleavable near-rings.
9.145 can be extended to f.o. cleavable near-rings.

(c) $R_o[x]$ with the antilexicographic order ($a_k x^k+\dots+a_r x^r>0$: <=> $a_k>0$). is not a f.o. nr. (although a f.o. ring when we use multiplication instead of composition (Fuchs)), for e.g. in this ordering $-x+1>0$, $x+2>0$, but $(-x+1)\circ(x+2) = -x-1<0$.

(d) One can define in an ordered nr. N $n\in N$ to be monotone (antitone) if $\forall\, n',n''\in N$: $n'\leq n'' \Rightarrow nn'\leq nn''$ ($nn'\geq nn''$).
n is positive definite : <=> $\forall\, n'\in N$: $nn'\geq 0$.
See Pilz (1) for results on these concepts.

(e) (Pilz (8)). Let N be a nr. with $(N,+) = (N_0,+)\dot{+}(N_c,+)$ (cf. 1.13), where N_0 and N_c are f.o. n.r.'s (by P_o, P_c).
Then the f.o. of N_0 and N_c can be extended to a f.o. on N iff $\forall\, p_o\in P_o$ $\forall\, p_c\in P_c$ $\forall\, n_0\in N_0$: $p_o\cdot(n_0+p_c)\in P_o$.
In this case the order is the "lexicographic" one determined by $n_0+n_c\geq 0 \Leftrightarrow (n_c>0) \vee (n_c = 0 \wedge n_0\geq 0)$ (see 9.150).

(f) It is hard to get full orders in "non-degenerated" near-rings (9.141). But it is very natural to look for lattice-orders (i.e. such that $(N,\leq)$ is a lattice). For instance, $M(\Gamma)$, where Γ is a f.o. group, can be given a lattice order by

$$m\leq m' \;:\Leftrightarrow\; \forall\, \gamma\in\Gamma : m(\gamma)\leq m'(\gamma).$$

For details and connections to "F-near-rings" N (these are subnr.'s and sublattices of a direct product $\prod_{i\in I} N_i$ of f.o. nr.'s $\in\mathfrak{N}_c$, lattice-ordered by $(\dots,n_i,\dots)\leq(\dots,n_i',\dots)$: <=> $\forall\, i\in I$: $n_i\leq n_i'$) and to vector-near-rings (F-nr.'s, where N is a subdirect product of the N_i's) see Pilz (1).

(g) See also Küsel (1).

f) MISCELLANEOUS TOPICS

In this final section we intend to give brief descriptions of topics we didn't discuss in our journey through the "nr.-universe" until now. Again it should be noted that being in this section should not imply any discrimination of this subject (as being "less important"). We have to reach an end of this monograph - the reader might be tired.

9.153 DICKSON-NEAR-RINGS

The definitions of coupling maps, derived nr.'s and Dickson nr.'s can be found in 9.90. For a detailed study of these concepts see Maxson (8) and Timm (6),(7).
Of course, a Dickson near-ring (=: DNR) is abelian.
One may write a DNR as $(D,+,\cdot,\circ)$, where $(D,+,\cdot)$ is a ring and $(D,+,\circ)$ the derived nr. $(\circ = \circ_\phi)$.
Maxson shows e.g. in (8) that $(D,+,\circ)$ has an identity iff $(D,+,\cdot)$ has one and $\forall\ d\in D^*$: $\phi_d \neq \bar{o}$. A finite DNR with identity is a nf..
The ideal structure of a DNR is also considered by Pieper (3) in comparing the left ideals of $(D,+,\cdot)$ and $(D,+,\circ)$. The connection between homomorphisms of $(D,+,\cdot)$ and $(D,+,\circ)$ are studied in Maxson (13) and Pokropp (3).
Kerby (5) settles the question in which cases the nr. of quotients of $(D,+,\circ)$ is a Dickson one w.r.t. the ring of quotients of $(D,+,\cdot)$.
Aside from these considerations, Magill (2) also studies "changed multiplications".

9.154 REGULAR NEAR-RINGS

$x\in N$ is regular if $\exists\ y\in N$: $xyx = x$. xy is then idempotent. N itself is called regular if all $x\in N$ are regular.
Examples of regular nr.'s are: $M_o(\Gamma)$ (Beidleman (10)), constant nr.'s, direct products of near-fields (Ligh (7)).

Every nr. $\in \mathcal{N}_0$ can be embedded into a regular one (Beidleman (10)).
Integral regular nr.'s are near-fields and regular dgnr.'s are $\{0\}$ or fields (Beidleman (10), Ligh (2)).
A regular nr. in which every idempotent is central is a subdirect product of nf.'s, hence abelian (Ligh (7)).
For more details see Ligh (7), Heatherly (9) (regular nr.'s with DCC), Ligh (16), Ligh-Utumi (2), Subrahmanyam (1), Heatherly (8), Plasser (1) (connections to nr.'s without nilpotent elements and to IFP-near-rings), Oswald (1),(2), (4),(8) and Choudhari (1).

9.155 LOCAL NEAR-RINGS

$N \in \mathcal{N}_0 \cap \mathcal{N}_1$ is called local if $L := L(N) := \{x \in N \mid x$ has no left inverse$\} \leq_N N$. (this happens iff L is a subgroup).
Maxson (1),(3) shows:
A local nr. is indecomposable. Hence a 1-semisimple one with DCC is simple. A nr. N is local iff N has a unique maximal N-subgroup (namely L). L is qr. and if N is not 2-radical then N is local iff $\mathcal{J}_2(N) = L$. So $L \triangleleft N$.
If N is local then N/L is a nf., hence a simple nr. is local iff it is a nf.. A local nr. has only 0 and 1 as idempotents. The additive group of a finite local nr. is a p-group.
Maxson (6) goes on to determine all local nr.'s of order p and p^2. In (9) and (12) he presents local non-rings on non-cyclic abelian (p-)groups of order ≥ 5 and more results in this direction.
Other examples of local nr.'s are given by $M_{aff}(V)$ (Maxson (1),(3)), $F[x]$, where F is a field with $|F| \geq 3$ (Clay-Doi (2) - 7.98), and $E(\Gamma)$, $A(\Gamma)$ and $I(\Gamma)$, where Γ is a generalized quaternion group (Malone (7)).
See Karzel-Meissner (1), Pieper (1),(3),(4) and Armentrout-Hardy-Maxson (1) for applications of local nr.'s to geometry (coordinatisation).

9.156 COMMUTATORS, DISTRIBUTORS AND SOLVABILITY

Distributors are defined in 9.79. For a detailed study of these concepts see Esch (1) and confer H.D. Brown (3). Esch (1) also contains results due to Fröhlich (1),(2) on distributors and "weak distributivity" in dgnr.'s (cf. 6.16). See also Mason (1) and Maxson (1).
Nr.'s generated by the commutators of a (non-abelian) group are studied in Gupta (1). See also Curjel (1).

9.157 DISTRIBUTIVE NEAR-RINGS

This is the place where the theories of near-rings and semirings meet. We mentioned these nr.'s already in 1.15, 1.107 and 1.108. All of §6 is applicable. Taussky (1) also showed that in a distributive nr. N either each element is a zero divisor or N is a ring. A simple distributive nr. is also a ring (Ferrero (1), Ligh (13)).
For more details see Heatherly (7),(8), Heatherly-Ligh (1), Ligh (8),(15), Malone (7) and (a unifying presentation) Weinert (1),(2).
N is said to be <u>n-distributive</u> ($n\in\mathbb{N}$) if $(N^2,+)$ is abelian and $\forall\ x,y_1,\dots,y_n,z_1,\dots,z_n\in N$:
$x(\sum_{i=1}^{n} y_iz_i) = \sum_{i=1}^{n} xy_iz_i$. N is <u>pseudo-distributive</u> if N is n-distributive for all $n\in\mathbb{N}$.
If one considers the nxn-matrices $M_n(N)$ with entries from some nr. N together with the usual addition and multiplication then (Heatherly (11), Ligh (20)) $M_n(N)$ is a nr. iff N is n-distributive. Also, one can study polynomials, formal power series, group near-rings and "Gaussian near-rings $N(i)$". These sets are (under the usual operations) always near-rings iff N is pseudo-distributive (see Heatherly-Ligh (1) for this and many other results concerning pseudo-distributive near-rings). Confer also Beidleman (1) and Gupta (1), as well as 9.160.

9.158 C-Z-TRANSITIVE AND C-Z-DECOMPOSABLE NEAR-RINGS

N is "C-Z-transitive" if $\forall\, n_c \varepsilon N_c^* \ \forall\, n_c' \varepsilon N_c \ \exists\, n_o \varepsilon N_o$: $n_o n_c = n_c'$.
In this case, ${}_{N_o}N_c$ is strongly monogenic. N is "C-Z-decomposable" if $N_c \trianglelefteq N$ (these nr.'s are closely related to a.a.n.r.'s !).
Heatherly (3) developes an ideal theory for these near-rings. Cf. also Pilz (1),(6).

9.159 N-SYSTEMS

A nr. $N \varepsilon \mathcal{N}_o$ with right cancellation law and a "halvable idempotent $e \neq 0$" (i.e. $\exists\, h \varepsilon N$: $h+h = e$) is called N-system.
Every N-system is abelian (see the proof of 9.13(b)) and integral (so 9b)2) is at hand). A finite N-system is a near-field, but there do exist infinite N-systems which are neither rings nor near-fields (see Ligh-Malone (1), Ligh-McQuarrie-Slotterbeck (1) and McQuarrie (1),(3)).
If $N \leq M_o(\Gamma)$ and N is an N-system containing id_N then every function of N is odd (cf. 9.152(b)).

9.160 SYLOW-TYPE THEOREMS; p-SINGULAR NEAR-RINGS

Ferrero (1),(2) shows that $|N| = m$, p^k/m but p^{k+1}/m and $N = N_d$ implies the existence of a two-sided invariant subgroup of N of order p^k. If $|N| = p \cdot q$ ($p,q \varepsilon \mathbb{P}$, $p<q$) and N is not abelian then N has no subnear-ring of order p.
If N is finite and $p \varepsilon \mathbb{P}$, N is called p-singular if p properly divides $|N|$, but N has no subnear-ring whose order is divisible by p. So p-singular nr.'s are "minimal for not fulfilling the Sylow-theorems". A p-singular nr. N is $\varepsilon \mathcal{N}_o$ and ${}_N N$ is strongly monogenic.
See Ferrero (4), (5), (7), (18) and (19).

9.161 CONDITIONS FOR N TO BE FINITE

Ligh (1) has shown that if N contains n right zero divisors (at least one of them εN_d) then $|N| \leq n^2$, hence N is

finite. See also Ligh-Malone (1).
For rings, the DCC and ACC on subrings force the ring to be finite. Bell-Ligh (1) extended this result to dgnr.'s and obtained similar other finiteness conditions (mainly for dgnr.'s).

9.162 ASSOCIATED RINGS

Let N be abelian and $A(N)$ the subnear-ring of $M((N,+))$ generated by all $h_n: \begin{array}{l} N \to N \\ m \to mn \end{array}$. Then $A(N)$ is a ring (the "ring associated to N") and was investigated in Williams (1). N and $A(N)$ are closely related.

9.163 NEAR-VECTOR SPACES

It seems not be quite clear how to define a near-vector space. Beidleman (1) defined it as a 2-semisimple N-group (N a nf.), and developed a kind of "nearly-linear" algebra.
Another approach to this concept is made by André (4).

9.164 SHEAF REPRESENTATION OF NEAR-RINGS

Dauns-Hofmann-type sheaf representations of near-rings are considered in Betsch (7) and Szeto (6),(8). See also Szeto-Wong (1).

9.165 VALUATION THEORY ON NEAR-RINGS

This is developed in Zemmer (4),(5) and (for near-fields) in Wefelscheid (6),(7).

9.166 TOPOLOGY IN NEAR-RINGS

The starting point was Beidleman-Cox (1) which contains definitions and structural properties of topological near-rings.
Topological nr.'s on relatively free groups were considered by Tharmaratnam (3) (see 6.35(f)).
Betsch (3) considers topological spaces induced by ν-primitive ideals (ν = 1,2).

Nr.'s of continuous mappings on topological groups (totally disconnected topological groups, Banach-spaces, real numbers,...) were considered by Betsch (3), Magill (1)-(3), Hofer (1)-(4), Yamamuro (5), Yamamuro-Palmer (1), Blackett (4)-(6), Su (1),(2) and Holcombe (3),(4).
For instance, Yamamuro obtains the following result in (5): Let B be a real Banach-space of dimension ≥ 2, and let N be a nr. of continuous mappings $B \to B$, containing $M_{aff}(B)$. Then every automorphism of B is inner. This implies that if B_1, N_1 and B_2, N_2 are two couples as above and $N_1 \cong N_2$ then N_1 and N_2 are also topologically isomorphic (homeomorphic).
See Sörensen (1) and Wefelscheid (1),(2) and (7) for topological near-fields.
See Neuberger (1),(2) for applications of nr.'s in functional analysis.

9.167 NEAR-RINGS IN ALGEBRAIC TOPOLOGY

In decomposing polyhedras one meets near-rings as structures which annihilate homology groups (see Curjel (1)).
Curjel (2) contains (among others) the following results: Let A be a finite complex, ΣA the reduced suspension of A and $N(\Sigma A) =: N$ the near-ring (with identity) of homotopy classes of base-point preserving selfmaps of ΣA. Using the induced endomorphisms of $H_*(\Sigma A)$, the following assertions can be shown to be equivalent:

(a) $\forall$ $m,n \in N$: $mn-nm$ is of finite additive order.

(b) The group of invertible elements in the monoid $(N,\cdot)$ (= its group kernel) is finite.

(c) $\forall$ $n \in N$: n nilpotent $\Rightarrow$ n is of finite additive order.

If the Betti-numbers of ΣA are known, one can decide whether or not N has these properties by a mechanical application of Hilton's formula for the homotopy groups of a union of spheres. Also,

$\{n \varepsilon N | n$ has finite additive order$\} \trianglelefteq N$.
$\mathfrak{J}_2(N)$ is nilpotent. The finiteness of $\mathfrak{J}_2(N)$ can again be seen by looking at the Betti-numbers of ΣA.

9.168 EXTENSIONS AND HOMOLOGY

Maxson (1), Choudhari (1), Seth-Tewari (1), Mason (3),(4) and Prehn (1), consider exact sequences of N-groups, injectivity, projectivity and the connections to semi-simplicity (see 5.49, 5.50 and 9.155).
Steinegger (1) describes extensions of near-rings by sets of functions (similar to the ring case).
For dgnr.'s these investigations were carried out by Fröhlich (5)-(8) ("non-abelian homological algebra") and Lausch (1),(3).

Finally:

9.169 NEAR-RINGS AND CATEGORIES

Let $\mathcal{C}$ be a category with finite products and a final object. Let $X \varepsilon \mathcal{C}$ be a group object. Then $Mor(X,X) = M(X)$ (cf. 1.4(a)) is a nr. with the obvious operations (Holcombe (3),(4)). Holcombe studies these near-rings in various categories. Homology and cohomology groups can be viewed as certain N-groups for some nr. N.
Similar considerations (in additive categories) can be found in Huq (1).
A categorical treatment of dgnr.'s was carried through by Fröhlich in (4)-(8).
A categorical investigation to radical theory is in Holcombe (8).

Now we are through. Thanks be to God.

A P P E N D I X

NEAR-RINGS OF LOW ORDER

Now we give a description of all near-rings of order ≤ 7 and of several classes of near-rings of order 8. The whole discussion is due to Clay (2), (4), (7), (8) and (9). Because of 9.92, this ammounts to the description of all mappings $\phi: \Gamma \to \text{End}\,\Gamma$, $\gamma \to \phi_\gamma$ with the property mentioned in 9.92, where Γ is a group of "small" order. The multiplication $\cdot_\phi$ in Γ is then given by $\gamma \cdot_\phi \delta = \phi_\delta(\gamma)$. This will be done in the following way:

(a) If $\Gamma = \{\gamma_1,\ldots,\gamma_n\}$ $(n \leq 8)$, we list the endomorphisms $\alpha_1,\ldots,\alpha_k$ of Γ.

(b) Every isomorphism class of near-rings of order n is determined by the n-tuple $(k_1,\ldots,k_n)$ of elements of $\mathbb{N}_0$, where $\phi_{\gamma_i} = \alpha_{k_i}$. So $\gamma_i \cdot_\phi \gamma_j = \phi_{\gamma_j}(\gamma_i) = \alpha_{k_j}(\gamma_i)$.

(c) The numbers following this n-tuple denote the numbers of those automorphisms of (a) which yield isomorphic near-rings on Γ.

(d) In A)-J), "Z" means that the near-ring considered is zero-symmetric, "D" means that it is distributive, "C" indicates that the near-ring is commutative and "I=η ($\varepsilon\Gamma$)" means that η is an identity.

Example: The near-ring

13) (0,7,13,9); 1,2,3;

on Klein's four group {0,a,b,c} means that ϕ is given by $\phi_0=\alpha_0$, $\phi_a=\alpha_7$, $\phi_b=\alpha_{13}$ and $\phi_c=\alpha_9$.

The multiplication table is then the following:

$\cdot_\phi$	0	a	b	c
0	$\phi_0(0)$	$\phi_a(0)$	$\phi_b(0)$	$\phi_c(0)$
a	$\phi_0(a)$	$\phi_a(a)$	$\phi_b(a)$	$\phi_c(a)$
b	$\phi_0(b)$	$\phi_a(b)$	$\phi_b(b)$	$\phi_c(b)$
c	$\phi_0(c)$	$\phi_a(c)$	$\phi_b(c)$	$\phi_c(c)$

=

$\cdot_\phi$	0	a	b	c
0	0	0	0	0
a	0	a	b	c
b	0	0	0	0
c	0	a	b	c

$\alpha_1, \alpha_2, \alpha_3$ are the automorphisms giving isomorphic near-rings. The near-ring is distributive.

A) $\mathbb{Z}_1 = \{0\}$: This case is trivial.

B) $\mathbb{Z}_2 = \{0,1\}$:

+	0	1
0	0	1
1	1	0

	0	1
α_0	0	0
α_1	0	1

1) (0,0); 1; Z,D,C
2) (0,1); 1; Z,D,C; I=1
3) (1,1); 1

C) $\mathbb{Z}_3 = \{0,1,2\}$:

+	0	1	2
0	0	1	2
1	1	2	0
2	2	0	1

	0	1	2
α_0	0	0	0
α_1	0	1	2
α_2	0	2	1

1) (0,0,0); 1,2; Z,D,C
2) (0,0,1); 1; Z
3) (0,1,0); 1; Z
4) (0,1,1); 1,2; Z
5) (1,1,1); 1,2
6) (0,1,2); 1,2; Z,D,C
7) (0,2,1); 1,2; Z,D,C

D) $\mathbb{Z}_4 = \{0,1,2,3\}$:

+	0	1	2	3
0	0	1	2	3
1	1	2	3	0
2	2	3	0	1
3	3	0	1	2

	0	1	2	3
α_0	0	0	0	0
α_1	0	1	2	3
α_2	0	3	2	1
α_3	0	2	0	2

1) (0,0,0,0); 1; Z,D,C
2) (0,1,0,0); 1,2; Z
3) (0,3,0,0); 1,2; Z
4) (0,0,1,0); 1; Z
5) (0,1,1,0); 1,2; Z
6) (0,1,0,1); 1; Z
7) (0,2,0,1); 1,2; Z
8) (0,1,1,1); 1; Z
9) (1,1,1,1); 1
10) (0,1,3,1); 1; Z
11) (0,1,3,2); 1,2; Z,D,C,I=1
12) (0,3,0,3); 1; Z,D,C

E) Klein's four group {0,a,b,c}:

+	0	a	b	c
0	0	a	b	c
a	a	0	c	b
b	b	c	0	a
c	c	b	a	0

	0	a	b	c
α_0	0	0	0	0
α_1	0	a	b	c
α_2	0	a	c	b
α_3	0	b	a	c
α_4	0	b	c	a
α_5	0	c	a	b

	0	a	b	c
α_6	0	c	b	a
α_7	0	a	0	a
α_8	0	a	a	0
α_9	0	c	0	c
α_{10}	0	0	c	c
α_{11}	0	0	b	b

	0	a	b	c
α_{12}	0	b	b	0
α_{13}	0	b	0	b
α_{14}	0	0	a	a
α_{15}	0	c	c	0

1) (0,1,1,1);1;Z
2) (0,14,1,1);1,3,4;Z
3) (0,0,1 1);1,3,4;Z,D
4) (0,14,2,1);1,2,3,4,5,6; Z,D,C; I=c
5) (0,0,2,1);1,2,3,4,5,6;Z
6) (0,4,5,1);1,2,5;Z,D,C;I=c
7) (0,7,11,1);1,2,5;Z,D,C;I=c
8) (0,14,11,1);1,2,3,4,5,6;Z; I=c
9) (0,7,0,1);1,2,3,4,5,6;Z
10) (0,0,0,1);1,2,5;Z
11) (0,7,13,7);1,2,3,4,5,6;Z
12) (0,7,0,7);1,2,3,4,5,6; Z,D,C
13) (0,7,13,9);1,2,3;Z,D
14) (0,7,0,9);1,2,3;Z
15) (0,13,0,13);1,2,3;Z,D,C
16) (0,0,0,14);1,2,3,4,5,6; Z
17) (0,0,0,0);1;Z,D,C
18) (7,7,1,1);1,2,3,4,5,6
19) (7,7,7,1);1,2,3,4,5,6
20) (7,8,1,2);1,2,3,4,5,6; I=b
21) (7,7,1,7);1,2,3,4,5,6
22) (7,7,7,7);1,2,3,4,5,6
23) (1,1,1,1);1

F) $\mathbb{Z}_5$ = {0,1,2,3,4}: Addition is modulo 5.

	0	1	2	3	4
α_0	0	0	0	0	0
α_1	0	1	2	3	4
α_2	0	2	4	1	3
α_3	0	3	1	4	2
α_4	0	4	3	2	1

1) (0,0,0,0,0);1;Z,D,C
2) (0,1,0,0,0);1,2,3,4;Z
3) (0,1,1,0,0);1,2,3,4;Z
4) (0,0,1,1,0);1,2;Z
5) (0,1,1,1,0);1,2,3,4;Z
6) (0,0,4,1,0);1,2,3,4;Z
7) (0,1,4,1,4);1,2,3,4;Z
8) (0,1,1,1,1);1;Z
9) (1,1,1,1,1);1
10) (0,1,2,3,4);1,2,3,4;Z,D, C;I=1

G) $\mathbb{Z}_6$ = {0,1,2,3,4,5}: Addition is modulo 6.

	0	1	2	3	4	5
α_0	0	0	0	0	0	0
α_1	0	1	2	3	4	5
α_2	0	2	4	0	2	4
α_3	0	3	0	3	0	3
α_4	0	4	2	0	4	2
α_5	0	5	4	3	2	1

1) (0,1,0,0,0,0);1,5;Z
2) (0,0,1,0,0,0);1,5;Z
3) (0,1,1;0,0,0);1,5;Z
4) (0,1,0,1,0,0);1,5;Z
5) (0,0,1,1,0,0);1,5;Z
6) (0,1,1,1,0,0);1,5;Z
7) (0,1,0,0,1,0);1,5;Z
8) (0,1,1,0,1,0);1,5;Z
9) (0,0,5,0,1,0);1,5;Z
10) (0,1,0,1,1,0);1,5;Z
11) (0,1,1,1,1,0);1,5;Z
12) (0,1,0,0,4,0);1,5;Z
13) (0,4,0,0,4,0);1,5;Z
14) (0,5,0,0,0,1);1,5;Z
15) (0,1,1,0,0,1);1,5;Z
16) (0,1,1,1,0,1);1,5;Z
17) (0,3,0,3,0,1);1,5;Z
18) (0,5,0,3,0,1);1,5;Z
19) (0,3,4,3,0,1);1,5;Z
20) (0,5,5,0,1,1);1,5;Z
21) (3,3,1,3,1,1);1,5
22) (3,1,3,3,1,1);1,5
23) (3,3,3,3,1,1);1,5
24) (3,5,5,3,1,1);1,5
25) (0,2,4,0,2,1);1,5;Z
26) (0,5,4,0,2,1);1,5;Z
27) (0,1,2,3,4,5);1,5;Z,D,C;I=1
28) (3,3,1,3,3,1);1,5
29) (3,3,3,3,3,1);1,5
30) (3,5,3,3,3,1);1,5
31) (0,4,4,0,4,1);1;5;Z
32) (4,4,4,1,4,1);1;5
33) (4,4,4,4,4,1);1;5
34) (0,5,1,0,5,1);1;5;Z
35) (3,5,1,3,5,1);1;5
36) (0,4,2,0,4,2);1;5;Z,D,C
37) (3,3,3,3,1,3);1;5
38) (3,3,5,3,1,3);1;5
39) (0,0,0,0,0,0);1;Z,D,C
40) (0,0,0,1,0,0);1;Z
41) (0,0,1,0,1,0);1;Z
42) (0,0,1,1,1,0);1;Z
43) (0,1,0,0,0,1);1;Z
44) (0,1,0,1,0,1);1;Z
45) (0,1,0,3,0,1);1;Z
46) (0,1,1,0,1,1);1;Z
47) (0,1,1,1,1,1);1;Z
48) (1,1,1,1,1,1);1
49) (3,1,1,3,1,1);1
50) (3,1,3,3,3,1);1
51) (0,1,4,0,4,1);1;Z
52) (4,1,4,1,4,1);1
53) (0,1,4,3,4,1);1;Z
54) (4,1,4,4,4,1);1
55) (0,3,0,3,0,3);1;Z,D,C
56) (3,3,1,3,1,3);1
57) (3,3,3,3,3,3);1
58) (0,4,4,0,4,4);1;Z
59) (4,4,4,1,4,4);1
60) (4,4,4,4,4,4);1

<u>H) S_3 = {0,a,b,c,x,y}:</u>

+	0	a	b	c	x	y
0	0	a	b	c	x	y
a	a	0	y	x	c	b
b	b	x	0	y	a	c
c	c	y	x	0	b	a
x	x	b	c	a	y	0
y	y	c	a	b	0	x

	0	a	b	c	x	y
α_0	0	0	0	0	0	0
α_1	0	a	b	c	x	y
α_2	0	a	c	b	y	x
α_3	0	b	a	c	y	x
α_4	0	b	c	a	x	y

	0	a	b	c	x	y
α_5	0	c	a	b	x	y
α_6	0	c	b	a	y	x
α_7	0	a	a	a	0	0
α_8	0	b	b	b	0	0
α_9	0	c	c	c	0	0

1) (0,0,2,1,2,1);1,2,3,4,5,6;Z
2) (0,0,1,2,2,1);1,2,3,4,5,6;Z
3) (0,0,0,0,2,1);1,2,3,4,5,6;Z
4) (0,0,1,1,0,1);1,2,3,4,5,6;Z
5) (0,0,0,1,0,1);1,2,3,4,5,6;Z
6) (0,7,2,1,0,0);1,2,3,4,5,6;Z
7) (0,0,2,1,0,0);1,2,3,4,5,6;Z
8) (0,7,7,1,0,0);1,2,3,4,5,6;Z
9) (0,7,8,7,0,0);1,2,3,4,5,6;Z
10) (7,7,1,1,1,7);1,2,3,4,5,6
11) (7,7,2,1,1,2);1,2,3,4,5,6
12) (7,7,7,1,1,1);1,2,3,4,5,6
13) (7,7,7,1,1,7);1,2,3,4,5,6
14) (7,7,1,2,1,2);1,2,3,4,5,6
15) (7,7,1,7,1,7);1,2,3,4,5,6
16) (7,7,7,7,1,2);1,2,3,4,5,6
17) (7,7,7,7,1,7);1,2,3,4,5,6
18) (7,7,2,1,7,7);1,2,3,4,5,6
19) (7,7,7,1,7,7);1,2,3,4,5,6
20) (0,1,1,1,0,1);1,2;Z
21)(0,0,0,0,0,1);1,2;Z
22)(0,0,1,1,1,1);1,3,5;Z
23)(0,0,0,1,1,1);1,2,4;Z
24)(0,7,1,1,0,0);1,3,5;Z
25)(0,0,1,1,0,0);1,3,5;Z
26)(0,4,5,1,0,0);1,2,4;Z
27)(0,7,8,1,0,0);1,2,4;Z
28)(0,0,0,1,0,0);1,2,4;Z
29)(0,7,7,7,0,0);1,3,5;Z,D,C
30)(7,7,1,1,1,1);1,3,5
31)(7,7,7,7,1,1);1,3,5
32)(7,7,1,1,7,7);1,3,5
33)(7,7,7,7,7,7);1,3,5
34)(0,1,1,1,1,1);1;Z
35)(0,0,0,0,1,1);1;Z
36)(0,1,1,1,0,0);1;Z
37)(0,7,8,9,0,0);1;Z
38)(0,0,0,0,0,0);1;Z,D,C
39)(1,1,1,1,1,1);1

I) $\mathbb{Z}_7$ = {0,1,2,3,4,5,6}: Addition is modulo 7.

	0	1	2	3	4	5	6
α_0	0	0	0	0	0	0	0
α_1	0	1	2	3	4	5	6
α_2	0	2	4	6	1	3	5
α_3	0	3	6	2	5	1	4

	0	1	2	3	4	5	6
α_4	0	4	1	5	2	6	3
α_5	0	5	3	1	6	4	2
α_6	0	6	5	4	3	2	1

1)(0,0,0,0,0,0,0);1;Z,D,C
2)(0,1,0,0,0,0,0);1,2,3,4,5,6;Z
3)(0,1,1,0,0,0,0);1,2,3,4,5,6;Z
4)(0,1,0,1,0,0,0);1,2,3,4,5,6;Z
5)(0,1,1,1,0,0,0);1,2,3,4,5,6;Z
6)(0,1,1,0,1,0,0);1,3;Z
7)(0,2,4,0,1,0,0);1,2,3,4,5,6;Z
8)(0,0,0,1,1,0,0);1,2,3;Z
9)(0,1,0,1,1,0,0);1,2,3,4,5,6;Z
10)(0,0,1,1,1,0,0);1,2,3,4,5,6;Z
11)(0,1,1,1,1,0,0);1,2,3,4,5,6;Z
12)(0,0,0,6,1,0,0);1,2,3,4,5,6;Z
13)(0,1,1,1,0,1,0);1,2,3,4,5,6;Z
14)(0,0,1,1,1,1,0);1,2,3;Z
15)(0,1,1,1,1,1,0);1-6;Z
16)(0,0,6,6,1,1,0);1-6;Z
17)(0,0,6,1,6,1,0);1-6;Z
18)(0,1,1,1,1,1,1);1;Z
19)(1,1,1,1,1,1,1);1
20)(0,6,6,6,1,1,1);1-6;Z
21)(0,6,6,1,6,1,1);1,3;Z
22)(0,2,4,4,1,2,1);1-6;Z
23)(0,1,2,3,4,5,6);1-6;Z,D,C,I=1
24)(0,1,2,4,4,2,1);1,2,3;Z

J) $\mathbb{Z}_8$ = {0,1,2,3,4,5,6,7}: Addition is modulo 8.

	0	1	2	3	4	5	6	7
α_0	0	0	0	0	0	0	0	0
α_1	0	1	2	3	4	5	6	7
α_2	0	2	4	6	0	2	4	6
α_3	0	3	6	1	4	7	2	5
α_4	0	4	0	4	0	4	0	4
α_5	0	5	2	7	4	1	6	3
α_6	0	6	4	2	0	6	4	2
α_7	0	7	6	5	4	3	2	1

1)(0,0,0,0,0,0,0,1);1,3,5,7;Z
2)(0,0,0,0,0,0,0,4);1,3,5,7;Z
3)(0,0,0,0,0,0,0,0);1,3,5,7;Z
4)(0,0,0,0,0,0,4,4);1,3,5,7;Z
5)(0,0,0,0,0,1,0,3);1,3,5,7;Z
6)(0,0,0,0,0,1,1,0);1,3,5,7;Z
7)(0,0,0,0,0,1,1,1);1,3,5,7;Z
8)(0,0,0,0,0,4,4,0);1,3,5,7;Z
9)(0,0,0,0,0,4,4,4);1,3,5,7;Z
10)(0,0,0,0,1,0,0,1);1,3,5,7;Z
11)(0,0,0,0,1,0,1,1);1,3,5,7;Z
12)(0,0,0,0,1,1,1,0);1,3,5,7;Z
13)(0,0,0,0,1,1,1,1);1,3,5,7;Z
14)(0,0,0,1,0,0,0,5);1,3,5,7;Z
15)(0,0,0,1,0,1,0,1);1,3,5,7;Z
16)(0,0,0,1,0,1,1,0);1,3,5,7;Z
17)(0,0,0,1,0,1,1,1);1,3,5,7;Z
18)(0,0,0,1,0,7,0,0);1,3,5,7;Z
19)(0,0,0,1,1,1,0,1);1,3,5,7;Z
20)(0,0,0,1,1,1,1,0);1,3,5,7;Z
21)(0,0,0,1,1,1,1,1);1,3,5,7;Z
22)(0,0,0,2,0,4,4,2);1,3,5,7;Z
23)(0,0,0,4,0,4,0,4);1,3,5,7;Z
24)(0,0,0,4,0,4,4,0);1,3,5,7;Z
25)(0,0,0,4,0,4,4,4);1,3,5,7;Z
26)(0,0,1,0,0,0,1,1);1,3,5,7;Z
27)(0,0,1,0,0,1,3,3);1,3,5,7;Z
28)(0,0,1,0,0,3,3,1);1,3,5,7;Z
29)(0,0,1,0,1,0,1,1);1,3,5,7;Z
30)(0,0,1,1,0,1,0,1);1,3,5,7;Z
31)(0,0,1,1,0,1,1,1);1,3,5,7;Z
32)(0,0,1,1,0,7,7,0);1,3,5,7;Z
33)(0,0,1,1,1,1,0,1);1,3,5,7;Z
34)(0,0,1,1,1,1,1,1);1,3,5,7;Z
35)(0,0,1,7,0,1,7,0);1,3,5,7;Z
36)(0,0,4,0,0,0,4,4);1,3,5,7;Z
37)(0,0,4,4,0,4,0,4);1,3,5,7;Z
38)(0,0,4,4,0,4,4,4);1,3,5,7;Z
39)(0,0,4,6,0,4,0,6);1,3,5,7;Z
40)(0,1,0,1,4,5,4,5);1,3,5,7;Z
41)(0,1,0,3,0,5,0,7);1,3,5,7;Z
42)(0,1,0,3,4,5,0,7);1,3,5,7;Z
43)(0,1,0,5,4,5,4,1);1,3,5,7;Z
44)(0,1,1,1,0,7,7,7);1,3,5,7;Z
45)(0,1,1,3,0,3,3,1);1,3,5,7;Z
46)(0,1,2,3,4,5,6,7);1,3,5,7;Z
D,C;I=1
47)(0,1,4,3,4,5,4,7);1,3,5,7;Z
48)(0,2,4,2,0,2,4,6);1,3,5,7;Z
49)(0,2,4,6,0,6,4,6);1,3,5,7;Z
50)(0,0,0,0,0,0,1,0);1,5;Z
51)(0,0,0,0,0,0,4,0);1,5;Z
52)(0,0,0,0,0,1,0,1);1,3;Z
53)(0,0,0,0,0,4,0,4);1,3;Z
54)(0,0,0,0,1,0,1,0);1,5;Z
55)(0,0,0,0,1,1,0,1);1,3;Z
56)(0,0,0,1,0,0,0,1);1,5;Z
57)(0,0,0,1,0,0,1,1);1,5;Z
58)(0,0,0,1,0,1,0,0);1,7;Z
59)(0,0,0,1,1,0,0,1);1,5;Z
60)(0,0,0,1,1,0,1,1);1,5;Z
61)(0,0,0,1,1,1,0,0);1,7;Z
62)(0,0,0,2,0,0,4,2);1,5;Z
63)(0,0,0,4,0,0,0,4);1,5;Z
64)(0,0,0,4,0,0,4,4);1,5;Z
65)(0,0,0,4,0,4,0,0);1,7;Z
66)(0,0,1,0,0,0,3,0);1,5;Z
67)(0,0,1,0,0,0,7,0);1,5;Z
68)(0,0,1,0,0,1,1,1);1,3;Z
69)(0,0,1,0,1,1,1,1);1,3;Z
70)(0,0,1,1,0,0,0,1);1,5;Z
71)(0,0,1,1,0,0,1,1);1,5;Z
72)(0,0,1,1,0,1,1,0);1,7;Z
73)(0,0,1,1,1,0,0,1);1,5;Z
74)(0,0,1,1,1,0,1,1);1,5;Z
75)(0,0,1,1,1,1,1,0);1,7;Z
76)(0,0,4,0,0,4,4,4);1,3;Z
77)(0,0,4,4,0,0,0,4);1,5;Z
78)(0,0,4,4,0,0,4,4);1,5;Z
79)(0,0,4,4,0,4,4,0);1,7;Z
80)(0,0,4,6,0,0,0,6);1,5;Z
81)(0,1,0,1,0,1,1,1);1,5;Z
82)(0,1,0,1,0,5,0,5);1,3;Z
83)(0,1,0,1,0,7,0,7);1,3;Z
84)(0,1,0,1,1,1,1,1);1,5;Z
85)(0,1,0,1,4,1,4,1);1,5;Z
86)(0,1,0,1,4,5,0,5);1,3;Z
87)(0,1,0,1,4,7,0,7);1,3;Z
88)(0,1,0,3,0,1,0,3);1,5;Z
89)(0,1,0,3,0,3,0,1);1,7;Z
90)(0,1,0,3,4,1,0,3);1,5;Z
91)(0,1,0,3,4,3,0,1);1,7;Z
92)(0,1,0,5,0,5,0,1);1,7;Z
93)(0,1,0,5,4,5,0,1);1,7;Z
94)(0,1,0,7,0,1,0,7);1,5;Z
95)(0,1,0,7,4,1,0,7);1,5;Z
96)(0,1,1,3,0,1,3,3);1,5;Z
97)(0,1,1,7,0,1,7,7);1,5;Z
98)(0,1,2,1,4,5,2,5);1,3;Z
99)(0,1,2,3,4,1,6,3);1,5;Z
100)(0,1,2,5,4,5,2,1);1,7;Z
101)(0,1,2,7,4,1,6,7);1,5;Z
102)(0,1,3,3,0,1,1,3);1,5;Z
103)(0,1,4,1,4,5,4,5);1,3;Z
104)(0,1,4,1,4,7,4,7);1,3;Z
105)(0,1,4,3,4,1,4,3);1,5;Z
106)(0,1,4,3,4,3,4,1);1,7;Z
107)(0,1,4,5,4,5,4,1);1,7;Z
108)(0,1,4,7,4,1,4,7);1,5;Z
109)(0,1,6,1,4,5,6,5);1,3;Z
110)(0,1,6,5,4,5,6,1);1,7;Z
111)(0,1,7,7,0,1,1,7);1,5;Z

112)(0,2,4,4,0,2,0,4);1,5;Z
113)(0,2,4,6,0,2,4,6);1,5;Z,D,C
114)(0,2,4,2,0,6,4,6);1,3;Z
115)(0,2,4,6,0,6,4,2);1,7
116)(0,4,0,4,0,4,4,4);1,5
117)(0,4,4,6,0,4,0,6);1,5
118)(1,1,1,1,1,1,1,1);1
119)(0,0,0,0,0,0,0,0);1;Z,D,C
120)(0,0,0,0,1,0,0,0);1;Z
121)(0,0,1,0,0,0,1,0);1;Z
122)(0,0,1,0,1,0,1,0);1;Z
123)(0,0,4,0,0,0,4,0);1;Z
124)(0,1,0,1,0,1,0,1);1;Z
125)(0,1,0,1,1,1,0,1);1;Z
126)(0,1,0,1,4,1,0,1);1;Z
127)(0,1,1,1,0,1,1,1);1;Z
128)(0,1,1,1,1,1,1,1);1;Z
129)(0,1,2,1,4,1,2,1);1;Z
130)(0,1,4,1,4,1,4,1);1;Z
131)(0,1,6,1,4,1,6,1);1;Z
132)(0,2,4,2,0,2,4,2);1;Z
133)(0,4,0,4,0,4,0,4);1;Z,D,C
134)(0,4,4,4,0,4,4,4);1;Z
135)(0,6,4,6,0,6,4,6);1;Z

K) The dihedral group D_8 = {0,a,2a,3a,b,a+b,2a+b,3a+b}:

	0	a	2a	3a	b	a+b	2a+b	3a+b
α_1	0	a	2a	3a	b	a+b	2a+b	3a+b
α_2	0	a	2a	3a	a+b	2a+b	3a+b	b
α_3	0	a	2a	3a	2a+b	3a+b	b	a+b
α_4	0	a	2a	3a	3a+b	b	a+b	2a+b
α_5	0	3a	2a	a	b	3a+b	2a+b	a+b
α_6	0	3a	2a	a	a+b	b	3a+b	2a+b
α_7	0	3a	2a	a	2a+b	a+b	b	3a+b
α_8	0	3a	2a	a	3a+b	2a+b	a+b	b
α_9	0	0	0	0	2a	2a	2a	2a
α_{10}	0	0	0	0	b	b	b	b
α_{11}	0	0	0	0	a+b	a+b	a+b	a+b
α_{12}	0	0	0	0	2a+b	2a+b	2a+b	2a+b
α_{13}	0	0	0	0	3a+b	3a+b	3a+b	3a+b
α_{14}	0	2a	0	2a	2a	0	2a	0
α_{15}	0	b	0	b	b	0	b	0
α_{16}	0	a+b	0	a+b	a+b	0	a+b	0
α_{17}	0	2a+b	0	2a+b	2a+b	0	2a+b	0
α_{18}	0	3a+b	0	3a+b	3a+b	0	3a+b	0
α_{19}	0	2a	0	2a	0	2a	0	2a
α_{20}	0	b	0	b	0	b	0	b
α_{21}	0	a+b	0	a+b	0	a+b	0	a+b
α_{22}	0	2a+b	0	2a+b	0	2a+b	0	2a+b
α_{23}	0	3a+b	0	3a+b	0	3a+b	0	3a+b
α_{24}	0	2a	0	2a	b	2a+b	b	2a+b
α_{25}	0	2a	0	2a	a+b	3a+b	a+b	3a+b
α_{26}	0	2a	0	2a	2a+b	b	2a+b	b
α_{27}	0	2a	0	2a	3a+b	a+b	3a+b	a+b
α_{28}	0	a+b	0	a+b	2a	3a+b	2a	3a+b
α_{29}	0	a+b	0	a+b	3a+b	2a	3a+b	2a
α_{30}	0	3a+b	0	3a+b	2a	a+b	2a	a+b
α_{31}	0	3a+b	0	3a+b	a+b	2a	a+b	2a
α_{32}	0	b	0	b	2a	2a+b	2a	2a+b
α_{33}	0	b	0	b	2a+b	2a	2a+b	2a
α_{34}	0	2a+b	0	2a+b	2a	b	2a	b
α_{35}	0	2a+b	0	2a+b	b	2a	b	2a
α_{36}	0	0	0	0	0	0	0	0

+	0	a	2a	3a	b	a+b	2a+b	3a+b
0	0	a	2a	3a	b	a+b	2a+b	3a+b
a	a	2a	3a	0	a+b	2a+b	3a+b	b
2a	2a	3a	0	a	2a+b	3a+b	b	a+b
3a	3a	0	a	2a	3a+b	b	a+b	2a+b
b	b	3a+b	2a+b	a+b	0	3a	2a	a
a+b	a+b	b	3a+b	2a+b	a	0	3a	2a
2a+b	2a+b	a+b	b	3a+b	2a	a	0	3a
3a+b	3a+b	2a+b	a+b	b	3a	2a	a	0

Near-rings with identity I=a on D_8:

1) (36, 1,14, 5,15,16,17,18);1,2,5,6
2) (36, 1,14, 5,15,21,17,23);1,2,5,6
3) (36, 1,14, 7,15,16,17,18);1,2,7,8
4) (36, 1,14, 7,15,21,17,23);1,4,6,7
5) (36, 1,14, 5,15,16,35,18);1,2,3,4,5,6,7,8
6) (36, 1,14, 7,15,16,17,31);1,2,3,4,5,6,7,8
7) (36, 1,14, 7,15,21,17,30);1,2,3,4,5,6,7,8

Non-zero-symmetric near-rings on D_8:

Class A (8 isomorphic copies under $\alpha_1,\alpha_2,\alpha_3,\alpha_4,\alpha_5,\alpha_6,\alpha_7,\alpha_8$)

1)(10, 1, 1, 1,10, 1, 1,10)
2)(10, 1, 1, 1,10, 1,10,10)
3)(10, 1, 1,10,10, 1, 1, 1)
4)(10, 1, 1,10,10, 1, 1,10)
5)(10, 1, 1,10,10, 1,10, 1)
6)(10, 1, 1,10,10, 1,10,10)
7)(10, 1, 1,10,10,10, 1, 1)
8)(10, 1, 1,10,10,10, 1,10)
9)(10, 1, 1,10,10,10,10, 1)
10)(10, 1, 1,10,10,10,10,10)
11)(10, 1,10, 1,10, 1, 1,10)
12)(10, 1,10, 1,10, 1,10,10)
13)(10, 1,10, 5,10, 1,10, 5)
14)(10, 1,10, 5,10, 5,10, 1)
15)(10, 1,10, 5,10,10,10,10)
16)(10, 1,10,10,10, 1, 1, 1)
17)(10, 1,10,10,10, 1, 1,10)
18)(10, 1,10,10,10, 1,10, 1)
19)(10, 1,10,10,10, 1,10,10)
20)(10, 1,10,10,10,10, 1, 1)
21)(10, 1,10,10,10,10, 1,10)
22)(10, 1,10,10,10,10,10, 1)
23)(10, 1,10,10,10,10,10,10)
24)(10, 1,24, 1,10,10,10,24)
25)(10, 1,24, 5,10, 1,25, 5)
26)(10, 1,24, 5,10, 5,24, 1)
27)(10, 1,24, 5,10,10,10,10)
28)(15, 1,15, 1,15, 1,15,15)
29)(15, 1,15, 1,15,15,35,35)
30)(15, 1,15, 5,15, 1,15, 5)
31)(15, 1,15, 5,15, 5,15, 1)
32)(15, 1,15, 5,15,15,15,15)
33)(15, 1,15, 5,15,15,35,15)
34)(10,10,24, 5,10,24,10,24)
35)(10,10, 1,10,10, 1, 1,10)
36)(10,10, 1,10,10, 1,10,10)
37)(10,10,10,10,10, 1, 1,10)
38)(10,10,10,10,10, 1,10, 5)
39)(10,10,10,10,10, 1,10,10)
40)(10,10,10,10,10, 1,24, 5)
41)(10,10,10,10,10,10,10,10)
42)(10,10,10,10,10,10,10,24)
43)(10,10,10,24,10, 1,24, 1)
44)(10,10,10,24,10,10,10,10)
45)(10,10,10,24,10,10,10,24)
46)(10,10,10,24,10,24,10,10)
47)(10,10,10,24,10,24,10,24)
48)(10,24,10,24,10, 1,24, 5)
49)(10,24,10,24,10,10,10,24)
50)(15, 1, 1, 1,15, 1, 1,15)
51)(15, 1, 1, 1,15, 1,15,15)
52)(15, 1, 1,15,15, 1, 1, 1)
53)(15, 1, 1,15,15, 1, 1,15)
54)(15, 1, 1,15,15, 1,15, 1)
55)(15, 1, 1,15,15, 1,15,15)
56)(15, 1, 1,15,15,15, 1, 1)
57)(15, 1, 1,15,15,15, 1,15)
58)(15, 1, 1,15,15,15,15, 1)
59)(15, 1, 1,15,15,15,15,15)
60)(15, 1,15, 1,15, 1, 1,15)
61)(15, 1,35, 5,15, 5,35, 1)
62)(15,15, 1,15,15, 1, 1,15)
63)(15,15, 1,15,15, 1,15,15)
64)(15,15,15,15,15, 1, 1,15)

65)(15,15,15,15,15, 1,15,15)
66)(15,15,15,15,15, 1,15,15)
67)(15, 1,15, 5,15,35,35,35)
68)(15, 1,15,15,15, 1, 1, 1)
69)(15, 1,15,15,15, 1, 1,15)
70)(15, 1,15,15,15, 1,15, 1)
71)(15, 1,15,15,15, 1,15,15)
72)(15, 1,15,15,15,15, 1, 1)
73)(15, 1,15,15,15,15, 1,15)
74)(15, 1,15,15,15,15,15, 1)
75)(15, 1,15,15,15,15,15,15)
76)(15, 1,35, 5,15, 1,35, 5)
77)(15,15,15,15,15,15,15,35)
78)(15,15,15,35,15,15,15,15)
79)(15,15,15,35,15,15,15,35)
80)(15,15,15,35,15,35,15,15)
81)(15,15,15,35,15,35,15,35)
82)(15,15,35,15,15, 1,15, 5)
83)(15,15,35,35,15, 1,15, 1)
84)(15,35,15,35,15,15,15,35)
85)(15,35,35,35,15, 1,15, 5)

Class B (4 isomorphic copies)

86)(10, 1, 1, 1,10, 1, 1, 1)
87)(10, 1, 1, 1,10, 1,10, 1)
88)(10, 1, 1, 1,10,10, 1,10)
89)(10, 1, 1, 1,10,10,10,10)
90)(10, 1,10, 1,10, 1, 1, 1)
91)(10, 1,10, 1,10, 1,10, 1)
92)(10, 1,10, 1,10,10, 1,10)
93)(10, 1,10, 1,10,10,10,10)
94)(10, 1,24, 1,10, 1,24, 1)
95)(10, 1,24, 1,10,10,10,10)
96)(10,24,10,24,10, 1,24, 1)
97)(10,24,10,24,10,10,10,10)
98)(10,24,10,24,10,24,10,24)
99)(15, 1, 1, 1,15, 1, 1, 1)
100)(15, 1, 1, 1,15, 1,15, 1)
101)(15, 1, 1, 1,15,15, 1,15)
102)(15, 1, 1, 1,15,15,15,15)
103)(15, 1,15, 1,15, 1, 1, 1)
104)(15, 1,15, 1,15, 1,15, 1)
105)(15, 1,15, 1,15,15, 1,15)
106)(15, 1,15, 1,15,15,15,15)
107)(15, 1,15, 1,15,15,35,15)
108)(15, 1,15, 1,15,35,35,35)
109)(15, 1,35, 1,15, 1,35, 1)
110)(10, 1,24, 1,10,24,10,24)
111)(10,10, 1,10,10, 1, 1, 1)
112)(10,10, 1,10,10, 1,10, 1)
113)(10,10, 1,10,10,10, 1,10)
114)(10,10, 1,10,10,10,10,10)
115)(10,10,10,10,10, 1, 1, 1)
116)(10,10,10,10,10, 1,10, 1)
117)(10,10,10,10,10, 1,24, 1)
118)(10,10,10,10,10,10, 1,10)
119)(10,10,10,10,10,24,10,24)
120)(15,15, 1,15,15, 1, 1, 1)
121)(15,15, 1,15,15, 1,15, 1)
122)(15,15, 1,15,15,15, 1,15)
123)(15,15, 1,15,15,15,15,15)
124)(15,15,15,15,15, 1, 1, 1)
125)(15,15,15,15,15, 1,15, 1)
126)(15,15,15,15,15,15, 1,15)
127)(15,15,15,15,15,15,15,15)
128)(15,15,15,15,15,35,15,35)
129)(15,15,35,15,15, 1,15, 1)
130)(15,35,15,35,15,15,15,15)
131)(15,35,15,35,15,35,15,35)
132)(15,35,15,35,15, 1,15, 1)

Integral near-rings on D_8:

1)(36, 1, 1, 1, 1, 1, 1, 1)
2)(1, 1, 1, 1, 1, 1, 1, 1)

Boolean near-rings on D_8:

1) (10, 1, 1, 1,10, 1, 1, 1); 1,2,3,4
2) (15,11, 1, 1,15, 1, 1, 1); 1,2,3,4
3) (36, 1, 1, 1, 1, 1, 1, 1); 1
4) (1, 1, 1, 1, 1, 1, 1, 1); 1

L) The quaternion group Q = {0,a,2a,3a,b,a+b,2a+b,3a+b}:

+	0	a	2a	3a	b	a+b	2a+b	3a+b
0	0	a	2a	3a	b	a+b	2a+b	3a+b
a	a	2a	3a	0	a+b	2a+b	3a+b	b
2a	2a	3a	0	a	2a+b	3a+b	b	a+b
3a	3a	0	a	2a	3a+b	b	a+b	2a+b
b	b	3a+b	2a+b	a+b	2a	a	0	3a
a+b	a+b	b	3a+b	2a+b	3a	2a	a	0
2a+b	2a+b	a+b	b	3a+b	0	3a	2a	a
3a+b	3a+b	2a+b	a+b	b	a	0	3a	2a

	0	a	2a	3a	b	a+b	2a+b	3a+b
α_1	0	a	2a	3a	b	a+b	2a+b	3a+b
α_2	0	a	2a	3a	2a+b	3a+b	b	a+b
α_3	0	a	2a	3a	a+b	2a+b	3a+b	b
α_4	0	a	2a	3a	3a+b	b	a+b	2a+b
α_5	0	3a	2a	a	b	3a+b	2a+b	a+b
α_6	0	3a	2a	a	2a+b	a+b	b	3a+b
α_7	0	3a	2a	a	a+b	b	3a+b	2a+b
α_8	0	3a	2a	a	3a+b	2a+b	a+b	b
α_9	0	b	2a	2a+b	a	3a+b	3a	a+b
α_{10}	0	b	2a	2a+b	3a	a+b	a	3a+b
α_{11}	0	b	2a	2a+b	a+b	a	3a+b	3a
α_{12}	0	b	2a	2a+b	3a+b	3a	a+b	a
α_{13}	0	2a+b	2a	b	a	a+b	3a	3a+b
α_{14}	0	2a+b	2a	b	3a	3a+b	a	a+b
α_{15}	0	2a+b	2a	b	a+b	3a	3a+b	a
α_{16}	0	2a+b	2a	b	3a+b	a	a+b	3a
α_{17}	0	a+b	2a	3a+b	a	b	3a	2a+b
α_{18}	0	a+b	2a	3a+b	3a	2a+b	a	b
α_{19}	0	a+b	2a	3a+b	b	3a	2a+b	a
α_{20}	0	a+b	2a	3a+b	2a+b	a	b	3a
α_{21}	0	3a+b	2a	a+b	a	2a+b	3a	b
α_{22}	0	3a+b	2a	a+b	3a	b	a	2a+b
α_{23}	0	3a+b	2a	a+b	b	a	2a+b	3a
α_{24}	0	3a+b	2a	a+b	2a+b	3a	b	a
α_{25}	0	0	0	0	2a	2a	2a	2a
α_{26}	0	2a	0	2a	0	2a	0	2a
α_{27}	0	2a	0	2a	2a	0	2a	0
α_{28}	0	0	0	0	0	0	0	0

Near-rings with identities on Q:

According to 9.116 (c), there are no near-rings of this kind.

<u>Non-zero-symmetric near-rings on **Q**:</u>

Only one: (1,1,1,1,1,1,1,1)

<u>Integral near-rings on **Q**:</u>

1) (28,1,1,1,1,1,1,1) 2) (1,1,1,1,1,1,1,1)

<u>Boolean near-rings on **Q**:</u>

1) (28,1,1,1,1,1,1,1) 2) (1,1,1,1,1,1,1,1).

LIST OF OPEN PROBLEMS

1) Generally, determine the structure of our special classes of near-rings (radicals, complete reducibility, semisimplicity, primitivity,...). For instance, what can one say about the radicals of planar near-rings?
2) Are all restrictions to zero-symmetric near-rings in this book really necessary?
2) Study measure and integration in near-rings (this is motivated by the $M_{cont}(\Gamma)$-type near-rings).
3) Do 2.63, 2.85, 5.54 and 5.62 hold for arbitrary ideals?
4) Is $\bar{M}_0(\Gamma)$ a near-ring if it contains a subnear-ring of $M(\Gamma)$ which is dense in $\bar{M}_0(\Gamma)$?
5) Is $\mathcal{J}_2$ always hereditary (cf. 5.18 and 5.67(s))?
6) Which radicals $\mathcal{J}_\nu(N)$ are "idempotent" (Holcombe) in the sense that $\mathcal{J}_\nu(\mathcal{J}_\nu(N) = \mathcal{J}_\nu(N)$?
7) In 5.20: does equality hold in general?
8) Recall the 4 problems on page 178.
9) Determine the ideal structure in the polynomial near-rings. Which ones have the DCCL, DCCN, et cetera ?
10) Do there exist proper near-domains (i.e. those ones which are no near-fields - cf. p. 247) ?
11) Find some examples of infinite near-fields which are not Dickson near-fields.
12) Is 9.21 (a) correct without the assumption that N has an identity (Bell) ?
13) Study lattice-ordered near-rings, F-near-rings and vector-near-rings (cf. p. 329).
14) How can one characterize those near-rings which can be fully ordered ?
15) Which (partial) orders in a near-ring can be extended to a full order (cf. (Fuchs)) ?
16) Is 9.133 true without the assumption that N is abelian ?
17) Compute the radicals of our near-rings of low order.

BIBLIOGRAPHY

PLEASE NOTE:

The classification scheme for the papers is as follows:

Near-rings: A... Additive groups of near-rings, near-rings on given groups
A'.. Affine near-rings
B... Boolean near-rings and generalisations (p-near-rings, IFP-near-rings,...)
C... Constructions (Sums and products, subdirect products,...)
D... Distributively generated near-rings
D'.. Distributors, distributive elements and near-rings, commutators, solvability
D".. Dickson near-rings
E... Elementary, examples, axiomatics, chain conditions, lattice of ideals,...
E'.. Embeddings
E".. Endomorphism near-rings ($E(\Gamma)$, $A(\Gamma)$, $I(\Gamma)$)
F... Near-fields
F'.. Free near-rings and N-groups
G... Geometric interpretations (coordinatisation, incidence groups,...)
H... Homological and categorical aspects, extensions
I... Idempotents, biregular near-rings
I'.. Integral near-rings, near-integral domains and generalisations
L... Local near-rings
M... Modularity
N... Nilpotence and non-nilpotence
O... Ordered near-rings and near-fields
P... Primitive near-rings, N-groups of type ν
P'.. Prime (semiprime, completely prime,...) ideals
P".. Planarity
P^o.. Polynomial near-rings, near-rings of formal power series
Q... Quasi-regularity
Q'.. Near-rings of quotients
R... Radical theory
R'.. Regular near-rings
S... Simplicity and semisimplicity
S'.. Sylow-type topics
S".. Relations to sharply transitive groups
T... Transformation near-rings ($M(\Gamma)$, $M_o(\Gamma)$, $M_G(\Gamma)$)
T'.. Topological considerations
V... Valuations
W... Near-rings without nilpotent elements
X... Other topics

Structures related to near-rings:
Cr.. Composition rings (TO-Algebras)
Na.. Near-algebras
Nd.. Near-domains (in the sense of "non-associative near-fields")
Rs.. Other related structures (seminear-rings,...)
Ua.. Universal algebraic context

Combined classifications give more information on the paper; for instance:
P",F... Planar near-fields or
D',R... Radical theory for distributively generated near-rings

ADAMS, William B., 235 Elsinora Str. at Concorde, Massachusetts 01742, USA

1. *Near-integral domains and fixed-point-free automorphisms*, Doctoral Diss.,Boston Univ.,Boston,Mass.,USA — A,I
2. *Near-integral domains on non-abelian groups*, submitted. — A,I
3. *Near-integral domains on finite abelian groups*,submitted. — A,I

ADLER, Irving, RFD, North Bennington, Vermont 05257, USA

1. *Composition rings*, Duke Math.J.29 (1962), 607-625. — Cr,H

AIJAZ,Kulsoom,Univ.of Islamabad, Pakistan
SEE AIJAZ-HUQ

AIJAZ,Kulsoom and HUQ,S.A.

1. *Categorical investigation of Γ-graded Λ-algebras*, Portugaliae Math.28 (1969), 21-36 — H

ANDRE, Johannes, Fachber. Math.,Univ.d.Saarld., 6600 Saarbrücken, Germany

1. *Projektive Ebenen über Fastkörpern*, Math.Z.62 (1955), 137-160. — F,G,Rs
2. *Über eine Beziehung zwischen Zentrum und Kern endlicher Fastkörper*, Arch.Math. 14 (1963), 145-146. — F
3. *Über unvollständige Fastkörper und verallgemeinerte affine Räume*, Math.Z.119, (1971), 254-266. — F,G,P,Rs
4. *Lineare Algebra über Fastkörpern*, Math. Z. 136 (1974),295-313. — F,P", X
5. *On finite non-commutative affine spaces*, Math. Centre Tracts 55 (1974) 60-107. — F,G
6. *Affine Geometrien über Fastkörpern*, Mitteilungen aus dem Mathem. Seminar Gießen 114 (1975), 1-99. — F,G
7. *Bemerkungen über Fastvektorräume*, FU Berlin, Lenz-Festband (1976), 28-36. — F,G,X
8. *Some topics on linear algebra over near-fields*, Oberwolfach 1976 — F,P",X

ANSHEL, Michael, 1140 5th Ave.New York, N.Y. 10028, USA
SEE ANSHEL- CLAY

ANSHEL,Michael and CLAY, James R.

1. *Planarity in algebraic systems*, Bull.Amer.Math.Soc. 74 (1968), 746-748 — P",G,I',A,E
2. *Planar algebraic systems, some geometric interpretations*, J.Algebra 10 (1968),166-173. — P",G,I',A,G

ARMENTROUT, Nancy, Dept.Math.Texas A&M Univ.,College Station,Texas 77843,USA

1. *On near-rings associated with generalized affine planes*, M.A.Thesis,Texas A&M 1971. — G,L

SEE ALSO ARMENTROUT-HARDY-MAXSON.

ARMENTROUT, Nancy,Hardy,F. LANE and MAXSON, C.J.

1. *On generalized affine planes*, J.Geometry 4 (1974), 143-159 — G,L

ARNOLD, Hans Joachim, Math.Sem.,GHS Duisburg, Lotharstr.65,41 Duisburg,Germany

1. *Algebraische und geometrische Kennzeichnung der schwach affinen Vektorräume über Fastkörpern*, Abh.Math.Sem.Univ.Hamburg, 32 (1968), 73-88 — F,G

BACHMANN, O.

1. *über eine Klasse verallgemeinerter affiner Räume*, Monatsh. Math. 79 (1975), 285-297. — F,G
2. *über die Unterräume von Fastvektorräumen*, manuscript. — F,X

BANASCHEWSKI, B., Dept. Math. McMasters Univ., Hamilton, Ont., Canada

SEE BANASCHEWSKI-NELSON

BANASCHEWSKI, B. and NELSON, E.

1. *On the non-existence of injective near-ring modules*, to appear — D,H

BASKARAN, S.

1. *Remarks on a paper of S. Ligh's (Monatsh. Math. 76 (1972),317-322)*, Math. Student 42 (1974), 351-352. — I',A

BEAUMONT,Ross A., Dept.Math.Univ. of Washington, Seattle, Wash. 98195, USA.

1. *Generalized rings*, Proc.Amer.Math.Soc. 9 (1958), 876-880. — Rs,E

BEIDLEMAN, James C. Dept.Math.Univ. of Kentucky, Lexington, Kentucky 40506, USA.

1. *On near-rings and near-ring modules*, Doctoral dissertation, Pensylv. State University, 1964. — E,D,E",F I,M,N,P Q,R,S,X
2. *Quasi-regularity in near-rings*, Math.Z. 89 (1965), 224-229. — Q,R,E,D,N
3. *A radical for near-ring modules*, Michigan Math.J. 12(1965),377-383 — D,R,S,N
4. *Distributively generated near-rings with descending chain condition*, Math.Z. 91 (1966), 65-69. — E,D,D"
5. *On groups and their near-rings of functions*, Amer.Math.Monthly 73 (1966), 981-983. — T,E
6. *Nonsemi-simple distributively generated near-rings with minimum condition*, Math.Ann. 170 (1967), 206-213. — D,N,I,R
7. *Strictly prime distributively generated near-rings*, Math.Z.100 (1967), 97-105. — P',D,P,E" M
8. *On the theory of radicals in d.g. near-rings I. The primitive radical*, Math.Ann. 173 (1967), 89-101. — R,D,P,D' N,E
9. *On the theory of radicals in d.g. near-rings II. The nil radical*, Math.Ann. 173 (1967), 200-218. — D,N,R,Q, E'
10. *A note on regular near-rings*, J.Indian Math.Soc. 33 (1969), 207-210. — R',N,I,I' E',F'
11. *On the additive group of a finite near-ring*, Indian J.Math. 12 (1970), 95-106. — A,D',D,P,R

SEE ALSO BEIDLEMAN-COX.

BEIDLEMAN, James C. and COX, R.H.

1. *Topological near-rings*, Arch.Math. (Basel) 18 (1967), 485-492. — T',Q,R,N

BELL, Howard E., Math.Dept.Brock Univ., St.Catharines,Ontario,Canada.

1. *Near-rings in which each element is a power of itself*, Bull.Austral. Math.Soc. 2 (1970), 363-368. | B,A,D,I' W,P'
2. *Certain near-rings are rings*, J.London Math.Soc.II Ser.4(1971), 264-270. | B,D
3. *Infinite subrings of infinite rings and near-rings*, Pacific J. Math. 59 (1975), 345-358. | D'
4. *Certain near-rings are rings II*, to appear. | B,I',D
5. *A commutativity theorem for near-rings*, to appear. | B,I',D
6. *Commutativity theorems for distributively generated near-rings*, Oberwolfach 1976. | B,I',D

SEE ALSO BELL-LIGH

BELL, Howard E. and LIGH, Steve

1. *On finiteness conditions for near-rings*, Publ. Math. Debrecen 22 (1975), 35-40. | D,W,E,X

BENZ, Walter, Fac. Math. Univ. Waterloo, Waterloo, Ont., Canada

1. *Vorlesungen über Geometrie der Algebren*, Springer Verl., Berlin-Heidelberg-New York 1973. | G,S"

BERMAN,Gerald, Combin-Optimization Dept.,Univ.Waterloo,Waterloo,Ontario,Can.

SEE BERMAN-SILVERMAN.

BERMAN, Gerald and SILVERMAN, Robert J.

1. *Near-rings*, Amer.Math.Monthly 66 (1959), 23-34 | E,I,E'
2. *Simplicity of near-rings of transformations*, Proc.Amer.Math.Soc. 10 (1959), 456-459. | T,S
3. *Embedding of algebraic systems*, Pacific J.Math. 60 (1960),777-786. | E',Ua

BERTANI, Laura, Istituto di Matematica, Università, 43100 Parma, Italy

1. *Costruzione di desegni regolari*, Riv. Mat. Univ. Parma (3) 1 (1972) | Rs,G,P"
2. *Costruzione di spazi di Steiner regolari*, Boll. della Unione Matematica Italiana (4) 11 (1975), 370-374. | Rs,G,P"

BETSCH, Gerhard,Math.Inst.Univ.Tübingen, Auf der Morgenstelle 10, D-7400 Tübingen, Germany.

1. *Fastringe*, Zulassungsarbeit, 1959. | E,F,D,S,
2. *Ein Radikal für Fastringe*, Math.Z. 78 (1962), 86-90. | R,F,S
3. *Struktursätze für Fastringe*,Diss. Univ.Tübingen,1963 | E,P,R,S, I,N,T,M
4. *Ein Satz über 2-primitive Fastringe*, Oberwolfach, 1968 | P,T
5. *Primitive near-rings*, Math.Z. 130 (1973), 351-361. | P,T,E'
6. *Some structure theorems on 2-primitive near-rings*, Colloquia Mathematica Societatis Janus Bolyai 6,Rings, modules, and radicals,Keszthely (Hungary), 1971. (North Holland 1973, 73-102). | P,T,I,D'
7. *Sheaf representation of near-rings*, Oberwolfach, 1972. | X
8. *Near-rings of group mappings*, Oberwolfach 1976. | T

SEE ALSO BETSCH-CLAY

BETSCH, Gerhard and CLAY, James R.

1. *Block designs from Frobenius groups and planar near-rings*, Proc. Conf. finite groups (Park City, Utah), Acad. Press 1976, 473-502. | P"

BLACKETT, Donald W., Math.Dept., Coll. of Lib.Arts, Boston Univ., Charles River Campus, Boston, Mass. 02215, USA

1. *Simple and semi-simple near-rings*, Doctoral dissertation, Princeton University, 1950. — S,I,P
2. *Simple and semi-simple near-rings*, Proc.Amer.Math.Soc. 4 (1953), 772-785. — S,I,P
3. *The near-ring of affine transformations*, Proc.Amer.Math.Soc. 7 (1956), 517-519. — A'
4. *Simple near-rings of differentiable transformations*, Proc.Amer.Math. Soc. 7 (1956), 599-606. — E,S,T'
5. *A countable near-ring dense in the near-ring of continous transformations* $R^n R$, Research Report, Dept. Math.Boston Univ., 1971 — E,T'
6. *Some near-rings dense in the near-ring of continous maps* $R^n \to R$, Research Report, Dept. Math., Boston University, 1972. — E,T'

BOCCIONI, Domenico, Via Ospedale 3, 35100 Padova, Italy

1. *Indipendenza delle condizioni di distributività*, Rend.Sem.Mat. Univ. Padova 28 (1958), 1-30. — R_s,E
2. *Indipendenza delle condizione mutua distributività*, Rend.Sem.Mat. Univ.Padova 28 (1958), 40-49. — Rs,E
3. *Dipendenza delle condizioni di mutua distributività nei bisistemi di ordine 3*, Rend. Sem. Mat. Univ. Padova 28 (1958), 50-67. — R_s,E
4. *Condizioni di distributività ed associadività unilaterali*, Rend.Sem. Mat. Univ. Padova 30 (1960), 178-193. — Rs,E
5. *Condizioni di distributività con almeno uno operazione commutativita*, Rend. Sem. Mat. Univ. Padova 31(1961), 87-103. — Rs,E
6. *Condizioni di autodistributività*, Rend. Sem. Mat. Univ. Padova 31 (1961), 171-197. — Rs,E
7. *Indipendenza delle condizioni di doppia e di tripla distributività*, Rend. Sem. Mat. Univ. Padova 32 (1962). — Rs,E
8. *Condizioni di mutua distributività con ripetizioni*, Rend. Sem. Mat. Univ. Padova 32 (1962). — Rs,E
9. *Condizioni indipendenti ed equivalenti a quelle di mutua distributività*, Rend. Sem. Mat. Univ. Padova 32 (1962). — Rs,E
10. *Simmetrizzazione di una operazione n-aria*, Rend. Sem. Mat. Univ. Padova 35 (1965), 82-106. — Rs,E
11. *Caratterizzazione di una classe di anelli generalizzati*, Rend. Sem. Mat. Univ. Padova 35 (1965), 116-127. — Rs,E

BOTERO DE MEZA, M.M., Dept. de Mat. Univ. de los Andes, Bogota, Columbia
SEE BOTERO DE MEZA-WEINERT

BOTERO DE MEZA, M.M. and WEINERT, H.J.

1. *Erweiterungen topologischer Halbringe durch Quotienten- und Differenzenbildung*, Jahresber. Deutsch. Math. Ver. 73 (1971), 60-85. — Rs,Q',T'

BRENNER, Joel L., 10 Phillips Rd.,Palo Alto, Calif. 94303, USA

1. *Maximal ideals in the near-ring of polynomials mod 2*, Pacific J.Math. 52 (1974), 595-600. — P^0

BROWN, Ferdinand L., Notre Dame, Adm. Bldg., Office 202, Notre Dame, Indiana, 46556, USA.

1. *Remarks concerning tri-operational algebra*, Reports of a Math. Colloq., Issue 5-6, Notre Dame (1944), 11-15. | Cr,E
2. *The accessory postulates of tri-operational algebra*, Reports of a Math. Colloqu. Issue 7, Notre Dame (1946), 61-67. | Cr,E
3. *Remaks concerning tri-operational algebra III*, Rep. Math. Coll. (2), Issue 8 (1948), 61-67. | Cr,E

BROWN, Harold David, Serre House, Comp. Science Dpt., Stanford Univ., Stanford, Calif. 94305.

1. *Near algebras*, Doctoral dissertation, Ohio State University, 1966. | Na, D',S,T
2. *Near algebras*, Illinois J.Math. 12 (1968), 215-227. | Na, D',S,T
3. *Distributor theory in near algebras*, Comm. Pure App. Math. 21 (1968), 535-544. | Na, D',I,C

BRUCK, R.H., 642 Inner Drive, Madison, Wisconsin 53705, USA.

1. *A survey of binary systems*, Berlin/Göttingen/Heidelberg, 1958. | Rs,E

BURKE, John C.

1. *Remarks concerning tri-operational algebra*, Report of a Math. Colloqu., Issue 7, Notre Dame (1946), 68-72. | Cr,E

BURN, R.P.

1. *Doubly transitive sets of permutations characterizing projective planes*. Geometriae dedicta 2 (1973), 57-63. | F,G

CARMICHAEL, R.D.

1. *Introduction to the theory of groups of finite order*, Abdruck der 1.Auflage, 1956, 396-402. | F,D"

CARTAN, H.

1. *Theory of analytic functions*, Addison-Wesley, Reading, Massachusetts, 1963, 9-16. | P°

CHAN, G.H. Dept.Math.Nanyang Univ. Singapore 22, Singapore
SEE CHAN-CHEW.

CHAN, G.H. and CHEW, K.L.

1. *On extensions of near-rings*, Nanta Math. 5 (1971), 12-21 | Q',E'

CHANDY, Attupurathuvadakkethil J., 1289 Drift Road, Westport, Mass., 02790, USA.

1. *Rings generated by inner automorphisms of non-abelian groups*, Doctoral dissertation, Boston Univ., 1965. | E"
2. *Near-rings generated by the inner automorphisms of L-groups*, submitted. | E"
3. *Rings generated by inner automorphisms of non-abelian groups*. Proc. Amer. Math. Soc. 30 (1971), 59-60. | E"

CHEW, Kim L., Nanyang Univ. Library, Singapore 22, Singapore
SEE CHAN-CHEW.

CHOUDHARI, S.C.

1. *On near-rings and near-ring modules*, Diss. Indian Inst. of Technology, Kanpur, India (1972). E,B,M,N,P,F Q,R,R',S,X

SEE ALSO CHOUDHARI-TEWARI.

CHOUDHARI, S.C. and TEWARI, K.

1. *On strictly semisimple near-rings*, Abh.Math.Sem.Univ.Hamburg 40 (1974), 256-264. S,P
2. *N.I.-bounded property in near-rings*, to appear.
3. *G-radical in near-rings*, to appear. R,Q,S,M

CLARK, John F., Jr.

1. *Rings associated with the rings of endomorphisms of finite groups*, J. Washington Acad. Sci. 40 (1950), 385-397. E,T

CLAY, James R. Dept.Math. Univ.Arizona, Tucson, Arizona, 85721, USA.

1. *The near-rings on a finite cylic group*, Amer. Math.Monthly 71 (1964), 47-50. A
2. *The near-rings definable on an arbitrary group and the group of left distributive multiplications definable on an abelian group*, Doctoral dissertation, University of Washington, 1966. E,A
3. *Imbedding an arbitrary ring in a non-trivial near-ring*, Amer.Math. Monthly 74 (1967), 406-407. E'
4. *The near-rings on groups of low order*, Math.Z. 104 (1968), 364-371. E
5. *Some geometric interpretations of planar near-rings*, Oberwolfach, 1968 P",G
6. *A note on integral domains that are not right distributive*, Elem. Math. 24 (1969), 40-41. I'
7. *The group of left distributive multiplications on an abelian group*, Acta Math. Sci.Hungar. 19 (1968), 221-227. A,E
8. *Research in near-rings using a digital computer*, Bit. 10, (1970), 249 - 265. A,E,I',B
9. *The near-rings on the cyclic group of order 8*; manuscript A
10. *Generating balanced incomplete block designs from planar near-rings*, J. Algebra 22 (1972), 319-331. P",I',A
11. *Some algebraic aspects of planarity*, Atti del Convegno di Geometrica Combinatoria e sue applicationi, Univ. degli studi, Perugia (1971), 163-172. P",I',G
12. *Generating balanced incomplete block designs from planar near-rings*, Oberwolfach 1972. P",A
13. *The structure of dilatation groups of generalized affine planes*, Journal of Geometry 6 (1975), 1-19. G,E,Q'
14. *The group of units of* $M_G(\Gamma)$, Oberwolfach 1976. T,X

SEE ALSO ANSHEL-CLAY, BETSCH-CLAY, CLAY-DOI, CLAY-LAWVER, CLAY-MALONE, CLAY-MAXSON.

CLAY, James R. and DOI, Donna K.

1. *Near-rings with identity on alternating groups*, Math. Scand. 23 (1968), 54-56. A
2. *Maximal ideals in the near-ring of polynomials over a field*, Colloqu. Math. Soc. Janus Bolyai 6 , Rings, Modules and Radicals, Keszthely (Hungary) 1971, North-Holland 1973, 117-133. P^0,S,R,G

CLAY, James R. and LAWVER, Donald A.

1. *Boolean near-rings*, Canad. Math. Bull. 12 (1969), 265-273. B

CLAY, James R. and MALONE, Joseph J.

1. *The near-rings with identities on certain finite groups*, Math. Scand. 19 (1966), 146-150. A

CLAY, James R. and MAXSON, Carlton J.

1. *The near-rings with identities on generalized quaternion groups*, Istituto Lombardo, Accademia di Science e lettere (A) 104 (1970) 525-530. A

COOPER, Charles

1. *Some properties of near-rings*, M.S. Thesis, McNeese State Univ.,1974 Rs

COX, R.H., Math. Dept. Univ. of Kentucky, Lexington, Kentucky 40506, USA
SEE BEIDLEMAN-COX.

CURJEL, Caspar R., Math. Dept. Univ. Washington, Seattle, Wash. 98195, USA

1. *On the homology decomposition of polyhedra*, Illinois J. Math. 7 (1963), 121-136. H
2. *Near-rings of homotopy classes*, manuscript. H,R,Q,N

DANCS-GROVES, Susan, Dept. Math. Inst. of Advanced Studies, Austral. Nat. Univ., Canberra A.C.T., 2600 P.O.Box 4, Australia

1. *The subnear-field structure of finite near-fields*, Bull. Austral. Math. Soc. 5 (1971), 275-280. F,D"
2. *On finite Dickson near-fields*, Abh. Math. Sem. Univ. Hamburg 37 (1972), 254-257. F,D"
3. *Locally finite near-fields*, Doctoral Diss. , Austral. National Univ., Canberra 1974. F,D"
4. *Locally finite near-fields*, Bull. Austral. Math. Soc. 11 (1974), 319-320. F,D"

DAVIS, Elwyn Herbert, 605 Oakcrest Drive, Pittsburgh, Kansas 66762, USA.

1. *Non-planar near-fiels*, Dissertation, University of Missouri at Columbia, 1969. F,G,D",P"
2. *Incidence systems associated with non-planar near-fields*, Canad. J.Math. 22 (1970) 939-952. F,G,D",P"
3. *A homomorphism of a pseudo-plane onto a projective plane*, Rocky Mountain J. Math. 3 (1973), 515-519. F,G

DEAN, Burton Victor, Operations Research Dept., Case Western Reserve,Cleveland, Ohio 44106, USA.

1. *Near-rings and their isotopes*,Doctoral dissertation, University of Illinois, 1952. X

DEMBOWSKI, Peter

1. *Finite geometries*, Springer 1968 (Ergebnisse der Mathematik,vol.44) F,G

DESKINS, Wilbur E., Dept.Math.Univ. Pittsburgh, Pennsylv. 15213,USA .

1. *A radical for near-rings*, Proc.Amer.Math.Soc. 5 (1954), 825-827. R,S
2. *A note on the system generated by a set of endomorphisms of a group*, Michigan Math. J. 6 (1959), 45-49.

DICKSON, Leonard E.

1. *Definitions of a group and a field by independent postulates*, Trans. Amer. Math Soc. 6 (1905), 198-204. — E,F,D"
2. *On finite algebras*, Nachr. Akad. Wiss. Göttingen (1905), 358-393. — E,F,D"

DOI, Donna K. (Watkins)

1. *Near-rings with identities on alternating groups and ideals in various near-rings*, Honors Thesis, University of Arizona, 1969. — E,A

SEE ALSO CLAY-DOI.

DOVER, Ronald E., Dept. Math. Lubbock Christian College 5601 W.19th Str. Lubbock, Texas 79407, USA.

1. *Quasi-rings*, M.S. Thesis, University of Texas at Arlington, 1968.

DU, Bau-Sen

1. *On regular near-rings*, Thesis, National Tsing Hua Univ. Taiwan, 1974 — R'

ELLERS, Erich W., Math. Dept. Univ. Toronto, Toronto 181, Ontario, Canada.

1. *Kommutative Inzidenzgruppen vom Exponenten n*, Abh. Math. Sem. Univ. Hamburg 31 (1967), 247-252. — F,G
2. *Eine Bemerkung über zweiseitige Inzidenzgruppen*, Abh. Math. Sem. Univ. Hamburg 33 (1969), 1-3. — F,G

SEE ALSO ELLERS-KARZEL.

ELLERS, Erich and KARZEL, Helmut

1. *Endliche Inzidenzgruppen*, Abh. Math. Sem. Univ. Hamburg 27 (1964), 250-264. — F,D",G

ESCH, Linda Sue, Math. Dept., Boston Univ., Charles River Campus, Boston, Mass. 02215, USA.

1. *Commutator and distributor theory in near-rings*, Doctoral dissertation, Boston University, 1974. — D'

EVANS Trevor, Math. Dept. Emory Univ. Atlanta, Georgia 30322, USA

SEE EVANS-NEFF

EVANS, Trevor and NEFF, M.F.

1. *Substitution algebras and near-rings I*, Notices Amer. Math. Soc. 11 (1964), November. — E

FAIN, Charles Gilbert, 1020 Aponi Rd., Vienna, Virginia 22180, USA

1. *Some structure theorems for near-rings*, Doctoral Dissertation, University of Oklahoma, 1968. — P,R,S,C,I E,F,M,N

FENZEL, William F.

1. *Regular near-rings*, M.S. Thesis, University of South Carolina, 1973. — R

FERRERO, Giovanni, Istituto Mat., Università, 43100 Parma, Italy.

1. *Sulla struttura aritmetica dei quasi-anelli finiti*, Atti Accad. Scienze Torino 97 (1963), 1-17. D',S'
2. *Sui problemi "tipo Sylow" relativi ai quasi anelli finiti*, Atti Accad. Scienze Torino 100 (1966), 643-657. S'
3. *Due generalizzazioni del concetto di anello e loro equivalenza nell' ambito degli "stems" finiti*, Riv.Mat.Univ. Parma 7 (1966),145-150. A,Rs
4. *Struttura degli "stems" p-singolari*, Riv.Mat.Univ. Parma 7 (1966), 243-254. S',S,A
5. *Classificazione e costruzione degli stems p-singolari*,Ist.Lombardo Accad. Sci. Lett. Rend. A. 102 (1968), 597-613. S',R,S
6. *Quasi anelli aritmeticamente notevoli*, Oberwolfach, 1968. S',D'
7. *Gli stems p-singolari con radicale proprio*,Ist. Lombardo Accad.Sci. Lett.A 104 (1970), 91-105. S',R
8. *Stems planari e BIB-disegni*, Riv.Math.Univ.Parma (2)11(1970),79-96 P",I',A
9. *Sui moltiplicatori (nel senso di Hall) e sui disegni ricchi di moltiplicatori*, Atti.Conv.Geo.Comb.Appl.,Perugia (1970), 233-237. P"
10. *Qualche disegno geometrico*, Le Matematiche (Catania), 26 (1971), 356-377. P",S'
11. *Applicazioni geometriche degli stems planari*, Oberwolfach, 1972. P"
12. *Su certe geometrie gruppali naturali*, Riv.Mat.Univ.Parma (3) 1 (1972) 97-111. P"
13. *Su una classe di nuovi disegni*, Ist. Lombardo Accad.Sci.Lett.Rend. A 106 (1972), 419-430. P"
14. *Osservazioni sugli elementi di prima categoria di un gruppo*, Riv. X,P"
15. *Deformazioni, raffinamenti e composizioni di funzioni di Steiner (I)*, Riv. Mat. Univ. Parma (3) 1 (1972). Rs,G,P"
16. *Gruppi di Steiner e sistemi finiti*, Le Matematiche 27 (1972), Fasc.1. Rs,G,P"
17. *Sui gruppi che ammettono funzioni di Steiner*, Rend. Ist. di Matem. Univ. Trieste 4 (1972), Fasc. II. Rs,G,P"
18. *Sul radicale degli stems p-singolari*, Atti Accad. Sci. Torino Cl. Sci. Fis. Mat. Natur. 107 (1973), 349-369. S',R
19. *Sul gruppo additivo di uno stem p-singolare*, Atti Accad. Sci. Torino Cl. Sci. Fis. Mat. Natur. 108 (1973/74), I:353-366; II:689-697. A,S'
20. *Su un problema relativo ai sistemi di Steiner disgiunti*, Rend Ist. di Mat. Univ. Trieste 7 (1975), Fasc. I. Rs,G,P"
21. *On a geometrical interpretation of distributivity*,Oberwolfach 1976 Rs,G

SEE ALSO FERRERO-SUPPA

FERRERO, Giovanni and SUPPA, Alberta

1. *Sistemi, anelloidi e funzioni di Steiner*, Atti del Seminario Matematico e Fisico dell' Università di Modena 20 (1972), Fasc. II. Rs,G,P"

FERRERO-COTTI, C., Istituto Matem. Università, 43100 Parma, Italy

1. *Una condizione di debole commutatività per gli anelli*, Riv. Mat. Univ. Parma (2) 10 (1969), 165-170. Rs
2. *Sugli stems in cui il prodotto è distributivo rispetto a se stesso*, Oberwolfach, 1972. B,S,D'
3. *Sugli stems in cui il prodotto è distributivo rispetto a se stesso*, Riv. Mat. Univ. Parma (3) 1 (1972), 203-220. B,S,D'
4. *On near-rings containing a ring with an involution*, Oberwolfach, 1976 B,Rs

FITTING, Hans

1. *Die Theorie der Automorphismenringe abelscher Gruppen und ihr Analagon bei nicht kommutativen Gruppen*, Math. Ann. 107 (1932), 514-542. [E'']

FOULSER, David A., Dept.Math. Univ. of Illinois, Chicago Circle, Il 60680, USA.

1. *On finite affine planes and their collineation groups*, Thesis, Univ. of Michigan, Ann Arbour, Michigan, 1962. [F,D'',G]

FREIBERGER, Helene,

1. *Fastringe*, Hausarbeit, Techn.Univ.Wien, Austria, 1975 [E]

FREIDMAN, Pavel Abramovic, Sverdlovskii Univ. ul.Libknechta 9a,Sverdlovsk, USSR.

1. *Distributively solvable near-rings*, Proceedings of the Riga Seminar on Algebra (Russian), 297-309,Latv.Gos.Univ.Riga, 1969. [D',N,R]

FRÖHLICH,Albrecht,Dept.Math.Univ.of London, Kings College, Strand London WC 2R, 2LS, London, England.

1. *Distributively generated near-rings I. Ideal theory*, Proc. London Math.Soc. 8 (1958), 76-94. [D,D',E,N]
2. *Distributively generated near-rings II. Representation theory*, Proc.London Math.Soc. 8 (1958), 95-108. [D,I]
3. *The near-ring generated by the inner automorphisms of a finite simple group*, J. London Math. Soc. 33 (1958), 95-107. [E'',D]
4. *On groups over a d.g. near-ring I. Sum constructions and free R-groups*, Quart.J.Math. Oxford Ser. 11 (1960), 193-210. [D,C,F',H]
5. *On groups over a d.g. near-ring II. Categories and functors*, Quart.J.Math.Oxford Ser.11 (1960), 211-228. [D,H]
6. *Non-abelian homological algebra I. Derived functors and satellites*, Proc.London Math. Soc. 11 (1961), 239-275. [H,D]
7. *Non-abelian homological algebra II. Varieties*, Proc.London Math. Soc. 12 (1962), 1-28. [H,D]
8. *Non-abelian homological algebra III. The functors EXT and TOR* Proc.London Math.Soc. 12 (1962), 739-768. [H,D]
9. *Some examples of near-rings*, Oberwolfach, 1968 [P^{o},X]

FURTWÄNGLER, Philipp

SEE FURTWÄNGLER-TAUSSKI

FURTWÄNGLER, Ph. and TAUSSKY, O.

1. *Über Schiefringe*, Sitzber. Akad.Wiss.Wien, Math.Nat.Klasse,Abt.IIA 145 (1936), 525. [D',A]

GANESAN, N., No.1, Umayal Lane, Chidambaram 1, Tamil Nadu, 608001 India

1. *Finite near-rings with zero divisors and regular elements*, Notices of the Amer. Math. Soc., August 1970, 70T-A168. [A,C,D',E]
2. *A study of finite rings and near-rings*, Doctoral Dissertation, Annamalai Univ., Tamil Nadu (India), 1971. [A,C,D',E]

GILBERT, Michael D.

1. *Commutativity in rings and near-rings*, M.S.Thesis, University of Southwestern Louisiana, Lafayette, 1972.

GORTON, R.

1. *λ-complete near-rings*, Fundamenta Math. 87 (1975), 73-78 | P,R,S

GOJAN, I.M., pr. Lenina 1, Kišinev 612, Moldov.SSR, USSR

1. *The Baer radical for near-rings*, Bull.Akad.Stiince RSS Moldoven 4 (1966), 32-38. | R,S,D

GONSHOR, Harry, Dept.Math.Rutgers Univ., New Brunswick, N.J. 08903, USA

1. *On abstract affine near-rings*, Pacific J.Math. 14 (1964), 1237-1240. | A'

GRAVES, James A.

1. *Near domains*, Doctoral dissertation, Texas A&M University, 1971. | I',Q',E'

SEE ALSO GRAVES-MALONE

GRAVES, James A. and MALONE, Joseph J.

1. *Embedding near domains*, Bull. Austral. Math. Soc. 9 (1973), 33-42. | I',Q',E'
2. *Euclidean near domains*, submitted | I',D,P^o
3. *Near domains as generalizations of D-rings*, Amer.Math.Monthly 82 (1975), 491-493. | I',P^o

GRAY, Mary W. Dept.Math.American Univ, Washington D.C. 20016, USA

1. *Rings, radicals and categories*, Addison-Wesley, Reading, Massach., 1969, Chapter 6, 120-122. | R,S,I

GUERCIA, Liana, Istituto di Matematica, Università, 73100 Lecce, Italy

SEE GUERCIA-LENZI

GUERCIA, Liana and LENZI, Domenico

1. *Su una generalizzazione del concetto di quasi-corpo associativo*, Atti Accad. Naz. Lincei Rend. Cl. Sci. Fis. Mat. Nat. (8) 57 (1974), 311-315. | Rs

GUPTA, N.D., Dept. Math. Univ. of Manitoba, Winnipeg 19, Canada

1. *Commutation near-rings of a group*, J. Austral. Math. Soc. 7 (1967), 135-140. | D'

GUTHRIE, Edgar R.

1. *The endomorphism near-ring on D_8*, M.S. Thesis, Texas A&M Univ., Coll. Station, 1969. | E",A

HALDER, H.-R., Lehrstuhl für Geometrie, Techn. Univ. München, 8 München 2, Barerstr. 23, Germany

SEE HALDER-HEISE

HALDER, H.-R. and HEISE, W.

1. *Einführung in die Kombinatorik*, Carl Hauser Verlag, München 1976 | G,S"

HALL, Marshall, Jr. Dept.Math.Caltech. Pasadena, Calif. 91109,USA

1. *The theory of groups*. New York, Macmillan 1959, 382-392. F,G,S",D"

HARDY, F.Lane, Math.Dept. Chicago State College, Chicago, Illinois 60621,USA

1. *Groups and near-rings*, unpublished.

SEE ALSO ARMENTROUT-HARDY-MAXSON

HARTNEY, J.F.T.

1. *On the radical theory of near-rings*, M.S .Thesis, University of Nottingham, 1968. P,R,S,D
2. *On the radical theory of a distributively generated near-ring*, Math.Scand. 23 (1968), 214-220. P,R,S,D

HAVEL, Vâclav, Math. Inst. Techn. Univ., Hilleko 6/I, 602-00 Brno, CSSR

1. *One characterization of special translation planes*, Archivum Mathematicum (Brno) 3 (1967), 157-160. S",G
2. *Über angeordnete Ternärgruppoide*, Časopis pro pestovani matematiky 9 (1969), 15-20. Nd,Rs,O
3. *Über schwach 2-Strukturen*, Abh. Math. Sem. Univ. Hamburg 33 (1969), 225-230. Rs
4. *Near-domains as linear pseudo ternaries*, Czech. Math. J. 21 (96) (1971), 344-347. Nd
5. *Geometrical interpretation of 2-fold transitive permutation groups* (Czech), Techn. Univ. Brno, 1973, 27p. S",Nd,G
6. *Nichtinjektive Homomorphismen von Viergeweben*, Abh. Math. Sem. Univ. Hamburg 37 (1972), 6-19. P",F,G

HEATHERLY, Henry E., Dept. Math. Univ. of Southwestern Louisiana, Lafayette, Louisiana 70501, USA.

1. *Embedding of near-rings*, Doctoral dissertation, Texas A&M University, College Station, 1968. T,E',S,A,D Rs
2. *Near-rings on simple groups*, manuscript. A
3. *C-2 transitivity and C-2 decomposable near-rings*, J.Algebra 19 (1971), 496-508. E,A',R,S,A
4. *One-sided ideals in near-rings of transformations*, J.Austral.Math. Soc. 13 (1972), 171-179. T,S
5. *Matrix near-rings*, J. London Math.Soc. (2) 7 (1973), 355-356. D'
6. *Near domains of composite characteristic*, Elem.Math. 28 (1973), 151-152. I',A
7. *Distributive near-rings*, Quart.J.Math.Oxford Ser. (2) 24 (1973), 63-70. D',N,A
8. *Near-rings without nilpotent elements*, Publ.Math.Debrecen 20 (1973), 201-205. W,R',I',B,P^{O}
9. *Regular near-rings*, J. Indian Math. Soc. 38 (1974), 345-354. R'
10. *Monogenic algebras*, submitted.
11. *Semiring multiplications on commutative monoids*, Publ.Math.Debrecen 21 (1974), 119-123. Rs,D',A

SEE ALSO HEATHERLY-LIGH, HEATHERLY-MALONE, HEATHERLY-OLIVIER, HEATHERLY-YEARBY.

HEATHERLY, Henry E. and LIGH, Steve,
1. *Pseudo-distributive near-rings*, Bull.Austral.Math.Soc. 12 (1975), 449-456. D',P°, R;A

HEATHERLY, Henry E. and MALONE, Joseph J., Jr.
1. *Some near-ring embeddings*, Quart. J.Math.Oxford Ser. 20 (1969), 81-85. E',D
2. *Some near-ring embeddings II*, Quart.J.Math.Oxford Ser. 21 (1970), 445-448. E',D

HEATHERLY, Henry E. and OLIVIER, Horace
1. *Near integral domains*, Monatsh. Math. 78 (1974), 215-222. I',A
2. *Near integral domains II*, Monatsh. Math. 80 (1975), 85-92. I',A
3. *H-monogenic near-rings*, submitted. I',D,A

HEATHERLY, Henry E. and YEARBY, Robert
1. *Distributive near-rings II*, submitted. D'

HEEREMA, Nickolas, Math.Dept. Florida State Univ., Tallahassee, Flor. 32306,USA
1. *Sums of normal endomorphisms*, Trans.Amer.Math. Soc. 84 (1957), 137-143. E"

HEISE, W., Lehrstuhl für Geometrie, Techn. Univ. München, 8 München 2, Barerstr. 23, Germany
1. *A combinatorial characterization of* $PGL(2,2^q)$, Abh. Math. Sem. Univ. Hamburg 44 (1975). S"
2. *On some Steiner systems*, Oberwolfach, 1976. Rs,G

SEE ALSO HALDER-HEISE, HEISE-KARZEL, HEISE-SÖRENSEN

HEISE, W. and KARZEL, Helmut
1. *Symmetrische Minkowski-Ebenen*, J. of Geometry 3 (1973), 5-20. G,S"
2. *Vollkommene fanosche Minkowski-Ebenen*, J. of Geometry 3 (1973),21-29 G,S"

HEISE, W. and SÖRENSEN, Kay
1. *Scharf n-fach transitive Permutationsmengen*, Abh. Math. Sem. Univ. Hamburg 44 (1975). S"

HELLER, Isidor
1. *On generalized polynomials*, Rep.Math.Colloq. 2 (1948), 58-60. P°, Cr

HILBERT, David
1. *Über den Zahlbegriff*, Jahresber.Dt.Math.Ver. 8 (1899), 180-184. E,A,D'

HOFER, Robert D., Math.Dept. State Univ. of New York,Coll. of Arts and Sciences, Plattsburgh, N.Y. 12901, USA.
1. *Restrictive semigroups of continous self-maps on arcwise connected spaces*, Proc. London Math. Soc. 25 (1972), 358-384. T',E
2. *Restrictive semigroups of continous functions on 0-dimensional spaces*, Canad. J.Math. 24 (1972), 598-611. T',E
3. *Simplicity of near-rings of continous functions on topological groups*, Oberwolfach, 1972. S,T,T'
4. *Simplicity of right distributive systems of functions on groupoids*, manuscript. Rs,S

HOGEWIJS, H.

1. *Semi-near-ring embedding*, Med. Konink. Acad. Wetensch. Lett. Schone Kunst Belgie Kl. Wetensch. 32 (1970), 3-11. Rs,T

HOLCOMBE, Wm. Michael Lloyd, Dept.of Pure Math., Queens Univ. of Belfast, BT-7 Inn, Northern Ireland.

1. *A class of 0-primitive near-rings*, Oberwolfach, 1968. P,T
2. *Primitive near-rings*, Doctoral dissertation, University of Leeds,1970 P,T,R
3. *Endomorphism near-rings in general categories*, Oberwolfach, 1972. H,F,T
4. *Endomorphism near-rings in categories I. Categorical representation*, J. London Math. Soc., to appear. H,F,T
5. *A class of 0-primitive near-rings*, Math.Z. 131 (1973), 251-268. P,T,R
6. *Representations of 2-primitive near-rings and the theory of near-algebras*, Proc. Royal Irish Acad. Sect. A 73 (1973), 169-177. P,T,R,Rs
7. *Near-rings of quotients of endomorphism near-rings*, Proc. Edinburgh Math.Soc. (2) 19 (1974/75), 345-352. Q',E"
8. *Special radical functors*, Oberwolfach, 1976. R,H

HULE, Harald, Dept. de Mat. Univ. de Brasilia, Brasilia D.F., Brasil.

1. *Polynome über universalen Algebren*, Monatsh. Math. 73 (1969), 329-340. P^o,S,Ua

SEE ALSO HULE-MÜLLER.

HULE, Harald and MÜLLER, Winfried

1. *On the compatibility of algebraic equations with extensions*, J. Austr. Math. Soc., to appear. Ua,P^o,X

HUQ, Syed A., Dept. Math. Univ. of Montreal, Montreal, Canada.

1. *Right abelian categories*, Rend. Sc. Fis. Mat. e Nat. Lincei 50 (1971) 284-289. H

SEE ALSO AIJAZ-HUQ.

ISTINGER, M., Inst. f. Algebra, Techn. Univ. Wien, Argentinierstr. 8, 1040 Wien, Austria.

SEE ISTINGER-KAISER.

ISTINGER, M. and KAISER, Hans K.

1. *A characterization of polynomially complete algebras*, submitted. X

JACOBSON, Nathan, Dept. Math. Yale Univ. New Heaven, Connecticut 06520,USA.

1. *The theory of rings*, Math. Surveys 2, New York, 1942, page 1. E

JACOBSON, Richard A., Dept.Math. Houghton College, Houghton, N.Y. 14744,USA.

1. *The structure of near-rings on a group of prime order*, Amer.Math. Monthly 73 (1966), 59-61. A

JOHNSON, Marjory J. Dept.Math. Univ. of South Carolina, South Carolina,92208,USA	
1. *Ideal and submodule structure of transformation near-rings*, Doctoral dissertation, University of Iowa, 1970.	T,E",R,S, D
2. *Radicals of endomorphism near-rings*, Rocky Mountain J.Math. 3 (1973), 1-7.	E",R,N
3. *Right ideals and right submodules of transformation near-rings*, J.Algebra 24 (1973),386-391.	E,T
4. *Near-rings with identities on dihedral groups*, Proc. Edingburgh Math.Soc. (2) 18 (1973), 219-228.	A
5. *Maximal right ideals of transformation near-rings*, J.Austral.Math. Soc. 19 (1975), 41o-412.	T
6. *Radicals of regular near-rings*, Monatsh. Math. 80 (1975), 331-341	R,R',P,S,T F
7. *Chain conditions on regular near-rings*, Univ.of South Carolina Math.Technical report No. 16A76-2, 1974	R',E
JONES, Patrica, Dept.Math.,Univ.of Southw. Louisiana, Lafayette, La 70501, USA.	
1. *Near-rings with involution*, Thesis, University of Southwestern Louisiana, 1976.	A
JONSSON, W.J.	
1. *Doubly transitive groups, near-fields and geometry*,Arch.Math. 17 (1966), 83-88.	F,T",G
JORDAN, Elfriede, Römerstr. 20, 4020 Linz, Austria.	
1. *Fastalgebren*, Thesis, Univ. Linz, 1976.	Na,E,E',D',F S,R,T'
JORDAN, Pascual, Isestr. 123, 2000 Hamburg 20, Germany.	
1. *Über polynomiale Fastringe*, Akad. Wiss. Mainz, Math.-Nat. Kl. (1951), 337-340.	P^o,E
JOUSSEN, Jakob, Kleine Schwerterstr. 44, 4600 Dortmund-Aplerbeck, Germany	
1. *Zum Transitivitätsverhalten der Projektionsgruppen endlicher Fastkörperebenen*, Abh. Math. Sem. Univ. Hamburg 35 (1971), 230-241.	F,G,D",P",S"
KAARLI, K.	
1. *A note on near-rings with identity*, (Russian; English and Estonian summaries), Tartu Riikl. Ül. Toimetised Vik. 336 (1974), 234-242.	A
2. *Minimal ideals in near-rings*, (Russian; English and Estonian summaries), Tartu Riikl. Ül. Toimetised Vik. 336 (1975), 105-142.	E,P,S,T,N
KAISER, Hans K., Inst. f. Algebra, Techn. Univ. Wien, Argentinierstr. 8, 1040 Wien, Austria. SEE ISTINGER-KAISER.	
KALLAHER, Michael J.	
1. *A survey of weak rank 3 affine plans*, Proc.internat. Conf. projective Planes, Washington State Univ. 1973, 121-143.	F,G

KALSCHEUER, Franz

1. *Die Bestimmung aller stetigen Fastkörper über dem Körper der reellen Zahlen als Grundkörper*, Abh. Math.Sem.Univ.Hamburg 13 (1940), 413-435. | F,T',V

KARZEL, Helmut, Lehrst. für Geometrie,Techn.Univ.,8 München 2,Barerstr.23,Germany

1. *Kommutative Inzidenzgruppen*, Arch.Math. 13 (1962), 535-538. | F,G
2. *Ebene Inzidenzgruppen*, Arch.Math. 15 (1964), 10-17. | F,G
3. *Bericht über projektive Inzidenzgruppen*, Jahresber. Dt.Math.Ver. 67 (1965), 58-92. | F,G,D"
4. *Projektive Räume mit einer kommutativen transitiven Kollineationsgruppe*, Math.Z. 87 (1965), 74-77. | F,G,S"
5. *Normale Fastkörper mit kommutativer Inzidenzgruppe*, Abh.Math.Sem. Univ.Hamburg 28 (1965), 124-132. | F,G
6. *Unendliche Dicksonsche Fastkörper*, Arch.Math. 16 (1965), 247-256. | F,D"
7. *Zweiseitige Inzidenzgruppen*, Abh.Math.Sem.Univ.Hamburg 29 (1965), 118-136. | F,G
8. *Zusammenhänge zwischen Fastbereichen, scharf zweifach transitiven Permutationsgruppen und 2-Strukturen mit Rechtecksaxiom*, Abh.Math. Sem.Univ.Hamburg 32 (1968), 191-206. | Nd,S",G
9. *Symmetrische Permutationsmengen*, submitted. | S"
10. *Permutation sets, near-domains and derived geometrical structures*, Oberwolfach, 1976. | Nd,S",G

SEE ALSO ELLERS-KARZEL, HEISE-KARZEL, KARZEL-MEISSNER, KARZEL-PIEPER, KARZEL-SÖRENSEN-WINDELBERG, KARZEL-STANIK-WÄHLING.

KARZEL, Helmut and MEISSNER, Hartwig

1. *Geschlitzte Inzidenzgruppen und normale Fastmoduln*, Abh.Math,Sem. Univ.Hamburg 31 (1967), 69-88. | G,F,L,Na

KARZEL, Helmut and PIEPER, Irene

1. *Bericht über geschlitzte Inzidenzgruppen*, Jahresber.d.Dt.Math.Ver. 72 (1970), 70-114. | G,F,L,Na,T

KARZEL, Helmut, SÖRENSEN, Kay and WINDELBERG, Dirk

1. *Einführung in die Geometrie*, Vandenhoeck & Ruprecht, Göttingen 1973 | G,F,P"

KARZEL, Helmut, STANIK, Rotraud and WÄHLING, Heinz

1. *Zum Anordnungsbegriff in affinen Geometrien*, Abh.Math.Sem.Univ. Hamburg 44 (1975), 24-31 | G,F

KAUTSCHITSCH, Hermann, Math.Inst. Univ.Klagenfurt, A-9010, Klagenfurt, Austria

1. *Über Vollideale in Potenzreihenringen*, submitted. | Cr,P°

KERBY, William E., Math.Sem.Univ.Hamburg, Bundesstr.55, 2000 Hamburg, Germany

1. *Anordnungsfragen in Fastkörpern*, Diss.Univ.Hamburg, 1966. F,O
2. *Projektive und nicht-projektive Fastkörper*, Abh.Math.Sem.Univ. Hamburg 32, (1968), 20-24. P'',F,P°, G
3. *Angeordnete Fastkörper*, Abh.Math.Sem.Univ.Hamburg 32 (1968), 135-146 O,F,P'',D''
4. *Quotientenbildung in Fastringen*, Oberwolfach, 1968. Q',D''
5. *Angeordnete Fastkörperebenen*, Abh.Math.Sem.Univ.Hamburg 33 (1969), 4-16. O,F,G
6. *Near domains and sharply 2-transitive permutation groups*, Oberwolfach, 1972. Nd,S''
7. *On infinite sharply multiply transitive groups*, Habilitationsschrift, Hannover 1972. Nd,F,Rs
8. *Automorphisms of homogenously generated algebras*, Abh.Math.Sem. Univ.Hamburg 39 (1973), 96-108. A',Na
9. *On infinite sharply multiply transitive groups*, Vandenhoeck and Ruprecht, Göttingen, 1974. Nd,F,Rs
10. *Sharply 3-transitive groups of characteristic ≡1 mod 3*, J.Algebra 32 (1974), 240-245. Nd,F,S''
11. *Sharply 2- and 3-transitive groups with kernels of finite index*, Aequationes Math. 14 (1976), 137-141. Nd,F,S''
12. *Near-domains and sharply n-transitive permutation groups*, Oberwolfach 1976. Nd,S''

SEE ALSO KERBY-WEFELSCHEID.

KERBY, William E. and WEFELSCHEID, Heinrich

1. *Über eine scharf 3-fach transitiven Gruppen zugeordnete algebraische Struktur*, Abh. Math.Sem. Univ.Hamburg 37 (1972), 225-235. F,D'',S'',Nd,F
2. *Conditions of finiteness on sharply 2-transitive groups*, Aequationes Math. 8 (1972), 287-290. Nd,S''
3. *Bemerkungen über Fastbereiche und scharf zweifach transitive Gruppen*, Abh.Math.Sem.Univ.Hamburg 37 (1972), 20-29. Nd,S'',F
4. *Ein Unterscheidungsmerkmal bei endlichen scharf 3-fach transitiven Gruppen*, Mitt.Math.Gesellsch. Hamburg 10 (1973), 81-87. S'',F,Nd
5. *Über eine Klasse von scharf 3-fach transitiven Gruppen*, J.Reine Angew.Math. 268/69 (1974), 17-26. S'',F,Nd,Rs D''
6. *The maximal subnear-fields of a near-domain*, J.Algebra 28 (1974), 319-325. Nd,F,S''

KING, Mary Katharine, Math.Dept. Texas A&M Univ., College Station, Texas 77843, USA.

1. *The endomorphism near-ring of the quaternion group*, M.S.Thesis, Texas A&M University, 1969. E''

KIST, Günter Peter, Lehrst. f. Geometrie, Techn. Univ. München, Barerstr. 23, 8 München 21, Germany.

1. *Punktiert affine Inzidenzgruppen und Fastkörpererweiterungen*, Abh. Math. Sem. Univ. Hamburg 44 (1975), 233-248. F,G

KRIMMEL, John Eric

1. *Conditions on near-rings with identity and the near-ring with identity on some metacyclic groups*, Doctoral Dissertation, Univ. of Arizona, Tucson, 1972. | A
2. *A condition on near-rings with identity*, Monatsh. Math. 77 (1973), 52-54. | A

KÖSEL, Joachim, Math. Sem. Univ. Hamburg, Bundesstr. 55, 2000 Hamburg 13, Germany.

1. *Archimedische Anordnung in Fastringen*, Diss. Univ. Bremen,1972. | O,F,P",A

KUZ'MIN, Ju. V.

1. *Representations of finite groups by automorphisms of nilpotent near-spaces and by automorphisms of nilpotent groups* (Russian), Sibirsk. Mat. Ž. 13 (1972), 107-117. | E',E"

LAUSCH, Hans, Dept.Math.,Monash Univ., Clayton, Victoria 3168, Australia.

1. *Kohomologie von distributiv erzeugten Fastringen I. Erweiterungen*, J. für Reine und Angewandte Mathematik 229 (1966), 137-146. | H,D
2. *Functions on groups with multiple operators*, J. London Math.Soc.42 (1967), 698-700. | P^o,Ua
3. *Kohomologie von distributiv erzeugten Fastringen II. Hindernistheorie für Erweiterungen*, J. für Reine und Angewandte Mathematik 231 (1968), 82-88. | H,D
4. *An application of a theorem of Gaschütz*, Bull.Austral.Math.Soc. 1, (1969), 381-384. | D,E
5. *Idempotents and blocks in Artinian d.g. near-rings with identity element*, Math. Annalen 188 (1970), 43-52. | D,I,N,E

SEE ALSO LAUSCH-NÖBAUER

LAUSCH, Hans and NÖBAUER, Winfried

1. *Algebra of polynomials*, North Holland/American Elsevier, Amsterdam, 1973 | P^o, D,I,N, R,S,Ua

LAUSCH, Hans, NÖBAUER, Winfried and SCHWEIGER, Fritz

1. *Poynompermutationen auf Gruppen*, Monatsh. Math. 69 (1965), 410-423. | E,X

LAWVER, Donald A., Dept.Math. Univ. of Arizona, Tucson, Arizona, 85721, USA

1. *Concerning nil groups for near-rings*, Acta Mathematica, Acad. Sci. Hungar. 22 (1972), 373-378. | A
2. *Existence of near-rings in special cases (near-rings on $Z(p^\infty)$)*, Oberwolfach, 1972. | A
3. *Cocyclic planar near-rings*, Acta Math. Acad.Sci.Hungar. 26 (1975), 87-90. | P",A,I

SEE ALSO CLAY-LAWVER

LAXTON, R.R., Dept.Math. Univ. of Nottingham, Univ.Park, Nottingh.,England

1. -------- *Doctoral dissertation*, University of London, 1961. — D,P,R,S,M, N,Q
2. *Primitive distributively generated near-rings*, Mathematika 8 (1961), 143-158. — P,D
3. *A radical and its theory for distributively generated near-rings*, J. London Math.Soc. 38 (1963), 40-49. — D,P,R,S,M,N Q
4. *Prime ideals and the ideal radical of a distributively generated near-ring*, Math.Z. 83 (1964), 8-17. — D,P,P',R,N
5. *A problem on free distributively generated near-rings*, Oberwolfach 1968. — D,F',D',E"
6. *Note on the radical of a near-ring*, J. London Math.Soc. (2) 6 (1972), 12-14. — D,R,N

SEE ALSO LAXTON-MACHIN

LAXTON, R.R. and MACHIN, Alan W.

1. *On the decomposition of near-rings*, Abh. Math. Sem.Univ. Hamburg 38 (1972), 221-230. — D,R

LENZI, Domenico, Istituto di Matematica, Università, 73100 Lecce, Italy

SEE GUERCIA-LENZI.

LEVI, F.W.

1. *Finite geometrical systems*, Calcutta, 1942. — F,G

LIGH, Steve, Dept.Math.Univ. of Southwestern Louisiana, Lafayette, Louisiana 70501, USA.

1. *On distributively generated near-rings*, Proc. Edingburgh Math.Soc. 16 (1969), 239-243. — D,F
2. *On division near-rings*, Canad.J.Math. 21 (1969), 1366-1371. — F,D,A
3. *Near-rings with descending chain condition*, Composito Mathematica 21 (1969), 162-166. — E,D,D',F,A
4. *On certain classes of near-rings*, Doctoral dissertation, Texas A&M University, College Station, 1969. — F,A,I,R',Q,D D'
5. *On Boolean near-rings*, Bull.Austral. Math.Soc. 1 (1969), 375-379. — B,D
6. *A generalization of a theorem of Zassenhaus*, Canad.Math.Bull. 12 (1969), 677-678. — A,F
7. *On regular near-rings*, Math. Japon. 15 (1970), 7-13. — R,I,A,S,F,B D'
8. *On the commutativity of near-rings*, Kyungpook Math.J. 10 (1970), 105-106. — B,W,D'
9. *Near-rings with identities on certain groups*, Monatsh.Math. 75 (1971), 38-43. — A
10. *D.G. near-rings on certain groups*, Monatsh.Math. 75 (1971), 244-249. — A,D,B
11. *On the commutativity of near-rings II*, Kyungpook Math.J. 11 (1971), 159-163. — B,D,A,W
12. *On the commutativity of near-rings III*, Bull. Austral. Math.Soc. 6 (1972), 459-464. — I',B,D,A
13. *On the additive groups of finite near integral domains and simple d.g. near-rings*, Monatsh. Math. 76 (1972), 317-322. — I', A,D,S
14. *The structure of a special class of near-rings*, J.Austral.Math. Soc. 13 (1972), 141-146. — B

15. *Some commutativity theorems for near-rings*, Kyungpook Math.J. 13 (1973), 165-170. D,D',B,A

16. *A special class of near-rings*, J. Austral.Math. Soc., to appear. B,D,R',W,.

17. *A note on matrix near-rings*, J.London Math.Soc. (2) 11 (1975) 383-384. X,D',A

18. *The structure of certain classes of rings and near-rings*, J.London Math. Soc. (2) 12 (1975). B,I',W,N

SEE ALSO BELL-LIGH, HEATHERLY-LIGH, LIGH-LUH, LIGH-MALONE, LIGH-McQUARRIE-SLOTTERBECK, LIGH-NEAL, LIGH-RAMAKOTAIAH-REDDY, LIGH-UTUMI.

LIGH, Steve and MALONE, Joseph J.,Jr.

1. *Zero divisors and finite near-rings*, J. Austral. Math.Soc. 11 (1970), 374-378. I',B,F,A,X

LIGH, Steve, MCQUARRIE, Bruce and SLOTTERBECK, Oberta

1. *On near-fields*, J. London Math.Soc. 5 (1972), 87-90. A,F,P^{o}

LIGH, Steve and LUH, Jiang

1. *Some commutativity theorems for rings and near-rings*, submitted. B,D,I',W

LIGH, Steve and NEAL, Larry

1. *A note on Mersenne numbers*, Math.Mag. 47 (1974), 231-233. F

LIGH, Steve, RAMAKOTAIAH, D. and REDDY, Venkatesvara Y.

1. *Near-rings with chain conditions*, Monatsh. Math. 80 (1975),119-130 A,E

LIGH, Steve and UTUMI, Yuzo

1. *Some generalizations of strongly regular near-rings*, submitted. R',B,I',I

LIGHTSTONE, A.H.

1. *A remark concerning the definition of a field*, Math. Mag. 37 (1964), 12-13. F

LUEDER, Kenneth

1. *Derivability in the irregular near-field planes*, Proc. intern. Conf. projective Planes, Washington State Univ., 1973, 181-189. F,G

LUH, Jiang, Math.Dept. N.Carolina State Univ., Raleigh, North Carolina 27607,USA

SEE LIGH-LUH.

LÜNEBURG, Heinz, Fachber. Math., Univ. Trier, 6750 Kaiserslautern, Postfach 1049, Pfaffenbergstr. 95, Germany.

1. *Über die Anzahl der Dickson'schen Fastkörper gegebener Ordnung*, Atti del Convegno di Geometria Combinatoria e sue Applicazioni, Ist. Mat. Univ. Perugia, Perugia, Italy, 1971, 319-322. D''

LYONS, Carter George, Math. Dept. Madison College, Harrisonburg, Virginia, 22801, USA.

1. *Endomorphism near-rings on the non-commutative group of order six*, M.S. Thesis, Texas A&M Univ., College Station, 1968. E",I
2. *Endomorphism near-rings*, Doctoral Dissertation, Texas A&M Univ., College Station, 1971. E",I
3. *Endomorphism near-rings*, Oberwolfach, 1972. E",I
4. *On decompositions of E(G)*, Rocky Mountain J. Math. 3 (1973),575-582. D,I,E",E'
5. *Finite groups with semisimple endomorphism rings*, Proc. Amer. Math. Soc. 53 (1975), 51-52. E",S
6. *A characterization of the radical of E(G) in terms of G*, Oberwolfach 1976. E",D,R

SEE ALSO LYONS-MALONE.

LYONS, Carter George and MALONE, Joseph J.

1. *Endomorphism near-rings*, Proc. Edingburgh Math.Soc.17 (1970),71-78 E",D,I
2. *Finite dihedral groups and d.g. near-rings I*, Compositio Mathematica 24 (1972), 305-312. E",R,A,E'
3. *Finite dihedral groups and d.g. near-rings II*, Compositio Mathematica 26 (1973), 249-259. E",I,R

MACHIN, Alan W.

1. *Right representation of a class of distributively generated near-rings*, Oberwolfach, 1968. D,R,P,P'
2. *On a class of near-rings*, Doctoral dissertation, University of Nottingham, 1971.

SEE ALSO LAXTON-MACHIN.

MAGILL, Kenneth D., Jr. Dept.Math. State Univ. of New York at Buffalo,4246 Ridge Lea Rd., Amherst, N.Y. 14226, USA.

1. *Automorphisms of the semigroup of all differentiable functions*, Glasgow Math. J. 8 (1967), 63-66. T'
2. *Semigroup structures for families of functions II*, J.Austral.Math. Soc. 7 (1967), 95-107. T'
3. *Semigroup structures for families of functions III*, J.Austral. Math. Soc. 7 (1967), 524-538. T'
4. *Near-rings of continous functions*, Oberwolfach, 1968. T'

MALONE, Joseph J., Math. Dept., Worcester Polytechnic Institute, Worcester, Mass. 01609, USA.

1. *Near-rings automorphisms*, Doctoral dissertation, Saint Louis University, 1962. E
2. *An additional remark concerning the definition of a field*, Math. Mag., 38 (1965), 94. F
3. *Near-rings with trivial multiplications*, Amer. Math.Soc. Monthly 74 (1967), 1111-1112. A
4. *Near-ring homomorphisms*, Canad.Math.Bull. 11 (1968), 35-41. E
5. *Automorphisms of abstract affine near-rings*,Math.Scand. 25 (1969) 128-132. E,A'
6. *A near-ring analoque of a ring embedding theorem*,J. Algebra 16 (1970), 237-238. E',E",D

7. *Generalized quaternion groups and distributively generated near-rings*, Proc.Edinburgh Math. Soc. 18 (1973), 235-238. — E",R,L,I,A
8. *Dg.near-rings on the infinite dihedral group*, submitted. — A,D
9. *Fraudee's groups and E(G)*, Oberwolfach, 1976. — E"

SEE ALSO CLAY-MALONE, GRAVES-MALONE, HEATHERLY-MALONE, LIGH-MALONE, LYONS-MALONE, MALONE-McQUARRIE.

MALONE, Joseph J., and McQUARRIE, Bruce

1. *Endomorphism rings of non-abelian groups*, Bull.Austral.Math.Soc. 3 (1970), 349-352. — E",A,T

MANNOS, Murray, Mitre, Bedford, Mass. 01703, USA.

1. *Ideals in tri-operational algebra I*, Reports of a Math.Colloq., Second Series, Issue 7, Notre Dame 1946, 73-79. — Cr

MARCHIONNA, Ermanno, Viale Abruzzi 44, 20131, Italy.

1. *Osservazioni sulla struttura aritmetica degli anelloidi finiti*, Boll. Unione Mat. Ital. 17 (1962), 289-319. — Rs

MARIN, V.G., Mat. Inst. Univ. Kishinew, USSR

1. *Near-algebras without nilpotent elements (Russian)*, Mat.Issled 6, Nr.4 (22), 1971, 123-139. — Na,W,I'

MARSCHOUN, A.

1. *Zur Theorie der dreioperationalen Algebren*, Doctoral dissertation University of Vienna, 1964. — Cr

MASON, Gordon, Dept.Math.Univ. of New Brunswick, Fredericton, N.Brunsw.,Can.

1. *Solvable and nilpotent near-ring modules*, Proc.Amer.Math.Soc. 40 (1973), 351-357. — D',D
2. *W-groups and near-ring modules*, Canad. Math. Bull. 18 (1975) — D',X
3. *Injective and projective near-ring modules*, Compositio Math., to appear. — D,S,H
4. *Injective and projective near-ring modules*, Oberwolfach 1976. — D,S,H

MAXSON, Carlton J., Math.Dept. Texas A&M, College Station, Texas 77843,USA

1. *On near-rings and near-ring modules*, Doctoral dissertation, Suny at Buffalo, 1967. — E,D,D',H,F' L,N,A,P',P^{0},Q
2. *On finite near-rings with identity*, Amer.Math.Monthly, 74 (1967), 1228-1230. — A
3. *On local near-rings*, Math.Z. 106 (1968), 197-205. — L,S,R,Q,I,A S,D',P^{0},A,F'
4. *A new characterization of finite prime fields*, Canad.Math.Bull. 11 (1968), 381-382. — A,S
5. *Dickson near-rings*, Oberwolfach 1968. — D"
6. *Local near-rings of cardinality p-square*, Canad,Math.Bull. 11 (1968), 555-561. — L,A
7. *On imbedding fields in non-trivial near-fields*,Amer.Math.Monthly 76 (1969), 275-276. — E',F
8. *Dickson near-rings*, J. Algebra 14 (1970), 152-169. — D",P^{0},I',R,S

9. *On the construction of finite local near-rings I. On non-cyclic abelian p-groups*, Quart.J.Math.(Oxford) (2) 21 (1970), 449-457. | L,A
10. *On the dimension of Veblen-Wedderburn systems*, Glasgow Math. J. 11 (1970), 114-116. | P',F,D",P^o
11. *On well ordered groups and near-rings*, Compositio Mat. 22 (1970), 241-244. | O
12. *On the construction of finite local near-rings II. On non-abelian p-groups*, Quart.J.Math.,Oxford Ser. (2) 22 (1971), 65-72. | L,A
13. *On morphisms of Dickson near-rings*, J. Algebra 17 (1971), 404-411. | D",P^o
14. *On groups and endomorphism rings*. Math.Z. 122 (1971), 294-298. | E",A,M,P

SEE ALSO ARMENTROUT-HARDY-MAXSON, CLAY-MAXSON.

MAZZOLA, Guerino, Math.Inst.Univ.,Freie Straße 36, CH-8032 Zürich,Switzerland

1. *Diophantische Gleichungen und die universelle Eigenschaft Finslerscher Zahlen*, Math.Ann. 202 (1973), 137-148. | X,H

MCQUARRIE, Bruce C. Dept.Math.,Worcester Polytechnic Institute, Worcester, Mass. 01609, USA.

1. *N-systems and related near-rings*, Doctoral Dissertation, Boston Univ., 1971. | E,I,A
2. *A non-abelian near-ring in which (-1) r=r implies r=0*, Canad.Bull. Math., to appear. | E,I
3. *Near-rings that are N-systems*, Oberwolfach, 1972. | I,A
4. *Correction to "A non-abelian near-ring in which (-1) r=r implies r=0"*, Canad.Math.Bull. 17 (1974), 425. | E,I

SEE ALSO LIGH-McQUARRIE-SLOTTERBECK, MALONE-McQUARRIE

MEISSNER, Hartwig, Pädagog.Hochschule Münster, 44 Münster,Fliednerstr.26,Germany

SEE KARZEL-MEISSNER.

MELDRUM, John D.P., Math. Dept. Univ. of Edinburgh, Mayfield Rd., Edinburgh EH9 3JZ, Scotland.

1. *Varieties and d.g. near-rings*, Proc.Edinburgh Math.Soc. 17 (Series II) (1971), 271-274. | E',E",D,T,U_a
2. *Representation theory of d.g. near-rings*, Oberwolfach, 1972. | D,U_a,E'
3. *The representation of d.g. near-rings*, J.Austral. Math.Soc. 16 (1973), 467-480. | D,F',E'
4. *The group d.g. near-ring*, Proc.London Math.Soc. (3) 32 (1976), 323-346. | C,D,F'
5. *Structure theorems for morphism near-rings*, Oberwolfach, 1976. | P,R,E',D
6. *The endomorphism near-ring of an infinite dihedral group*, Proc.Royal Soc. Edinburgh, to appear. | D,E"
7. *The structure of morphism near-rings*, submitted. | D,E",E

MENGER, Karl, Illinois Institute of Technology, Chicago, Illinois 60616,USA

1. *Algebra of analysis*, Notre Dame Mathematical Lectures, No.3,1944 | Cr,E,X
2. *Tri-operational algebra*, Reports of a Math.Colloq. Second Series Issue 5-6, Notre Dame, 1944, 3-10. | Cr,P^o,X
3. *General algebra of analysis*, Reports of a Math.Colloq., Second Series, Issue 7, Notre Dame, 1946, 46-60. | Cr
4. *Gulliver in a land without 1, 2, 3*, Math.Gaz. 43 (1959),241-250. | (-)
5. *The algebra of functions; past, present and future*, Rend.Mat.20 (1961), 409-430. | Cr

MILGRAM, A.N.

1. *Saturated polynomials*, Reports of a Math.Colloq. Second Series, Issue 7, Notre Dame, 1946, 65-67. [P^{o}, Cr]

MISFELD, Jürgen, 3000 Hannover-Herrenhausen,Osteroderweg 10, Germany

1. *Beziehungen zwischen topologisch-algebraischen und topologisch-geometrischen Strukturen*, Oberwolfach 1968 [F,T',G]
2. *Zur Konstruktion topologischer Fastkörper*, Oberwolfach, 1972. [F,D",T']

SEE ALSO MISFELD-TIMM.

MISFELD, Jürgen and TIMM, Jürgen

1. *Topologische Dicksonsche Fastkörper*, Abh.Math.Sem.Univ.Hamburg 37 (1972), 60-67 [F,D",T']

MITSCH, Heinz,Math.Inst. Univ. Strudlhofg. 4, A-1090 Wien, Austria.

1. *Trioperationale Algebren auf Verbänden*, Doctoral dissertation, University of Vienna, 1967. [T,O,Cr,Rs]
2. *über Polynome und Polynomfunktionen auf Verbänden*, Monatsh.Math. 74 (1970), 238-243. [P^{o},Rs]
3. *Rechtsverbandshalbgruppen*, J. Reine Angew.Math. 264 (1973),173-181 [T,O,Rs]
4. *Ideale in Rechtsverbandshalbgruppen*, Monatsh. f.Math.78 (1974), 360-378. [Rs]
5. *Kongruenzen in Rechtsverbandshalbgruppen*, J. reine u.angew.Math., to appear. [Rs]

MLITZ, Rainer, III.Inst. f.Math., Techn.Univ. Wien, A-1040 Wien, Gußhausstr. 27-29, Austria

1. *Ein Radikal für universale Algebren und seine Anwendung auf Polynomringe mit Komposition*, Monatsh. Math. 75 (1971), 144-152. [R,Ua,P^{o}]
2. *Verallgemeinerte Jacobson-Radikale in Polynomkompositionsfastringen*, Oberwolfach, 1972. [R,P^{o}]
3. *Jacobson-Radikale in Fastringen mit einseitiger Null*, Math.Nachr. 63 (1974), 49-65. [P,R,S,M,P^{o} Ua]
4. *Modules and radicals of universal algebras* (Russian), submitted. [Ua,R,M,P]
5. *Cyclic radicals of universal algebras*, submitted. [P,R,S,Ua]
6. *The application of some ideas of the near-ring radical theory to universal algebra*, Oberwolfach, 1976. [Ua,R]

MÖLLER, Winfried, Math. Inst. Univ. Klagenfurt, 9010 Klagenfurt, Austria.

1. *Eindeutige Abbildungen mit Summen-, Produkt- und Kettenregel im Polynomring*, Monatsh. Math. 73 (1969), 354-367. [Cr,P^{o},X]
2. *El Álgebra de Derivaciones*, An. Acad. Brasil Ciênc. 45 (1973),339-343 [P^{o},X,Cr]
3. *über die Kettenregel in Fastringen*, Abh. Math. Sem. Univ. Hamburg 48 [X,P^{o}]
4. *über die Abhängigkeit von Summen-, Produkt- und Kettenregel im rationalen Funktionenkörper*, Sitzber. Österr. Akad. Wiss. Math.-Natur. Kl. II, to appear. [Cr,P^{o},X,Q']
5. *Derivationen in Kompositionsalgebren*, Sitzber. Österr. Akad. Wiss. Math.-Natur. Kl. II, to appear. [P^{o},X,Cr,Na Ua]
6. *über Abbildungen mit Kettenregel in Fastringen*, Oberwolfach, 1976. [X]

SEE ALSO HULE-MÖLLER.

MURDOCH, David C., Dept.Math. of British Columbia, Vancouver 8, B.C.,Canada
See Murdoch-Ore.

MURDOCH, David C. and ORE,
1. *On generalized rings*, Amer.Math.63 (1941), 73-86. | Rs,E

NAUMANN, Herbert, 4010 Hilden, Immermannstr. 8, Germany
1. *Stufen der Begründung der ebenen affinen Geometrie*, Math. Z. 60 1954 , 120-141. | G,F,D"

NEAL, Larry, Dept.Math. Univ. of Southw. Louisiana, Lafayette, La. 70501, USA
SEE LIGH-NEAL.

NEFF, M.F., Math. Dept. Emory Univ., Atlanta, Georgia 30322, USA.
SEE EVANS-NEFF.

NELSON, Evelyn, Dept. Math. McMasters Univ., Hamilton, Ont., Canada.
SEE BANASCHEWSKI-NELSON.

NEUBERGER, John W., Dept. Math. Emory Univ., Atlanta, Georgia, 30322, USA.
1. *Toward a characterization of the identity component of rings and near-rings of continuous transformations*, Journ. Reine Angew. Math. 238 (1969), 100-104. | T,T'
2. *Differentiability of the exponential of a member of a near-ring*, Proc. Amer. Math. Soc. 48 (1975), 98-100. | T'

NEUMANN, B.H., FAA, Frs, Dept.Math. Institute of Advanced Studies, Australian Nat. Univ., P.O.Box 4, Canberra A.C.T. 2600.
1. *On the commutativity of addition*, J. London Math. Soc. 15 (1940), 203-208. | A,I
2. *Groups with automorphisms that leave only the neutral element fixed*, Archiv der Mathematik 7 (1956), 1-5. | A

NEUMANN, Hanna
1. *Near-rings connected with free groups*, Proc.International Conference, Amsterdam, II, (1954), 46-47. | E",D
2. *On varieties of groups and their associated near-rings*, Math.Z. 65 (1956), 36-69. | E",D,E

NÖBAUER, Wilfried, Inst.f.Algebra ,Techn.Univ. Wien, Argentinierstr. 8, A-1040 Wien, Austria.
1. *über die Operation des Einsetzens in Polynomringen*, Math.Ann. 134 (1958), 248-259. | Cr,P^o
2. *Die Operation des Einsetzens bei Polynomen in mehreren Unbestimmten*, J.Reine Angew.Math. 201 (1959), 207-220. | Cr,P^o
3. *Zur Theorie der Vollideale I*, Monatsh. Math. 64 (1960), 176-183. | Cr
4. *Zur Theorie der Vollideale II*, Monatsh. Math. 64 (1960), 335-348. | Cr
5. *über die Ableitungen der Vollideale*, Math. Z. 75 (1961), 14-21. | Cr
6. *Funktionen auf kommutativen Ringen*, Math. Ann. 147 (1962), 166-175 | Cr,P^o,S,X
7. *Die Operation des Einsetzens bei rationalen Funktionen*, Österr. Akad. Wiss. Math.-Natur. Kl. S.-B. II 170 (1962), 35-84. | Cr,E

8. *Über die Darstellung von universellen Algebren durch Funktionenalgebren*, Publ. Math. Debrecen 10 (1963), 151-154. — E,T

9. *Derivationssysteme mit Kettenregel*, Monatsh. Math. 67 (1963), 36-49. — Cr,X

10. *Transformationen von Teilalgebren und Kongruenzrelationen in allgemeinen Algebren*, J. Reine Angew. Math. 214/215 (1965), 412-418. — E,S,P^o,Ua

11. *Über die Automorphismen von Kompositionsalgebren*, Acta Math. Acad. Sci. Hungar. 26 (1975), 275-278. — E,Cr,Ua

SEE ALSO LAUSCH-NÖBAUER, LAUSCH-NÖBAUER-SCHWEIGER, NÖBAUER-PHILIPP.

NÖBAUER, Wilfried and PHILIPP, Walter

1. *Über die Einfachheit von Funktionenalgebren*, Monatsh. Math. 66 (1962), 441-452. — S,T,Ua

2. *Die Einfachheit der mehrdimensionalen Funktionenalgebren*, Arch. Math. 15 (1964), 1-5. — S,T,Ua

OLIVIER, Horace R., Dept. Math. Univ. of Southwestern Louisiana, Lafayette, Louisiana 70501, USA.

1. *Endomorphism near-rings on certain groups*, M.S. Thesis, University of Southwestern Lousiana, 1970. — E"

2. *Near integral domains*, Doctoral dissertation, University of Southwestern Louisiana, Lafayette, 1976. — I'

SEE ALSO HEATHERLY-OLIVIER

ORE, Oystein

1. *Linear equations in non-commutative fields*, Ann.of Math. 32 (1931) 463-477. — Q',E",R'

SEE ALSO MURDOCH-ORE.

OSWALD, A., Dept. Math. Teeside Polytechnic, Middlesbrough, Teeside, England.

1. *Some topics in the structure theory of near-rings*, Doctoral Dissertation, Univ. of York, 1973. — D,D',E,E',F',N P,P',Q,Q',R',R S,T,X

2. *Near-rings in which every N-subgroup is principal*, Proc. London Math. Soc. (3) 28 (1974), 67-88. — E,P',P,X,D,T'

3. *Completely reducible near-rings*, to appear. — E,S,R',W

4. *Semisimple near-rings have the maximum condition on N-subgroups*, J. London Math. Soc. (2) 11 (1975), 408-412. — S,R,E

5. *Conditions on near-rings which imply that nil N-subgroups are nilpotent*, Proc. Edinb. Math. Soc., to appear. — D,E,N

6. *A note on injective modules*, submitted. — E,X

7. *Near-rings of mappings of torsion-free S-sets*, manuscript. — E,Q'

8. *Completely reducible near-rings*, Oberwolfach, 1976. — E,S,R',W

OUBRE, Glenn J., Math. Dept. Univ. of Southwestern Louisiana, Lafayette, Louisiana 70501, USA.

1. *The Krull-Schmidt theorem for near-rings*, M.S. Thesis, Univ. of Southwestern Louisiana, 1970. — I,C

PALMER, O.

SEE PALMER-YAMAMURO

PALMER, O. and YAMAMURO, Sadayuki.

1. *A note on finite dimensional differentiable mappings*, J.Austral. Math.Soc. 9 (1969), 405-408. T'

PANKIN, Mark.D., Dept.Math. Univ. of Iowa, Iowa City, Iowa, USA.

1. *On finite planes of type 1-4 and planar division near-rings*, J.Algebra 27 (1973), 257-277. G,P",F

PASSMAN, Donald S., Dept.Math. Yale Univ. New Heaven Ct 06520, USA.

1. *Permutation groups*, Lecture Notes, Yale University, Benjamin, New York-Amsterdam, 1968. §§ 17-20. F,D",G

PATER, Z., Sf. Gheorghe, Romania

1. *über Funktionenalgebren*, Bull.Math.Soc. Sci.Math.R.S. Roumanie, Tome 9 (57), Nr. 2 (1965), 87-103. E,Cr,Rs

PENNER, Sidney, 3148 Grand Concourse, Bronx, New York 10458, USA

1. *Geometric axiomatics of substitution*, M.S. Thesis, Univ. of Chicago, 1958. G,Cr
2. *Bi- and tri-operational algebras of functions*, Doctoral dissertation, Illinois Institute of Technology, 1964. G,Rs

PHILIPP, Walter, Dept.Math. Univ. of Illinois, Urbana, Illinois 61801, USA

1. *über die Einfachheit von Funktionenalgebren über Verbänden*, Monatshefte Math. 67 (1963), 259-268. Rs,S

SEE ALSO NÖBAUER-PHILIPP.

PICKERT, Günter, Math.Inst.Justus Liebig Univ., Arnoldstr.2, 63 Gießen, Germany

1. *Projektive Ebenen*, Berlin, Göttingen, Heidelberg, Springer, 1955. G,F

PIEPER, Irene, 2050 Hamburg 80, Ladenbecker Furtweg 11a, Germany.

1. *Darstellung zweiseitiger geschlitzter Inzidenzgruppen*, Abh.Math. Univ.Hamburg 32 (1968), 97-126. F,G,L,Na
2. *über gekoppelte Abbildungen auf Fastringen*, Oberwolfach, 1968. D",E,M,N
3. *Zur Darstellung zweiseitiger affiner Inzidenzgruppen*, Abh.Math. Sem.Univ.Hamburg 35 (1970), 121-130. F,G,L,Na
4. *On a class of near-modules*, Oberwolfach, 1972. L,Na,D",F

SEE ALSO KARZEL-PIEPER.

PILZ, Günter, Inst.Math. d.Joh.Kepler-Univ. Linz, A-4045, Linz, Austria

1. *Ordnungstheorie in Kompositionsringen*, Doctoral dissertation, University of Vienna, 1967. G,O,E
2. *Ordnungstheorie in Fastringen*, Oberwolfach, 1968. O
3. *über geordnete Kompositionsringe*, Monatsh.Math. 73 (1969), 159-169. Cr,O
4. *Ω - groups with composition*, Publ.Math.Univ. Debrecen 17 (1970), 313-320. E,O,Ua
5. *Geordnete Fastringe*, Abh. Math.Sem.Univ.Hamburg 35 (1970), 83-89. O
6. *Parallelism in near-rings*, Rocky Mountain J. Math. 1 (1970), 483-487. G,O
7. *On direct sums of ordered near-rings*, J.Algebra 18 (1971), 340-342 O,S
8. *Zur Charakterisierung der Ordnungen in Fastringen*, Monatsh.Math. 76 (1972), 250-253. O

9. *On the construction of near-rings from a Z- and a C-near-ring*, Oberwolfach, 1972.	C,D,A',O
10. *A construction method for near-rings*, Acta Math. Acad. Sci. Hungar. 24 (1973), 97-105.	C
11. *Primitive near-rings with one-sided zero*, Institutsbericht No. 38, Math. Inst. Univ. Linz, 1976.	P,X
12. *Free near-rings and N-groups*, Institutsbericht No.39, Math.Inst. Univ.Linz, 1976.	F'
13. *Completely decomposable near-rings*, Institutsbericht No.40,Math. Inst. Univ. Linz, 1976.	E
14. *Radicals of related near-rings*, Institutsbericht No.41, Math.Inst. Univ.Linz, 1976.	R,M,P
15. *On the endomorphism near-rings* $E(\Gamma)$, $A(\Gamma)$ *and* $I(\Gamma)$, Institutsbericht No. 42, Math.Inst.Univ. Linz, 1976.	E",E,R,P
16. *Affine near-rings*, Institutsbericht Nr.43, Math.Inst.Univ.Linz,1976	A'
17. *Prime ideals in near-rings*, Institutsbericht No.47, Math.Inst. Univ.Linz, 1976.	P',R
18. *Constructing distributively generated near-rings*, Institutsbericht No.48, Math.Inst.Univ.Linz, 1976.	D,C,F'
19. *Modular left ideals of near-rings*, Institutsbericht No.49, Math. Inst. Univ.Linz. 1976.	M
20. *On the theory of near-ring radicals*, Oberwolfach, 1976.	R,P,X
21. *Near-rings*, North-Holland/American Elsevier, Amsterdam, 1976	All from A to X
PLASSER, Kurt, Math.Inst.Univ.Linz, A-4045 Linz, Austria	
1. *Subdirekte Darstellung von Ringen und Fastringen mit Booleschen Eigenschaften*, Diplomarbeit, Univ.Linz, Austria, 1974.	B,I,R,W
PLOTKIN, Boris Isakovič, Vysse Komand noinz.,Uc. b.Padomiu 5, Riga.USSR	
1. *Ω-semigroups, Ω-rings and representations*, Soviet Math. 4 (1963), 523-526. Doklady Akad. Nauk SSSR 149.	E",E',P,R
2. *Some questions on the general theory of representations of groups*, Amer.Math.Soc. Translations, Series 2, Vol. 52, pp. 171-200,1966.	E",E',P,R
3. *Group of automorphisms of algebraic systems*, Russian:Moskow 1966 English: Walters-Noordhoff Publ., Gröningen 1972.	Ua,R,E,E"
POKROPP, Fritz , Inst.f.Statistik u.Ök.d.Univ.,Von Melle-Park 5,2000 Hamburg 13	
1. *Dicksonsche Fastkörper*, Doctoral dissertation, University of Hamburg, 1965.	F,D"
2. *Dicksonsche Fastkörper*, Abh.Math.Sem. Univ. Hamburg 30 (1967), 188-219.	F,D"
3. *Isomorphe Gruppen und Fastkörperpaare*, Arch.Math. 18 (1967), 235-240.	F,D"
4. *Gekoppelte Abbildungen auf Gruppen*, Abh.Math.Sem.Univ.Hamburg 32 (1968) 147-159.	D",F
PFEHN, Renate, Hans-Scholz-Str. 28, 50 Erfurt, German Democratic Republic	
1. *Injektive Gruppen über Fastringen*, Publ. Math. Debrecen, to appear.	E,E',H,Ua

PROHASKA, Olaf

1. *The exceptional Lüneburg planes*, Proc.internat. Conf. projective planes, Washington State Univ., 1973, 275-279. — F,G

POLIN, S.V., Math.Inst. Univ.Moscow, USSR

1. *Primitive M-near-rings over multioperator groups*, Math.USSR Sbornic 13 (1971), 247-265. — P,T,Ua
2. *Radicals in m-Ω -near-rings I*, Izvestija vysš. učebn. Zaved., Mat. 1972, No.1(116), 64-75 (1972) (Russian). — R,M,N,P;Q,
3. *Radicals in m-Ω -near-rings II*, Izvestija vysš. učebn. Zaved., Mat. 1972, No.2(117), 63-71 (1972) (Russian). — P,R,S,D',Ua Q,N,M
4. *Generalized rings*, in: Bohut' - Kuźmin-Širšov (ed.), Rings II (Russian), 1973, Novosibirsk, Institut Matematiki, Sibir. AN, USSR.

RAMAKOTAIAH, D. Dept.Math. Andhra Univ. Postgraduate Center, Guntur-522005 (A.P.), India.

1. *Radicals for near-rings*, Math.Z. 97 (1967), 45-56. — R,S,P,M,N,Q
2. *Structure of 1-primitive near-rings*, Math.Z. 110 (1969), 15-26. — P',I,P,T
3. *Theory of near-rings*, Ph.D.dissertation, Andhra Univ., 1968.
4. *A radical for near-rings*, Arch.Math.(Basel) 23 (1972), 482-483. — R,S,Q
5. *Isomorphisms of near-rings of transformations*, J.London Math. Soc.9(1974), 272-278. — T,E"
6. *Structure theorems on 1-completely reducible N-groups*, manuscript. — R,S,P
7. *One-sided ideals in near-rings of transformations*, submitted. — T,T',P
8. *One-sided ideals in near-rings of transformations*, Oberwolfach, 1976 — T,T',P

SEE ALSO LIGH-RAMAKOTAIAH-REDDY, RAMAKOTAIAH-RAO, RAMAKOTAIAH-REDDY.

RAMAKOTAIAH, D. and RAO, G. Koteswara

1. *Topological formulation of density theorem for 0-primitive near-rings*, to appear. — P,T'
2. *A special class of near-rings*, to appear. — B,I;P;R;F

RAMAKOTAIAH, D., and REDDY, Venkateswara, Y.

1. *Zero divisors in near-rings*, to appear. — I',X,B

RAO, G. Koteswara, Dept. Math. Andhra Univ., Postgraduate Center, Guntur-5 A.P., India.

SEE RAMAKOTAIAH-RAO.

RAO, I.H.Nagaraja, Dept.Math. Andhra Univ., Postgraduate Center, Guntur, 522005 (A.P.), India.

1. *Sum constructions of N-groups*, Indian J.Math. 11 (1969), 75-82. — C,H,F'

RATLIFF, Ernest F., Jr., Math.Dept., Southwest Texas State Univ., San Marcos, Texas 78666, USA.

1. *Some results on p-near-rings and related near-rings*, Ph.D. Dissertation, University of Oklahoma, 1971. — B

REIDEMEISTER, K.

1. *Vorlesungen über Grundlagen der Geometrie*, Berlin, 1930, pages 41-43 and 110-128. — G,F,D"

REDDY, Venkateswara Y., Math.Dept. A.U.P.G.Centre, Guntur-522005 (A.P.),India
SEE LIGH-RAMAKOTAIAH-REDDY.

RIEDL, Christiane
1. *Radikale für Fastmoduln, Fastringe und Kompositionsringe*, Doctoral dissertation, University of Vienna, 1966. R,E,M,D,G Rs

ROBINSON, Daniel, A., Dept.Math.Atlanta Univ., Vienna, Georgia 30332, USA
1. *Sums of normal semi-endomorphisms*, Math.Monthly 70 (1963),537-539. E"

ROTH, R.J., P.O.Box 318, Montclair, New Jersey 07042, USA.
1. *The structure of near-rings and near-ring modules*, Doctoral dissertation, Duke University, 1962. E,C,S

SCHWEIGER, Fritz, Math. Inst. Univ. Salzburg, Porschestr. 1/I, 5020 Salzburg, Austria.
SEE LAUSCH-NÖBAUER-SCHWEIGER.

SCOTT, S.D., 10 Beacon Ave., Campbell's Bay, Auckland, New Zealand.
1. *Near-rings and near-ring modules*, Doctoral dissertation, Australian National University, 1970.
2. *Non-nilpotent ideals of near-rings with minimal condition*, Oberwolfach, 1972. N
3. *Formation radicals for near-rings*, Proc.London,Math.Soc. (3) 25 (1972), 441-464. R,N,I,E
4. *Idempotents in near-rings with minimal condition*, J. London Math. Soc. (2) 26 (1973), 464-466. I,N
5. *Minimal ideals of near-rings with minimal condition*, J. London Math.Soc. (2) 7 (1974), 8-12. E,S,N
6. *Near-rings with minimal condition on right N-subgroups*, submitted.
7. *Near-rings generated by fixed-point-free automorphisms*, Oberwolfach, 1976. E",D'

SCOTT, William, R., Dept.Math. Univ. Utah, Salt Lake City, Utah 84112, USA.
1. *Group theory*, Prentice Hall, 1964, pp. 270-277. F,G

SETH, Vibha, Dept.Math. Indian Institute of Technology,Kanpur, 208016,India
1. *Near-rings of quotients*, Doctoral dissertation, Indian Institute of Technology, 1974. Q'
SEE ALSO SETH-TEWARI.

SETH, Vibha and TEWARI, K.
1. *On injective near-ring modules*, Canad.Math.Bull. 17 (1974),137-141. D
2. *Near-rings of quotients*, to appear. Q'

SHODA, Kenjiro
1. *Daisugaku-Tsuron*, Kyoritsu-Shuppan, Tokyo, 1947, Cap. III.

SHRIVASTAVA, Krischna Kumar, Lucknow, India.
1. *Annihilators in near-rings*, Math.Balcanica 2 (1972), 215-218. E,N

SILVERMAN , Robert J., Dept.Math. Univ. of New Hampshire, Durham, New Hampshire 03824, USA.
SEE BERMAN-SILVERMAN.

SLOTTERBECK, Oberta, 1965 Merill Rd., Kent, Ohio 44240, USA.
SEE LIGH-McQUARRIE-SLOTTERBECK.

SO, Yong-Sian, Math.Inst. Univ. Linz, A-4o45 Linz, Austria.
1. *Polynom - Fastringe*, Doctoral Dissertation, Univ. Linz, 1976. P^o,R,E,G

SÖRENSEN, Kay, Math.Inst.Techn.Univ.München,Barerstr.23,8 München 21,Germany
1. *Topologische normale Fastmoduln und Inzidenzgruppen*, Oberwolfach, 1968. T',G
2. *Topologische Inzidenzgruppen*, Abh. Math. Sem. Univ. Hamburg 35 (1970), 75-82. T',G
SEE ALSO HEISE-SÖRENSEN, KARZEL-SÖRENSEN-WINDELBERG.

SPERNER, Emanuel, Hagener Allee 41, 2070 Ahrensburg i.H., Germany
1. *On non-Desarguan geometries*, Seminari dell' Istituto Nazionale di Alta Matematica 1962-63, 574-594. F,G

STANIK, Rotraud, Math.Inst.Techn.Univ.München, Barerstr. 23, 8 München 21,Germany
SEE KARZEL-STANIK-WÄHLING.

STEINEGGER, Günter, Waldeggstr. 91, 4020 Linz, Austria.
1. *Erweiterungstheorie von Fastringen*, Doctoral Dissertation, Univ. Salzburg, Austria, 1972. E,H,Cr

STENDER, Klaus, Math.Sem.Univ.Hamburg, Bundesstr. 55, 2000 Hamburg 13, Germany
1. *Normalitätsbedingungen in Fastbereichen*, to appear. S",Nd

STRAUS, E.G., Dept. Math. Univ. of Calif. at Los Angeles, Los Angeles, Calif. 90024, USA.
1. *Remarks on the paper "Ideals in near-rings of polynomials over a field"*, Pac. J. Math. 52 (1974), 601-603. P^o

STUEBEN, E.F.
1. *Ideals in two-place tri-operational algebras*, Monatsh.Math. 69 (1965), 177-182. Cr

SU, Li Pi, Math.Dept.Univ. of Oklahoma, Norman, Oklahoma 73069, USA.
1. *Homomorphisms of near-rings of continous functions*, Pacific J.Math. 38 (1971), 261-266. T',O,F
2. *Near-rings of continous functions*, Chinese University of Hong-Kong (1972), 141-150. T'

SUBRAHMANYAM, N.V., Dept.Math.,Andhra Univ., Waltair, India.
1. *Boolean semirings*, Math.Annalen 148 (1962) 395-401. B

SUVAK, John Alvin
1. *Full ideals and their ring groups for commutative rings with identity*, Doctoral dissertation, University of Arizona,Tucson, 1971. P^o,Cr
2. *Two classes of ring groups for* Z_n, submitted. P^o,Cr

SZETO, George, Math.Dept., Bradley Univ. Peoria, Illinois, 61606, USA.

1. *On a class of near-rings*, J.Austral.Math.Soc. 14 (1972), 17-19. — B
2. *The sub-semigroups excluding zero of near-ring*, Monatsh.Math. 77 (1973), 357-362. — I',B
3. *Planar and strongly uniform near-rings*, Proc.Amer.Math.Soc. 44 (1974), 269-274. — P'',X
4. *On regular near-rings with no non-zero nilpotent elements*, Math. Japan. 79 (1974) 65-70. — R',W
5. *Finite near-rings with trivial annihilators*, J.Austral.Math.Soc. 18 (1974), 194-199. — I',A,P''
6. *The Peirce sheaf representation of near-rings*, Oberwolfach, 1976. — X,B
7. *The automorphism group of a class of semigroups*, Monatsh. Math., to appear. — X
8. *On a sheaf representation of a class of near-rings*, J. Austral. Math. Soc., to appear. — X,B

SEE ALSO SZETO-WONG.

SZETO, George and WONG, Yuen-Fat

1. *On sheaf representations of near-algebras without nilpotent elements*, to appear. — Nd,X

TAUSSKY-TODD, Olga, Calif.Inst. of Techn.,Alfred Sloan Laboratory of Math. and Physics, Pasadena, Calif. 91109, USA.

1. *Rings with non-commutative addition*, Bull.Calcutta Math.Soc. 28 (1936), 245-246. — A

SEE ALSO FURTWÄNGLER-TAUSSKI.

TEWARI, K., Math.Dept. Indian Institute of Technology, Kanpur, 208016,India

1. *Quotient near-rings and near-rings modules*,Oberwolfach, 1972. — Q',D
2. *Radicals of near-rings*, to appear.

SEE ALSO CHOUDHARI-TEWARI, SETH-TEWARI.

THARMARATNAM, B., Dept. Math. Univ. of Sri Lanka, Sri Lanka, Colombo Camp., Colombo 3, Ceylon

1. -------------, *Doctoral dissertation*, University of London, 1964. — T,T',D,P,E''
2. *Complete primitive distributively generated near-rings*, Quart.J. Math.Oxford 18 (1967), 293-313. — T,T',D,P
3. *Endomorphism near-rings of relatively free groups*, Math. Z. 113 (1970), 119-135. — E'',T',D,P

TIMM, Jürgen,Fachs. Math., Univ. Bremen, 28 Bremen 33, Achterstr., Germany.

1. *Über das verallgemeinerte Dickson-Verfahren*, Oberwolfach, 1968. — D'',A
2. *Eine Klasse schwacher binärer Doppelstrukturen*, Abh.Math.Sem.Univ. Hamburg 33 (1969),102 -118. — E,D'',Rs
3. *Über die additiven Gruppen spezieller Fastringe*,J. Reine Angew. Math. 239/240 (1969), 47-54. — A,Rs
4. *Die Lösung eines Problems von Havel*, Arch.Math. (Brno) 6 (1970), 25-28. — D'',Rs
5. *Zur Theorie der nicht notwendig assoziativen Fastringe*, Abh.Sem. Univ. Hamburg 35, (1970), 14-32. — E,F,P'',A,Rs
6. *Zur Konstruktion von Fastringen, I*, Abh.Math.Sem.Univ.Hamburg 35 (1970), 57-73. — D'',Rs
7. *Zur Theorie der Fastringkonstruktionen II*, Abh. Math.Sem. Univ. Hamburg 36 (1971), 16-32. — D'',Rs
8. *Free near-algebras*, Oberwolfach, 1972. — Na,F', Ua,P°, O

SEE ALSO MISFELD-TIMM.

TITS, Jaques, Collège de France, 11 Place Marcelin-Berthelot,75231 Paris, France

1. *Sur les groupes doublement transitifs continus*, Comm.Math.Helv. 26 (1952), 203-224. — S", F, T', Rs
2. *Corrections et complements*, Comm.Math.Helv. 30 (1956), 234-240. — S"

UTUMI, Yuzo, Univ. of Osaka Prefecture 4-804 Mozuume-machi, Sàkai, 591,Japan
SEE LIGH-UTUMI.

VAN der WALT, A.P.J., Technological Univ., Delft, The Netherlands and Potchefstroom Univ., Potchefstroom, Republ. of South Africa.

1. *Prime ideals and nil radicals in near-rings*, Arch.Math. 15 (1964), 408-414. — P", R, N

VAN HOORN, Willy G., Math.Dept. Agricultural Univ. de Dreijen,Wageningen, Holland.

1. *Some generalizations of the Jacobson radical for seminear-rings*, Oberwolfach, 1968. — Rs, R, P
2. *Some generalizations of the Jacobson radical for semi-near-rings and semirings*, Math. Z. 118 (1970), 69-82. — Rs, P, R, M, N,

SEE ALSO VAN HOORN-VAN ROOTSELAAR.

VAN HOORN, Willy G. and VAN ROOTSELAAR, B.

1. *Fundamental notions in the theory of seminear-rings*, Compositio Math. 18 (1966), 65-78. — Rs

VAN ROOTSELAAR, B., Van Nijerodeweg 914, Amsterdam 1,11, The Netherlands

1. *Die Struktur der rekursiven Wortarithmetik des Herrn V.Vukovic*, Indag.Math. 24 (1962), 192-200. — Rs
2. *Algebraische Kennzeichnung freier Wortarithmetiken*, Compositio Math. 15 (1963), 156-168. — Rs
3. *Zum ALE-Fasthalbringbegriff*, Nieuw Archief voor Wiskunde 15(1967) 247-249. — Rs

SEE ALSO VAN HOORN-VAN ROOTSELAAR.

VEBLEN, O.
SEE VEBLEN-WEDDERBURN.

VEBLEN, O. and WEDDERBURN, J.H.M.

1. *Non-Desarguesian and non-Pascalian geometries*, Trans.Amer.Math.Soc. 8 (1907), 379-388. — G, Nf

VERONESI, Maria Luisa

1. *Sulla struttura arithmetica dei gruppi finiti con operatori*, Istituto Lombardo (Rend.Sc.) A 102 (1968), 3-11. — Rs

WÄHLING, Heinz, Math.Sem. Univ.Hamburg,Bundesstr. 55,2000 Hamburg 13, Germany.

1. *Darstellung zweiseitiger Inzidenzgruppen durch Divisionsalgebren*, 30 (1967), 220-240. — F,G
2. *Einige Sätze über Fastkörper*, Oberwolfach, 1968. — F,D''
3. *Invariante und vertauschbare Teilfastkörper*, Abh. Math.Sem. Univ. Hamburg, 33 (1969), 197-2o2. — F
4. *Automorphismen Dicksonscher Fastkörper*, Oberwolfach, 1972. — F,D''
5. *Ein Zassenhauskriterium für unendliche Fastkörper*, to appear. — F,D''
6. *Automorphismen Dicksonscher Fastkörperpaare mit kleiner Dicksongruppe*, Math. Z. 147 (1976), 65-78. — F,D''
7. *Normale Fastkörper mit kommutativer bzw.zweiseitiger Inzidenzgruppe*, to appear. — F,G
8. *Zur Theorie der Fastkörper*, Habilitationsschrift, Hamburg, 1972. — F,D'',G
9. *Bericht über Fastkörper*, Jahresbericht Dt.Math.Ver. 76 (1975), 41-103. — F,D'',G
10. *Automorphismen Dicksonscher Fastkörperpaare mit kleiner Dicksongruppe*, Abh.Math.Sem. Univ. Hamburg 44 (1975), 122-138. — F,D''

SEE ALSO KARZEL-STANIK-WÄHLING.

WEDDERBURN, H.H.M.

SEE VEBLEN-WEDDERBURN.

WEFELSCHEID, Heinrich, Fachber.Math., GHS Duisburg, Postfach 919, 4100 Duisburg 1, Germany.

1. *Vervollständigung topologisch-algebraischer Strukturen*, Doctoral Dissertation, Univ. Hamburg (Germany), 1966. — T',F,D'',Rs
2. *Vervollständigung topologischer Fastkörper*, Math. Z. 99 (1967), 279-298. — F,T',D''
3. *About a connection between order and valuation in near-fields*, Oberwolfach, 1968. — F,O,V
4. *Zur Konstruktion scharf 3-fach transitiver Permutationsgruppen mit Hilfe von Fastkörpern*, Oberwolfach, 1972. — Nd,S''
5. *Untersuchungen über Fastkörper und Fastbereiche*, Habilitationsschrift, Hamburg, 1972. — F,Nd,D''
6. *Zur Konstruktion bewerteter Fastkörper*, Abh. Math. Sem. Univ. Hamburg 38 (1972), 106-117. — V,F,D''
7. *Bewertung und Topologie in Fastkörpern*, Abh. Math. Sem. Univ. Hamburg 39 (1973), 130-146. — F,T',V,D''
8. *Über eine Orthogonalitätsbeziehung in Hyperbelstrukturen*, Abh. Math. Sem. Univ. Hamburg, to appear. — G,S'',Nd
9. *Zur Charakterisierung einer Klasse von Hyperbelstrukturen*, submitted. — G,S'',Nd
10. *Die Automorphismengruppe der Hyperbelstrukturen*, Proc. Symp. on Geometric Algebra, Duisburg, Germany, 1976, to appear. — G,S'',Nd,F
11. *Near-domains and sharply n-transitive permutation groups*, Oberwolfach, 1976. — Nd,S''

SEE ALSO KERBY-WEFELSCHEID.

WEINERT, Hans-Joachim, Math.Inst. Techn.Univ. Clausthal, Erzstr. 1, D-3392, Clausthal-Zellerfeld, Germany.

1. *Halbringe und Halbkörper I*, Acta Math. Acad. Sci.Hungar. 13 (1962) 365-378. — D',Rs
2. *Halbringe und Halbkörper II*, Acta. Math. Acad. Sci. Hungar., 14 (1963), 209-227. — D',Rs,Q'
3. *Ringe mit nichtkommutativer Addition I*, Jahresber. Dt.Math.Ver. 77 (1975), 10-27. — D',C,E',B,A
4. *Ringe mit nichtkommutativer Addition II*, Acta.Math.Acad. Sci. Hungar., 26(1975), 295-310. — D',E,H
5. *Related representation theorems for rings, semirings, near-rings and seminear-rings by partial transformations and partial endomorphisms*, Proc. Edinburgh Math. Soc., to appear. — E,T,Rs
6. *On distributive near-rings*, Oberwolfach, 1976. — D',E',H

SEE ALSO BOTERO DE MEZA-WEINERT.

WIELANDT, Helmut, Math.Inst. d. Univ., 74 Tübingen, Germany.

1. *über Bereiche aus Gruppenabbildungen*, Deutsche Mathematik 3 (1938), 9-10. — E,P,T
2. *Unveröffentlichte Manuskripte aus den Jahren 1937-1952.*
3. *How to single out function near-rings*, Oberwolfach, 1972. — E,T

WIESENBAUER, Johann, Institut für Algebra, Techn. Univ. Wien, Argentinierstr. 8, 1040 Wien, Austria.

1. *über die Klassifikation endlicher Kompositionsringe*, Diplomarbeit, Techn. Univ. Wien, 1971. — Cr,E
2. *über die Unabhängigkeit der Distributivitätsbedingungen*, Atti Accad. Naz. Lincei Cl. Sc. fis. mat. e nat. 58 (1975), 819-822. — Rs,E

WINDELBERG, Dirk, Math. Inst. Techn. Univ. Hannover, Hannover, Germany.

SEE KARZEL-SÖRENSEN-WINDELBERG.

WHITTINGTON, Robert J.

1. *Computer aided determination of near domains*, M.S.Thesis, Univ. of Southwestern Lousiana, Lafayette, 1973. — I',A

WILLHITE, Mary Lynn

1. *Distributively generated near-rings on the dihedral group of order eight*, M.S.Thesis, Texas A&M University, 1970. — D,A

WILLIAMS, Robert, E., Dept.Math. Kansas State Univ., Manhatten,Kansas 66504,USA.

1. *Simple near-rings and their associated rings*, Doctoral dissertation University of Missouri, 1965. — X,S,R
2. *A note on near-rings over vector spaces*, Amer.Math.Monthly 74 (1967), 173-175. — A,Na

WOLF, S.A.

1. *Spaces of constant curvature*, McGraw-Hill, New York 1967, §§ 5-6. — F,D"

WOLFSON, Kenneth G., Math.Dept. Rutgers Univ., New Bruncswick, N.J, 08903,USA

1. *Two sided ideals of the affine near-ring*, Amer.Math.Monthly 65 (1958), 29-30. — A'

WONG, Yuen-Fat, Dept. Math. DePaul Univ., Chicago, Illinois, USA.

1. *Sheaf representations of near-algebras*, Oberwolfach, 1976. — Nd,X

SEE ALSO SZETO-WONG.

WUYTACK, F.

1. *Boolean composition algebras*, Simon Stevin 37 (1963/64),97-125. — B,Rs

YAMAMURO, Sadayuki, Dept.Math. Inst. of Adv.Studies, Austral.Nat.Univ., Box 4, G.P.O., Canberra A.C.T., 2600 Australia.

1. *On near-algebras of mappings of Banach spaces*, Proc.Japan Acad. 41 (1965), 889-892. — Na,T'
2. *Ideals and homomorphisms in some near-algebras*, Proc.Japan Acad. 42 (1966), 427-432. — Na,T'
3. *A note on D-ideals in some near-algebras*, J. Austral. Math.Soc. 7 (1967), 129-134. — Na,T'
4. *On the spaces of mappings on Banach spaces*, J. Austral.Math.Soc. 7 (1967), 160-164. — Na,T'
5. *A note on near-rings of mappings*, J.Austral. Math.Soc. 16 (1973), 214-215. — T',T

SEE ALSO PALMER-YAMAMURO.

YEARBY, Robert Lee, Math.Dept. Grambling College, Grambling, Louisiana, USA

1. *A computer-aided investigation of near-rings on low order groups*, Doctoral Dissertation, Univ. of Southwestern Louisiana, Lafayette, 1973. — A

ZASSENHAUS, Hans, Math.Dept. Ohio State Univ., Columbus,Ohio, 43210,USA

1. *Über endliche Fastkörper*, Abh.Math.Sem. Univ. Hamburg 11 (1935/36), 187-220. — F,D",Rs
2. *Kennzeichnung endlicher linearer Gruppen als Permutationsgruppen*, Abh. Math.Sem. Univ. Hamburg 11 (1936), 17-40. — S",F
3. *Gruppentheorie*, Vandenhoek und Rupprecht 1958, pp. 105-107. — S",F

ZEAMER, R.W., Dept. Math. Univ. of Montreal, Montreal, Canada.

1. *Near-rings on free groups*, Oberwolfach, 1976. — F

ZEMMER, J.L., Dept. Math., Math. Science Building, Univ. of Missouri-Columbia, Columbia, Missouri, 65201 USA.

1. *Near-fields, planar and non-planar*, The Math. Student 31 (1964), 145-150. — F,P",A
2. *On a class of doubly transitive groups*, The Math.Student 31, (1964), 155-156. — S",F
3. *The additive group of an infinite near-field is abelian*, J.London Math.Soc. 44 (1969), 65-67. — F,A,S"
4. *Valuation near-rings*, Oberwolfach, 1972. — V,F,L,R
5. *Valuation near-rings*, Math.Z. 130 (1973), 175-188. — V,F

SUPPLEMENTARY WORKS

ACZÊL, J.: "Über die Gleichheit der Polynomfunktionen auf Ringen", Acta Sci. Math.(Szeged) 21 B (1960), 105-107.

ARTIN, E., NESBITT,C.J. and THRALL,R.M.:"Rings with minimum condition", Ann Arbour, Univ. of Michigan Press, 1955.

BAER, Reinhold: "Linear algebra and projective geometry", New York, 1952.

BEAUMONT, Ross A.:" Rings with additive group which is the direct sum of cyclic groups", Duke Math. J. 15 (1948), 367-369.

COCHRAN, W.G. and COX, G.M. :"Experimental designs", 2nd ed., Wiley and Sons, New York, 1957.

DEMBOWSKI, Peter:"Finite geometries", Springer Verl., 1968 (Ergebnisse der Mathematik, vol. 44).

DUBREIL,P. and DUBREIL-JACOTIN, M. :"Lectures on modern algebra", Hafner 1967, New York.

FUCHS, Lászlő :"Teilweise geordnete algebraische Strukturen", Vandenhoeck und Ruprecht, Göttingen, 1966.

GABOVICH, E. :" Partially ordered Ω-groups" (Russian), Uč. Sap. Tartusk. Univ. 102 (1961) 294-300.

GASCHÜTZ, W. :" Zu einem von B.H. und H. Neumann gestellten Problem", Math. Nachr. 14 (1955), 249-252.

GRÄTZER, George :"Universal algebra", Van Nostrand, 1968.

HALL, Marshal Jr. :"Combinatorial theory", Ginn/Blaisdell, Waltham, Mass., 1967.

HIGGINS, P.J. :"Groups with multiple operators", Proc. London Math. Soc. (3) 6 (1956), 366-416.

HOEHNKE, H.J. :"Radikale in allgemeinen Algebren", Math. Nachr. 32 (1964), 347-383.

HUPPERT, B. :"Endliche Gruppen I", Springer Verl., Berlin 1967.

JACOBSON, Nathan :"Structure of rings", Amer. Math. Soc. Colloquium Publ., vol 37, Rhode Island, 1968.

KERTÉSZ, Andor :" Vorlesungen über artinsche Ringe", Akademiai Kiadó, Budapest, 1968.

KUROSH, A.G. :"Lectures on general algebra", Chelsea Publ. Co., New York, 1963.

MAURER, W.D. and RHODES, J.L. :"A property of finite simple non-abelian groups", Proc. Amer. Math. Soc. 16 (1965), 552-554.

McCOY, N.H. :"The theory of rings", McMillan, 1969.

ROTMAN, J.J.:"The theory of groups", Allyn & Bacon, 1965.

SCOTT, W.R. :"Group theory", Prentice-Hall, 1964.

THIRRIN, G. :"On duo rings", Canad. Math. Bull. 3 (1960), 167-172.

THOMPSON, J.:"Finite groups with fixed-point-free automorphisms of prime order", Proc. Nat. Acad. of Science of USA 45 (1959), 578-581.

WIELANDT, H.:"Finite permutation groups", Academic Press, New York, 1964.

LIST OF SYMBOLS AND ABBREVIATIONS

I N D E X